Michael Springborg, Meijuan Zhou
Quantum Chemistry

Also of Interest

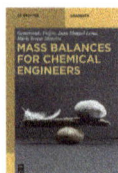

Mass Balances for Chemical Engineers
Gumersindo Feijoo, Juan Manuel Lema, Maria Teresa Moreira, 2020
ISBN 978-3-11-062428-1, e-ISBN (PDF) 978-3-11-062430-4,
e-ISBN (EPUB) 978-3-11-062431-1

Data Science in Chemistry
Artificial Intelligence, Big Data, Chemometrics and Quantum
Computing with Jupyter
Thorsten Gressling, 2021
ISBN 978-3-11-062939-2, e-ISBN (PDF) 978-3-11-062945-3,
e-ISBN (EPUB) 978-3-11-063053-4

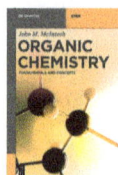

Organic Chemistry
Fundamentals and Concepts
John M. McIntosh, 2018
ISBN 978-3-11-056512-6, e-ISBN (PDF) 978-3-11-056514-0,
e-ISBN (EPUB) 978-3-11-056521-8

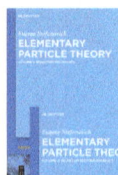

Elementary Particle Theory
Eugene Stefanovich, 2019

Volume 1 Quantum Mechanics
ISBN 978-3-11-049088-6, e-ISBN (PDF) 978-3-11-049213-2,
e-ISBN (EPUB) 978-3-11-049103-6

Volume 2 Quantum Electrodynamics
ISBN 978-3-11-049089-3, e-ISBN (PDF) 978-3-11-049320-7,
e-ISBN (EPUB) 978-3-11-049143-2

Volume 3 Relativistic Quantum Dynamics
ISBN 978-3-11-049090-9, e-ISBN (PDF) 978-3-11-049322-1,
e-ISBN (EPUB) 978-3-11-049139-5

Michael Springborg, Meijuan Zhou

Quantum Chemistry

An Introduction

DE GRUYTER

Authors

Prof. Dr. Michael Springborg
Physical and Theoretical Chemistry
University of Saarland
Campus B2.2
66123 Saarbrücken
Germany
m.springborg@mx.uni-saarland.de

Dr. Meijuan Zhou
Physical and Theoretical Chemistry
University of Saarland
Campus B2.2
66123 Saarbrücken
Germany
chemzhoumj@163.com

ISBN 978-3-11-074219-0
e-ISBN (PDF) 978-3-11-074220-6
e-ISBN (EPUB) 978-3-11-074223-7

Library of Congress Control Number: 2021938643

Bibliographic information published by the Deutsche Nationalbibliothek
The Deutsche Nationalbibliothek lists this publication in the Deutsche Nationalbibliografie;
detailed bibliographic data are available on the Internet at http://dnb.dnb.de.

© 2021 Walter de Gruyter GmbH, Berlin/Boston
Cover image: EzumeImages / iStock / Getty Images Plus
Typesetting: VTeX UAB, Lithuania
Printing and binding: CPI books GmbH, Leck

www.degruyter.com

Introduction

The quantum theory forms the basis for the understanding of the chemical bond and for spectroscopy. Because the manipulation of chemical bonds is at the heart of chemistry, a good understanding of chemical bonding is mandatory. This plus the great importance of spectroscopy in the characterization of materials and of chemical bonding in the rationalization of chemical findings, a good understanding of quantum theory is an important component of chemistry. Quantum theory also forms the basis for the area of chemical modeling, whereby properties, structures, reaction paths, etc. can be investigated with computer calculations. This was simultaneously the field of the research activities of AK Springborg at the University of Saarland.

The quantum theory is the theory which is to be used in dealing with very small systems. From this point of view, quantum theory seems to be of limited relevance to our everyday life, where we usually deal with much larger systems. Therefore, it may be surprising to know that an estimated 20 % of all produced goods are based on technological developments that would not exist without quantum theory.

Chemical modeling is also increasingly important for more experimental chemists. First of all, scientific research papers are rarely accepted by better chemical journals today if the experimental studies are not supported and supplemented by theoretical calculations. Therefore, for a doctoral thesis, which includes mainly experimental results, it is also important to include results of theoretical studies. Remembering that an unpublished scientific work is a meaningless work, one must recognize that publishing is an essential part of the doctoral thesis, implying that theoretical calculations are an important part of any chemical study.

Ultimately, theoretical calculations can replace a significant part of experimental studies. This can be illustrated through an example from our own experience. At the University of Tianjin in China, there is a working group dealing with the production of materials for the application in solar cells. The aim is to identify materials with a maximum yield (i. e., a maximum amount of electrical current that is generated from the solar radiation). For this purpose, about 50 members of the working group were investigating materials based on porphyrins (these form the building blocks of the so-called Grätzel cells, a type of solar cells). The porphyrins are modified by all the possible skills of organic chemistry. Each synthesis and subsequent study of the performance of a new compound takes one person about one year, but it is very often recognized that it does not meet the requirements. A theoretical calculation can be carried through in about one week, and the person who makes such a calculation can handle more than one compound in parallel (for example, 10 compounds). Therefore, about five hundreds compounds can be theoretically treated by a single theorist per year, a much larger number than the single compound that can be studied by the experimentalist. Even if the calculations are subject to certain inaccuracies, the results of such calculations are very relevant to identify optimal, promising compounds in

https://doi.org/10.1515/9783110742206-201

the laboratory. So, as can be seen, theoretical calculations can provide very useful information for experimentalists.

The aim of the present manuscript is to present the foundations of quantum theory as well as examples of its application in treating atomic and molecular systems. It is a course aimed at the chemistry students in their undergraduate studies at a university. It would be helpful for the readers to possess a first understanding of quantum theory, including having heard about orbitals, the Aufbau principle, the chemical bond, etc. But even without this prior knowledge, it should be, in principle, possible to adapt to the subject of this manuscript, although not without some effort. Sometimes examples are briefly discussed, which are discussed in more detail later in this manuscript and, therefore, this prior knowledge would be very helpful.

Saarbrücken, March 2021
Michael Springborg and Meijuan Zhou

Contents

1 What is quantum theory?

1.1 Classical physics

The quantum theory, together with the theory of relativity, is one of the two most important developments in physics during the 20th century. The quantum theory is important when considering objects that are very small (e. g., electrons and atoms), while the relativity theory becomes important when the objects move very fast. In both cases, phenomena occur that are different from what we are experiencing in our everyday life. The book of George Gamow, *Mr. Tompkins in Wonderland* can be recommended for the interested readers. This book describes what the world would look if either quantum effects or relativistic effects were noticeable in our everyday life. The book is written as an entertainment, and the author who also was a very capable scientist, has been able to present science to the educated layman.

Before these two theories were developed, the situation in the second half-part of the 19th century has often been compared with our present situation. One was largely convinced that all important physical laws were understood, and one needed "only" to apply these on all sorts of issues. That everything would change within few years was not considered possible. Whether we will soon experience another, similar revolution in the natural sciences is an open question.

Here, we shall briefly discuss some concepts of classical physics, which are not always valid with the introduction of quantum theory. "Not always" means that classical physics must be regarded as sufficiently accurate when it comes to macroscopic objects, e. g., when sending a rocket to the moon.

According to classical physics:

- Position and momentum coordinates are independent of each other and can have arbitrary values.
- If one knows the position and momentum of an object at a certain point in time, as well as all forces acting on the object, the position and momentum coordinates at any later time, in principle, can be computed with any precision.
- The energy of an object can take any value.
- A part of physics deals with bodies, while another part deals with waves. The two parts have little to do with each other.

We shall now see how experimental and theoretical results led to questioning these statements.

1.2 Black-body radiation

A black body with a certain temperature emits electromagnetic radiation (see Figure 1.1). This radiation is composed of radiation with all possible wavelengths λ and

https://doi.org/10.1515/9783110742206-001

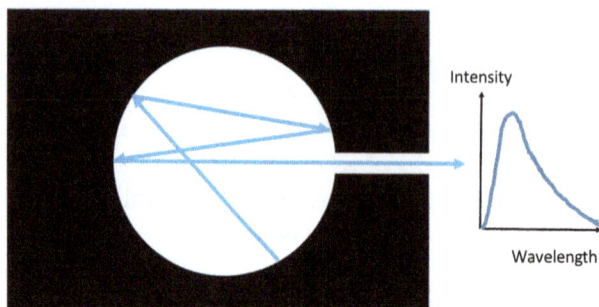

Figure 1.1: Schematic representation of the radiation of a black body. From the surface of the internal cavity of the body, radiation (photons) is emitted, which is brought into thermal equilibrium with the black body through multiple reflections at the surface before it emerges through a small opening. Ultimately, the intensity of the emitted radiation can be measured as a function of the wavelength.

the whole spectrum depends on the temperature of the body, as shown in Figure 1.2. This effect is already known from everyday life. When you turn on a hearth, it becomes increasingly hot, and thereby its color changes from black to red, and even later to orange and yellow.

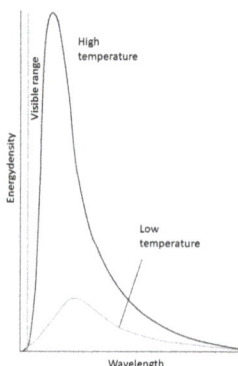

Figure 1.2: The radiation of a black body as measured experimentally at two different temperatures. Adapted from the book of Peter W. Atkins, *Kurzlehrbuch Physikalische Chemie*, Wiley-VCH, 2001.

According to Wilhelm Wien, the wavelength λ_{max}, at which the spectrum has a maximum, and the temperature T of the body obeys

$$\lambda_{max} \cdot T = \text{constant}. \tag{1.1}$$

Before knowing the origin of the radiation from the sun (as a result of nuclear reactions inside the sun), it was thought that the sun also represented such a black body. By examining the spectrum of the sun, one has then suggested that the temperature of the sun must be about 6000 K. In fact, this is not a bad approximation for the temperature at the surface of the sun, but far less than the several hundred million K that are found inside the sun.

Josef Stefan and Ludwig Boltzmann found empirically that the total radiated energy is proportional to T^4.

To explain the characteristics of such a black body, John William Strutt, better known as Baron Rayleigh, and James Jeans suggested that the radiation can be described by means of small oscillators (i. e., "something" vibrates and radiates). With this theory they obtained the spectrum in Figure 1.3. Compared to the experimental spectra in Figure 1.2, it is clear that the theory of Rayleigh and Jeans fails especially at small wavelengths: instead of approaching the value zero, the proposed spectrum diverges. This happens with smaller wavelengths than that of visible light and therefore this failure is called the **Ultraviolet catastrophe.**

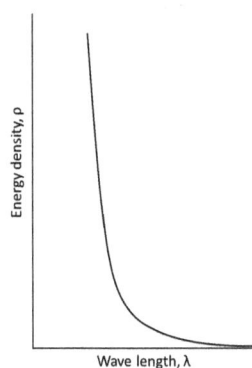

Figure 1.3: Radiation of black body according to the theory of Rayleigh and Jeans. Adapted from the book of Peter W. Atkins, *Kurzlehrbuch Physikalische Chemie*, Wiley-VCH, 2001.

In 1900, Max Planck presented a modified theory. Originally, he did not quite believe in the validity of this theory, but only that it showed that in principle the experimental spectrum can be reproduced, although he thought that different assumptions should be introduced rather than what he did. Max Planck modified the theory of Rayleigh and Jeans in such a way that he assumed that the oscillators could not emit any energy in the form of radiation, but radiation with a frequency v could only be emitted in multiples of v,

$$E = nhv. \tag{1.2}$$

n is an integer, and h is a constant—the one we nowadays call **Planck's constant.** With this only slightly modified theory, Max Planck obtained spectra like that in Figure 1.4. It is clear that the experimental spectra can be reproduced, and in particular that the ultraviolet catastrophe has disappeared.

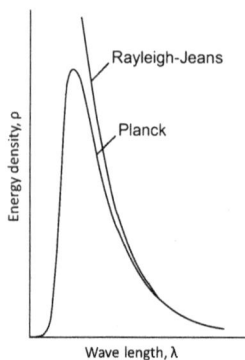

Figure 1.4: Radiation energy density of a black body according to the theory of Planck and the theory of Rayleigh and Jeans. Adapted from the book of Peter W. Atkins, *Kurzlehrbuch Physikalische Chemie*, Wiley-VCH, 2001.

The time of this proposal by Max Planck is often referred to as the birth time of quantum theory: Max Planck presented his theory at a meeting of the German Physical Society on December 14, 1900, in Berlin.

1.3 Heat capacities of solids

From the kinetic gas theory and, above all, the equipartition theorem it is known that the molar heat capacity C_V, of a solid consisting of only one type of atoms should be equal to $3R$ and independent of the temperature. In deriving this, it is assumed that the solid state can absorb energy only through vibrations.

According to experiment, however, something different is found; cf. Figure 1.5. C_V is not at all independent of the temperature and at low temperatures $C_V \to 0$. For some solids, it is also found that C_V becomes larger than $3R$ at higher temperatures. This latter can be explained by the fact that also the electrons contribute to C_V, which was not taken into account within the kinetic gas theory.

To explain the deviations at low temperatures, Albert Einstein proposed that, similar to the black body, the vibrations can not have any energy. According to his model, each atom of a solid can vibrate at a certain frequency, v_E, the Einstein frequency. By assuming the energy of the vibrations is quantized,

$$E = n \cdot h \cdot v_E \tag{1.3}$$

(n is an integer and h is a constant) he found a much improved description of the experimental results (see Figure 1.6). It is astonishing to see that the constant h has the same value as we have found above for the black body. This suggests that this constant is a universal natural constant.

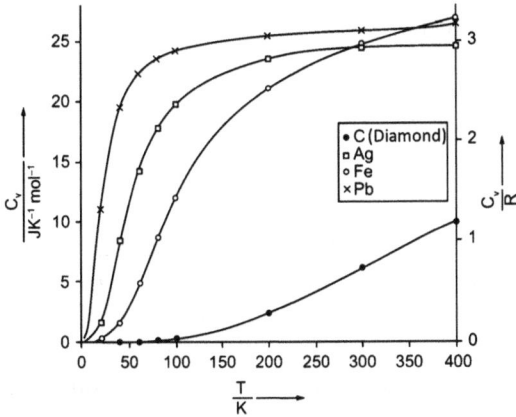

Figure 1.5: Molar heat capacity of different solids as a function of temperature. Adapted from the book, Gerd Wedler *Lehrbuch der Physikalichen Chemie*, Wiley-VCH, 2004.

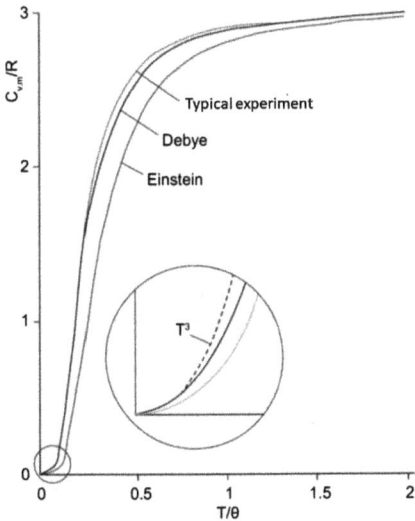

Figure 1.6: Molar heat capacity of a solid as a function of the temperature according to experiment, the theory of Einstein, and the theory of Debye. Adapted from the book Peter W. Atkins, *Kurzlehrbuch Physikalische Chemie*, Wiley-VCH, 2001.

Peter Debye later improved the model of Einstein by assuming that the frequencies of the vibrations of a solid can have all values up to a maximum value, v_D, i. e., the Debye frequency, and that the distribution of the frequencies has a certain form that shall not be discussed here. For each frequency v, a relation such as equation (1.3) is assumed

$$E = n \cdot h \cdot v. \tag{1.4}$$

Subsequently, the contributions of the individual oscillations are summed over all frequencies. Thereby the agreement with the experiment could be improved further; see Figure 1.6.

1.4 Photoelectric effect

When light falls on a metal plate, electrons can be kicked out from the metal; see Figure 1.7. This is the **photoelectric effect**. The electrons are charged so that the released electrons produce an electric current that can be measured. If this is done as a function of the frequency of the electromagnetic radiation, curves like the ones shown in Figure 1.8 are obtained. It is seen that the current is equal to zero up to a certain threshold frequency. It is astonishing that the same threshold is obtained when, e. g., increasing the intensity of the electromagnetic radiation; see Figure 1.8.

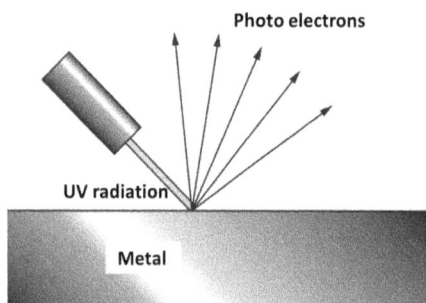

Figure 1.7: The photoelectric effect. Adapted from the book of Peter W. Atkins, *Kurzlehrbuch Physikalische Chemie*, Wiley-VCH, 2001.

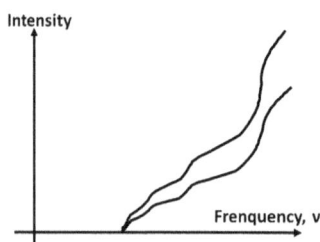

Figure 1.8: The photoelectric effect at two different intensities of the electromagnetic radiation.

In his so-called miraculous year, 1905, Albert Einstein explained this effect and later received the Nobel Prize in physics for this work. He assumed that the energy of the electromagnetic radiation is quantized, so that the radiation with the frequency v occurs in small packages with the energy hv. Part of this energy is used to pull out an electron from the metal, and the rest becomes the kinetic energy of the electron,

$$hv = \Phi + \frac{1}{2}mv^2. \tag{1.5}$$

Φ is the so-called **work function** and $\frac{1}{2}mv^2$ is the kinetic energy of the electron. Thus, only if the energy of the radiation is greater than the work function, electrons can leave the metal.

Through this interpretation, electromagnetic radiation occurs in small packages, called **photons**. Thus, a phenomenon normally treated as a wave also has a particle character.

1.5 The double-slit experiment

We first look at a long, narrow canal filled with water. In this, we can generate waves that propagate in parallel (so-called plane waves), as illustrated in the left parts of Figures 1.9 and 1.10. If we have installed barriers in the channel so that the waves can only propagate through one or two narrow slits, the behavior of the waves changes. With only one slit, circular waves are formed from the plane waves, as shown in Figure 1.9 This can be seen as a confirmation of the principle of Christiaan Huygens: each point of a wavefront can be considered as the starting point of a new spherical wavefront. With two slits, a fascinating pattern is created on the other side of these two slits, a so-called interference pattern, which is caused by the fact that the two spherical wave fronts are constructively or destructively added (Figure 1.10).

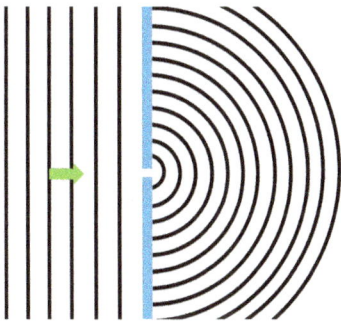

Figure 1.9: The single-slit experiment. According to the principle of Huygens, the plane waves propagate like spherical waves after a narrow slit.

An interference pattern can also be generated with other types of waves, e. g., light (i. e., electromagnetic waves). If one has not two slits but only one narrow slit, as shown in Figure 1.9, the electromagnetic wave propagates in all possible directions after the slit. For the two-slit case, we obtain spherical waves that propagate close to

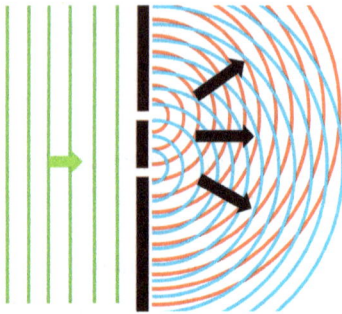

Figure 1.10: The double-slit experiment. According to the principle of Huygens, the plane waves propagate like a spherical wave after a narrow slit. For two narrow slits, there are certain directions in which an increased intensity occurs because the waves overlap constructively in these directions. Some of these directions are indicated by the arrows.

each other (see Figure 1.10). Then, the intensity of the light measured in a particular direction will depend on whether the light waves superimpose constructively or destructively. This creates the interference pattern.

We repeat the double-split experiment with electrons. This means that we have an electron source far away on the left side in Figure 1.11. This produces electrons that propagate from left to right. Also these will be separated by the two slits. If the electrons were small particles, at the right side we would see that these particles only pass through the two slits at two places (see Figure 1.11). The positions at which the charged electrons move after the two slits are indicated by the solid circles, whereas the motion of the charged electrons before the slits is shown by the colored arrows. In this case, we would see only two arrows to the right of the double slit. However, the experiment gives a pattern that closely resembles the interference pattern of the water waves.

This **double-slit experiment** shows that electrons behave not only like particles, but also like waves.

1.6 Compton diffraction

In an experiment, Arthur Holly Compton showed that electromagnetic radiation can also behave like particles. He used high-energy electromagnetic radiation (e. g., X-ray radiation) that is scattered by electrons (see Figure 1.12) whereby the radiation propagates in one direction, while the electron moves in another direction. The whole process can be described very well by assuming that the radiation consists of small particles (photons), and then treating the process as a collision process (as among billiard balls) between two particles, i. e., the photon and the electron.

In this experiment, we have seen that electromagnetic radiation not only possess wave properties but can also behave like a particle.

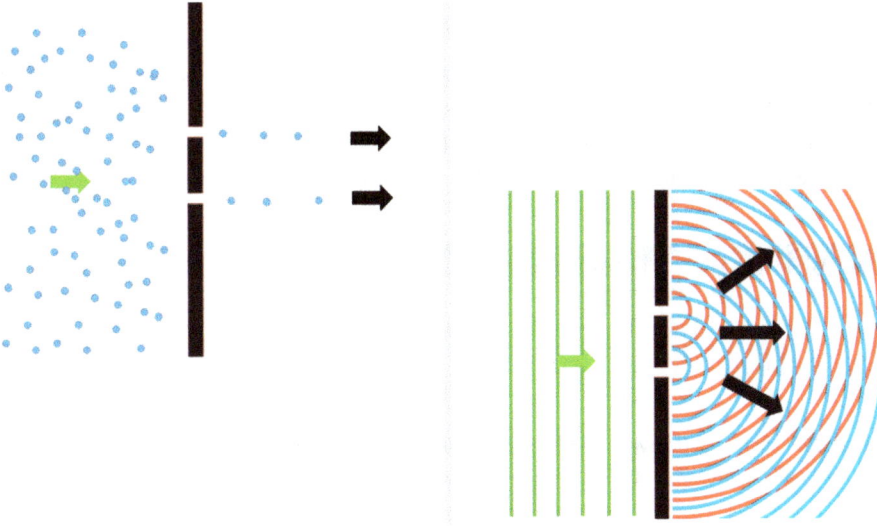

Figure 1.11: The double-slit experiment with electrons and the two possible results: the upper left part shows the results of the electrons behaving as small particles, while the results in the lower right part correspond to the case that the electrons behave like waves.

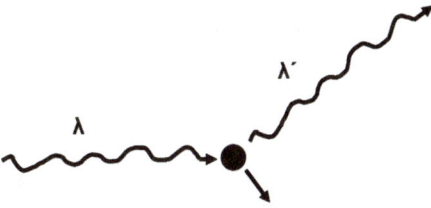

Figure 1.12: The experiment of Compton: an electromagnetic wave hits a particle, is scattered, and continues in another direction with a different wavelength.

1.7 Wave-particle duality

The examples above show that the separation into waves and particles can no longer be sustained in the atomic world. The objects (light, electrons, ...) sometimes behave like particles and sometimes as waves. This is called the **wave-particle duality**, or **particle-wave duality**.

In 1924, Louis de Broglie proposed that there is the following relation between the wave nature (described by the wavelength λ) and the particle character (described by the momentum p),

$$\lambda = \frac{h}{p}. \tag{1.6}$$

1.8 Spectra

As a final example, we show experimental spectra in Figure 1.13. In one case, Hg atoms are first excited (with the help of electromagnetic radiation). Subsequently, it is observed that the Hg atoms emit their absorbed energy again. The special feature is that the emitted energy can only assume certain discrete values.

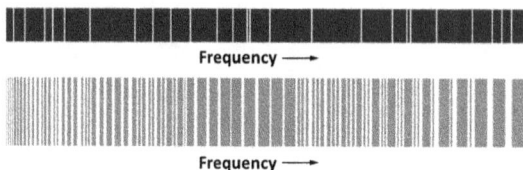

Frequency ⟶

Frequency ⟶

Figure 1.13: Example of (above) an emission spectrum and (below) an absorption spectrum. Adapted from the book of Peter W. Atkins, *Kurzlehrbuch Physikalische Chemie*, Wiley-VCH, 2001.

In the other case, the energy absorbed by ScF gas molecules is measured. In this case, it can be seen that the absorbed energy can only take certain discrete values, too.

In both cases, we have a situation that is interpreted as in Figure 1.14. A molecule or atom can only have certain energies. If the system is excited, it can emit the energy again by falling back from a level of higher energy to an energetically lower level. The energy that becomes free, ΔE, is radiated in the form of a photon, whose frequency v is given by

$$hv = \Delta E. \tag{1.7}$$

Conversely, the system can be excited by absorbing a photon whose frequency satisfies equation (1.7). Then ΔE is the energy difference between the final and the initial state.

This is another example of how energy is quantified. This quantization is specific to each system. Therefore, measuring the discrete energies can be used to characterize a system. This is the basis for all forms of spectroscopy.

We have not yet explained where these discrete energies are coming from. The basis for this is provided by the Schrödinger equation, which we shall discuss in the next chapter.

1.9 Problems with answers

1. **Problem:** Sketch the absorption spectrum of a system that can have the energies ϵ_0, $2\epsilon_0$, and $4\epsilon_0$, respectively.
   Answer: The absorption spectrum has maxima for energies, $0 < E = hv = h\frac{c}{\lambda}$, $E = E_n - E_m$. E_n and E_m are two possible energies of the system. In the present example, this is the case for $E = \epsilon_0$, $E = 2\epsilon_0$ and $E = 3\epsilon_0$, so the spectrum is as sketched in Figure 1.15.

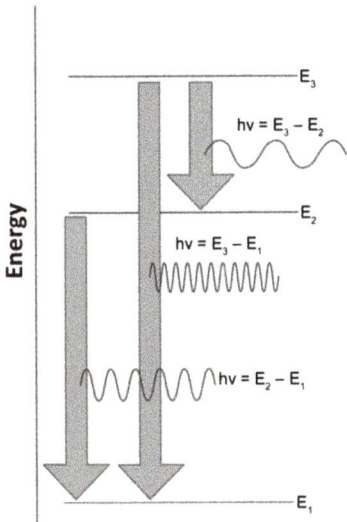

Figure 1.14: Schematic representation of the energy levels of an atom or molecule as well as the emitted radiation. Adapted from the book of Peter W. Atkins, *Kurzlehrbuch Physikalische Chemie*, Wiley-VCH, 2001.

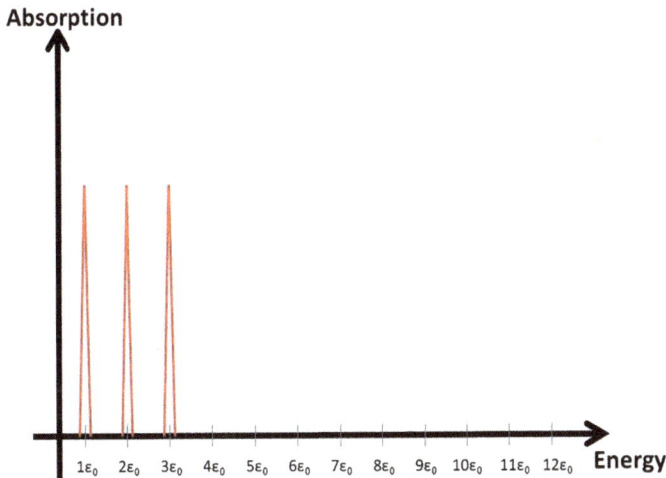

Figure 1.15: Illustration of the answer to question 1 in Section 1.9.

2. **Problem:** For a quantum system, there are the following possible energies: ϵ_0, $4\epsilon_0$, $9\epsilon_0$, $16\epsilon_0$,…. What does an absorption spectrum look like qualitatively for this system?

   Answer: An absorption spectrum has maxima for energies, $0 < E = h v = h\frac{c}{\lambda}$, with $E = E_n - E_m$ and E_n and E_m being two possible energies of the system. In the present example, for $E = (n^2 - m^2) \cdot \epsilon_0$ where $n > m \geq 1$. The smallest energies are $E =$

$3\epsilon_0, 5\epsilon_0, 7\epsilon_0, 8\epsilon_0, 9\epsilon_0, 11\epsilon_0, 12\epsilon_0, 13\epsilon_0, \ldots$, respectively. The part of the spectrum for the smallest energies is shown in Figure 1.16.

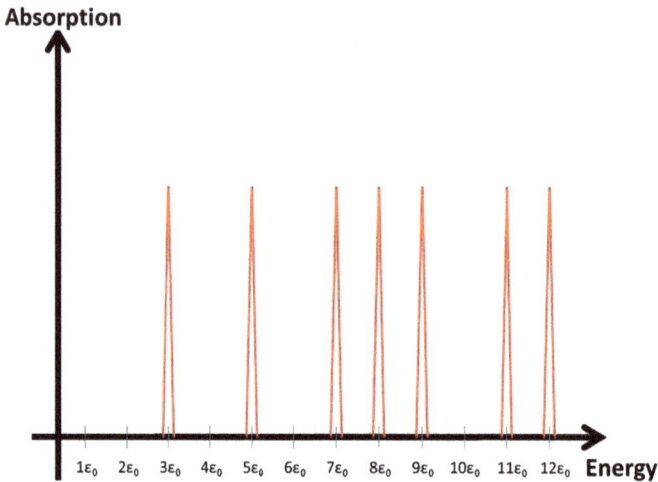

Figure 1.16: Illustration of the answer to question 2 in Section 1.9.

3. **Problem:** For a system consisting of many independent particles, the absorption peaks are located at the following energies, 1, 2, 8, 9, and 10 eV. The lowest energy is known to be 2 eV for this system. Determine the possible energies of the individual particles.

   Answer: For a system with N different energy levels, there are at most $\frac{N(N-1)}{2}$ different absorption energies. In the present case, we have 5 absorption energies, indicating that $N = 4$. For the absorption energies, ΔE, $\Delta E = E_i - E_j$, where E_i and E_j are the two energies of the system. Because the greatest excitation energy is equal to 10 eV, the highest energy of the system is 10 eV higher than the lowest. By experimenting, one can then see that energies of 0, 1, 2, and 10 eV fulfill this. But since the lowest energy equals 2 eV in this system, the energies must be 2, 3, 4, and 12 eV, respectively. There is another solution, i. e., the energies can be 2, 10, 11, and 12 eV, respectively. The information provided by the spectrum is not enough to decide which solution is right.

4. **Problem:** For a system consisting of many independent particles, the absorption peaks are located at the following energies, 1, 2, and 3 eV. The lowest energy is known to be 0 eV for this system. Use this information to determine the possible energies of the individual particles.

   Answer: As before, for a system with N different energy levels, there are at most $\frac{N(N-1)}{2}$ different absorption energies. In the present case, we have 3 absorption energies, indicating that $N = 3$. For the absorption energies, ΔE, $\Delta E = E_i - E_j$, where

E_i and E_j are two energies of the system. By testing, it can be seen that energies of 0, 1, and 3 eV fulfill this, whereby the lowest energy equals 0 eV. There is another solution, i. e., the energies can be 0, 2, and 3 eV, respectively. And finally, there is a third possible proposition, this time for $N = 4$: 0, 1, 2, and 3 eV. The information from the spectrum is not enough to decide which solution is the right one.

1.10 Problems

1. For a quantum system, the following energies are possible: $\frac{1}{2}\epsilon_0, \frac{3}{2}\epsilon_0, \frac{5}{2}\epsilon_0, \frac{7}{2}\epsilon_0, \ldots$ How does an absorption spectrum look qualitatively for this system?
2. For a quantum system, the following energies are possible: $-\epsilon_0, -\frac{1}{4}\epsilon_0, -\frac{1}{9}\epsilon_0, -\frac{1}{16}\epsilon_0, \ldots$ How does an absorption spectrum look qualitatively for this system?
3. Sketch the intensity of the radiation of a black body as a function of the wavelength. With the help of this sketch, explain the laws of Wien and of Stefan and Boltzmann as well as the ultraviolet catastrophe. Describe Planck's hypothesis, which explains the radiation of a black body.
4. Sketch the molar heat capacity of an elemental solid as a function of the temperature. Explain the prediction of the equipartition theorem (of the kinetic gas theory) by means of this sketch, as well as the predictions of the models of Einstein and Debye.
5. Explain briefly the Compton experiment, and why it could not be explained with classical physics.
6. Explain the concept of "photoelectric effect," including "work function" and Einstein's hypothesis.
7. Explain the concept "wave-particle duality," including the relationship of de Broglie.
8. Explain the double-slits experiment, including its relationship to quantum theory.
9. For a system consisting of many independent particles, the absorption spectrum has peaks located at the following energies: 1, 5, 7, 8, 12, and 13 eV. The lowest energy is known to be 5 eV for this system. Use this information to determine the possible energies of the individual particles.
10. For a system consisting of many independent particles, the absorption spectrum has peaks located at the following energies: 5, 10, and 15 eV. The lowest energy is known to be 4 eV for this system. Use this information to determine the possible energies of the individual particles.

2 Basics of quantum theory

2.1 The time-dependent Schrödinger equation

In 1926, Erwin Schrödinger and Werner Heisenberg presented two mathematically quite different theories, both of which provided a qualitative and quantitative description of the quantum effects discussed in the previous chapter. A short time later, it became clear that the two theories are equivalent. For our purposes, the formulation of Schrödinger is best suited and, therefore, we will concentrate on this. In this chapter, we will briefly introduce the basics. At first, some of this can be confusing and, therefore, we will discuss an example later (Chapter 4) in detail. In Chapter 3, we will also discuss in detail the relevant properties of operators, which are essential for quantum theory.

Without further introduction, we start directly by presenting the Schrödinger equation for a particle moving in an external (and possibly time-dependent) potential $V(\vec{r}, t)$ in the 3-dimensional space. The properties of this system can be determined through the time-dependent Schrödinger equation,

$$-\frac{\hbar^2}{2m}\nabla^2\tilde{\psi}(\vec{r}, t) + V(\vec{r}, t)\tilde{\psi}(\vec{r}, t) = i\hbar\frac{\partial}{\partial t}\tilde{\psi}(\vec{r}, t). \tag{2.1}$$

m is the mass of the particle,

$$\hbar = \frac{h}{2\pi} = 1.05459 \cdot 10^{-34}\,\text{J}\cdot\text{s}, \tag{2.2}$$

where h is the Planck constant mentioned in the previous chapter, and $\tilde{\psi}(\vec{r}, t)$ is the (time-dependent) wave function. For this, the probability of finding the particle in a volume element $d\vec{r}$ at $\vec{r}$ at time t is given by $|\tilde{\psi}(\vec{r}, t)|^2 d\vec{r} = \tilde{\psi}^*(\vec{r}, t)\tilde{\psi}(\vec{r}, t)d\vec{r}$ (for the reader who is unfamiliar with continuous distribution functions, please consult Section 16.1). Finally, in equation (2.2), ∇^2 is the Laplace operator. In Cartesian coordinates, this is

$$\nabla^2 = \frac{\partial^2}{\partial x^2} + \frac{\partial^2}{\partial y^2} + \frac{\partial^2}{\partial z^2}, \tag{2.3}$$

while in spherical coordinates

$$\nabla^2 = \frac{\partial^2}{\partial r^2} + \frac{2}{r}\frac{\partial}{\partial r} + \frac{1}{r^2\sin^2\theta}\frac{\partial^2}{\partial\phi^2} + \frac{1}{r^2\sin\theta}\frac{\partial}{\partial\theta}\sin\theta\frac{\partial}{\partial\theta}. \tag{2.4}$$

In many cases, the external potential $V(\vec{r}, t) = V(\vec{r})$ is static, that is, independent of time. This is, e. g., the case for an electron which is moving in the electrostatic field of atomic nuclei, which, on the other hand, do not move, while the assumption is no longer valid when the nuclei are moving, e. g., vibrating back and forth. Also for the case that the electron experiences an additional oscillating electromagnetic field, as

https://doi.org/10.1515/9783110742206-002

is the case for spectroscopy, the assumption of the static field is no longer valid. But even in these cases, the results obtained in the time-independent case are very useful.

In the following chapters, we shall discuss the Schrödinger equation in more detail. But we already now recognize that the equation can be written as

$$\hat{H}\tilde{\psi}(\vec{r},t) = i\hbar \frac{\partial}{\partial t}\tilde{\psi}(\vec{r},t) \tag{2.5}$$

Here,

$$\hat{H} = -\frac{\hbar^2}{2m}\nabla^2 + V(\vec{r},t), \tag{2.6}$$

is an operator (the Hamilton operator) that acts on the wave function. Operators are discussed in detail in the next chapter.

2.2 The time-independent Schrödinger equation

Stationary solutions are wave functions, which "always look the same." A (hypothetical) example is shown in Figure 2.1. If one compares this with Figure 2.2, where a nonstationary wave function is shown, the difference is (hopefully) clear: In the stationary case, the wave function can be obtained at any time from the wave function at any other time by scaling. Thus, for all $\vec{r}$

$$\tilde{\psi}(\vec{r},t_1) = \tilde{\psi}(\vec{r},t_2) \cdot a(t_1,t_2). \tag{2.7}$$

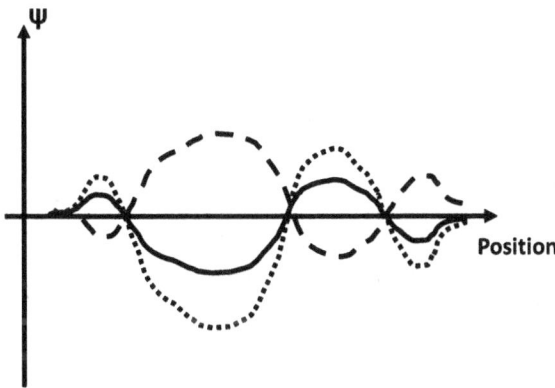

Figure 2.1: Schematic representation of a static wave function at three different times. The three different curves show the wave function at the three times. It can be seen that the three functions always look the same, so they differ pairwise only through a position-independent factor.

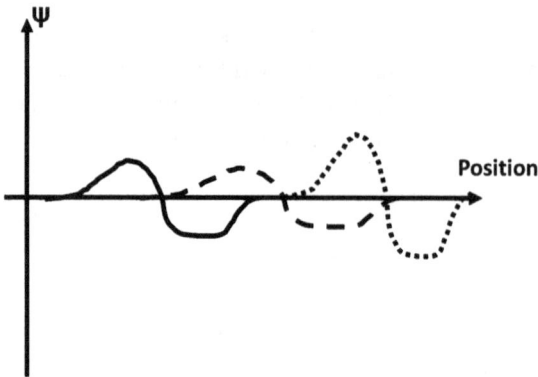

Figure 2.2: Schematic representation of a nonstatic wave function at three different times. The three different curves show the wave function at the three times. It can be seen that the three functions do not look the same, so they differ pairwise through a position-dependent factor.

The time-independent Schrödinger equation makes sense only if the Hamilton operator has no time dependency,

$$V(\vec{r}, t) = V(\vec{r}). \tag{2.8}$$

This is, e. g., the case for the electrons and nuclei of an isolated molecule without interaction with the environment, and when the molecule is not experiencing the field of electromagnetic radiation. The latter is, e. g., the case for spectroscopic investigations of the molecule, but it can be shown (see Section 9.8) that the information which describes the spectroscopic properties of the molecule can be obtained with the aid of the time-independent Schrödinger equation.

For stationary wave functions, equation (2.7) gives that

$$\tilde{\psi}(\vec{r}, t) = \psi(\vec{r}) \cdot A(t) \tag{2.9}$$

has to be valid. Inserting this into the time-dependent Schrödinger equation, equation (2.1), one finds that the stationary solutions must satisfy the equation

$$-\frac{\hbar^2}{2m} \nabla^2 \psi(\vec{r}) + V(\vec{r})\psi(\vec{r}) = E\psi(\vec{r}). \tag{2.10}$$

E is then the energy of the particle.

If $\hat{H}$ has no time dependence, equation (2.10) results as follows. We insert equation (2.9) into equation (2.1),

$$\hat{H}[\psi(\vec{r})A(t)] = i\hbar \frac{\partial[\psi(\vec{r})A(t)]}{\partial t}. \tag{2.11}$$

Then we use that $\hat{H}$ is a so-called linear operator (which is explained in more detail in the following chapter) and independent of t. From this,

$$A(t)\hat{H}\psi(\vec{r}) = i\hbar\psi(\vec{r})\frac{\partial A(t)}{\partial t}. \tag{2.12}$$

One divides by $A \cdot \psi$ (and ignores those points $(\vec{r}, t)$ where the product disappears)

$$\frac{\hat{H}\psi(\vec{r})}{\psi(\vec{r})} = i\hbar \frac{\frac{\partial A(t)}{\partial t}}{A(t)}. \tag{2.13}$$

It is now important to recognize that the left-hand side has no t dependence, and that the right-hand side has no $\vec{r}$ dependence. The two expressions can therefore be identical only if they are both equal to some constant. We will call this constant E.

This gives us two equations. The first equation is

$$i\hbar \frac{\partial A(t)}{\partial t} = E \cdot A(t), \tag{2.14}$$

that has the solution

$$A(t) = \exp\left[-i\frac{Et}{\hbar}\right], \tag{2.15}$$

by setting a constant prefactor equal to 1.

The second equation is

$$\hat{H}\psi = E\psi, \tag{2.16}$$

i. e., equation (2.10).

In order to derive the time-independent Schrödinger equation, one can formally proceed as follows (NB: this is not a mathematically correct derivation, but this "derivation" is given here because it illustrates various aspects). For the sake of simplicity, we consider a particle in one dimension that moves in the potential $V(x)$ (see Figure 2.3). According to classical mechanics, the total energy is the sum of the kinetic and the potential energy,

$$E = E_{\text{kin}} + V(x) = \frac{p^2}{2m} + V(x). \tag{2.17}$$

In order to obtain quantum mechanical expressions, one must know that in quantum theory, observables are expressed not by functions but by operators. This will be discussed in detail in the next chapter. Thereby, the momentum p becomes the operator

$$\hat{p} = -i\hbar \frac{d}{dx}, \tag{2.18}$$

from which we obtain that p^2 becomes the operator $-\hbar^2 \frac{d^2}{dx^2}$. Inserting this into equation (2.17), we obtain

$$E = -\frac{\hbar^2}{2m}\frac{d^2}{dx^2} + V(x) \tag{2.19}$$

V(x)

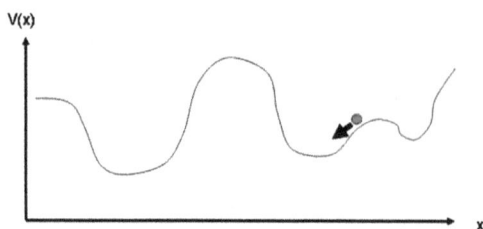

Figure 2.3: A particle in a 1-dimensional potential $V(x)$.

or by "multiplication" on both sides with "something," e. g., $\psi(x)$,

$$E\psi(x) = -\frac{\hbar^2}{2m}\frac{d^2}{dx^2}\psi(x) + V(x)\psi(x) = \hat{H}\psi(x). \tag{2.20}$$

Thus, we have obtained the Schrödinger equation without actually deducing it. We also recognize that quantum theory uses operators that can be derived from classical functions. For example, the Hamilton operator is the operator for energy.

2.3 The wave function

A wave function is generally complex, and thus takes complex, rather than real, function values. It has to obey the following:
–   It must be unique.
–   It cannot be infinite in a finite interval.
–   It must be continuous, even though it does not have to be differentiable everywhere.
–   The wave function can be nondifferentiable only in points where the potential diverges.

Examples of wave functions are given in Figure 2.4.

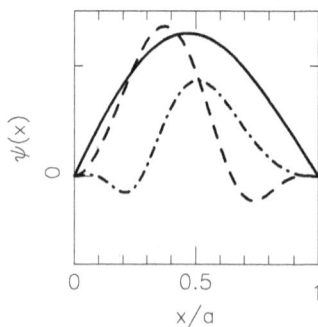

Figure 2.4: Examples of wave functions which are nonzero only for $0 \leq x \leq a$.

Max Born interpreted the wave function as follows:

$$|\psi(\vec{r})|^2 d\vec{r} = \psi^*(\vec{r})\psi(\vec{r})d\vec{r} = \psi^*(x,y,z)\psi(x,y,z)dx\,dy\,dz \tag{2.21}$$

is the probability of finding the particle in a volume element $d\vec{r}$ at the point $\vec{r}$; see Figure 2.5. Because of this interpretation, we must have

$$1 = \int |\psi(\vec{r})|^2\,d\vec{r}. \tag{2.22}$$

At the same time, this interpretation implies that one can consider $|\psi(\vec{r})|^2$ as a probability density (or, for short, only density). Then it becomes possible to determine expectation values for experimentally measurable quantities. An expectation value is the result which an experiment would give on average, if exactly the same experiment is carried out very, very many times under the same circumstances. Since in quantum theory, operators and not functions are used to describe experimentally measurable variables (this is discussed in more detail in Chapter 3), it is not initially clear how the expectation values are calculated. In fact, one uses

$$\langle Q \rangle = \int \psi^*(x)\hat{Q}\psi(x)\,dx = \int \psi^*(x)[\hat{Q}\psi(x)]\,dx, \tag{2.23}$$

where we again have assumed that the particle moves in only one dimension. $\hat{Q}$ is the quantum mechanical operator for the experimental quantity Q, that is, e. g., x, x^2, $-i\hbar\frac{d}{dx}$, and $-\hbar^2\frac{d^2}{dx^2}$ for the position coordinate, the square of the position coordinate, the momentum coordinate, and the square of the momentum coordinate, respectively.

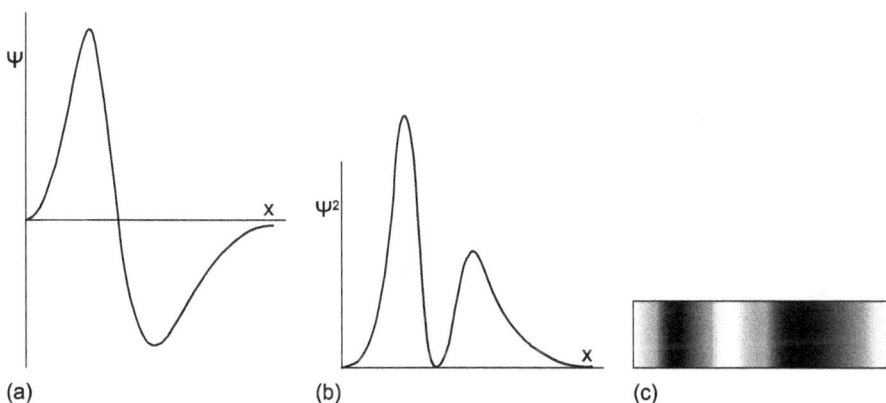

(a)　　　　　　　　(b)　　　　　　　　(c)

Figure 2.5: The interpretation of the wave function according to Born. (a) shows the wave function, which is real in this example. The square of this is shown in (b) and describes the probability of finding the particle in a small interval. If the position of the particle were repeatedly measured, a result as in (c) would be obtained. Adapted from the book of Peter W. Atkins, *Kurzlehrbuch Physikalische Chemie*, Wiley-VCH, 2001.

In equation (2.23), we have also emphasized how the integral is computed: First, we let $\hat{Q}$ operate on the wave function $\psi(x)$. The result is then multiplied by $\psi^*(x)$, and the result of this multiplication is ultimately integrated over the whole x space.

Furthermore, in equation (2.23) we have also used equation (2.22). In the more general case, for any wave function, $\psi(x)$, which does not necessarily obey equation (2.22)

$$\langle Q \rangle = \frac{\int \psi^*(x)\hat{Q}\psi(x)\,dx}{\int \psi^*(x)\psi(x)\,dx}. \tag{2.24}$$

From this, we also recognize that if the wave function is changed by a phase factor:

$$\psi(x) \rightarrow e^{i\theta}\psi(x) \tag{2.25}$$

with θ being equal to a real number, then the expectation values do not change. Therefore, phase factors can be chosen as desired.

2.4 Heisenberg's uncertainty principle

Werner Heisenberg presented the uncertainty principles that now carry his name. According to these (for a particle that moves in one dimension),

$$\Delta x \cdot \Delta p \geq \frac{\hbar}{2} \tag{2.26}$$

Here, Δx is the width (uncertainty) of the position coordinate and Δp that of the momentum. The width of the measured variables are defined as

$$\Delta s = \left(\langle s^2 \rangle - \langle s \rangle^2 \right)^{1/2}. \tag{2.27}$$

The larger the width, the less accurate is a measurement of this quantity (see Section 16.1). Therefore, the width is also-called **uncertainty**.

Equation (2.26) shows that position and momentum coordinates are not independent of each other. The more precisely one coordinate can be determined (i.e., the smaller the associated width), the other coordinate will be less accurate. This relationship between position and momentum coordinates departs from classical physics, as discussed in the previous chapter (Section 1.1). The fact that the position and momentum coordinates are no longer independent of each other can also be seen from the fact that the information on the properties of a particle in momentum space can be determined by means of the wave function in position space. That is, from equations (2.23) and equation (2.18), one obtains

$$\langle p^n \rangle = \int \psi^*(x) \left[\left(-i\hbar \frac{d}{dx} \right)^n \psi(x) \right] dx. \tag{2.28}$$

For particles moving in several (e. g., 3) dimensions, equation (2.26) is generalized to

$$\Delta x \cdot \Delta p_x \geq \frac{\hbar}{2}$$
$$\Delta y \cdot \Delta p_y \geq \frac{\hbar}{2}$$
$$\Delta z \cdot \Delta p_z \geq \frac{\hbar}{2}. \tag{2.29}$$

For combinations such as $\Delta x \cdot \Delta p_y$, there is no such relationship.

2.5 More particles

In the case that we have more particles, say $\mathcal{N}$, the wave function becomes a function of the coordinates of all particles. In the stationary case, the wave function becomes $\psi(\vec{r}_1, \vec{r}_2, \dots, \vec{r}_\mathcal{N})$. Equivalently, the Hamilton operator corresponds to the energy of the complete system.

Very often, the system of our interest is an atom or a molecule. In that case, the $\mathcal{N}$ particles are the nuclei and the electrons. The Hamilton operator contains then contributions from the individual particles like the kinetic energy and potential energy from external (for instance, electromagnetic) fields as well as contributions from all pairs of particles, i. e., electrostatic interactions between the charged particles. Other contributions may exist but these are not relevant here. Thus, the Hamilton operator can be written as

$$\hat{H} = \sum_{i=1}^{\mathcal{N}} \hat{h}_{1,i}(\vec{r}_i) + \sum_{i=1}^{\mathcal{N}-1} \sum_{j=i+1}^{\mathcal{N}} \hat{h}_{2,ij}(\vec{r}_i, \vec{r}_j), \tag{2.30}$$

where $\hat{h}_1$ are the operators that act on the coordinates of just a single particle, whereas $\hat{h}_2$ are those that act on pairs of particles.

A commonly employed simplification is the Born–Oppenheimer approximation (see Section 10.3) that is used for atomic and molecular systems. Thereby, only the electrons are treated using the quantum theory. Then $\mathcal{N} = N$, the number of electrons in the system. Moreover, as is known, the electrons are indistinguishable, a fact that we shall use later (see Section 10.7). This also means that the operators become independent of the particle indices, i. e.,

$$\hat{h}_{1,i}(\vec{r}_i) = \hat{h}_1(\vec{r}_i)$$
$$\hat{h}_{2,ij}(\vec{r}_i, \vec{r}_j) = \hat{h}_2(\vec{r}_i, \vec{r}_j). \tag{2.31}$$

An expectation value for an operator of equivalent single-electron contributions,

$$\hat{Q} = \sum_{i=1}^{N} \hat{q}(\vec{r}_i) \tag{2.32}$$

becomes then [cf. equation (2.24)]

$$\langle Q \rangle = \frac{\int \int \cdots \int \psi^*(\vec{r}_1, \vec{r}_2, \ldots, \vec{r}_N) \hat{Q} \psi(\vec{r}_1, \vec{r}_2, \ldots, \vec{r}_N) \, d\vec{r}_1 \, d\vec{r}_2 \, \ldots \, d\vec{r}_N}{\int \int \cdots \int \psi^*(\vec{r}_1, \vec{r}_2, \ldots, \vec{r}_N) Q \psi(\vec{r}_1, \vec{r}_2, \ldots, \vec{r}_N) \, d\vec{r}_1 \, d\vec{r}_2 \, \ldots \, d\vec{r}_N}$$
$$= \frac{N \int \int \cdots \int \psi^*(\vec{r}_1, \vec{r}_2, \ldots, \vec{r}_N) \hat{q}(\vec{r}_1) \psi(\vec{r}_1, \vec{r}_2, \ldots, \vec{r}_N) \, d\vec{r}_1 \, d\vec{r}_2 \, \ldots \, d\vec{r}_N}{\int \int \cdots \int \psi^*(\vec{r}_1, \vec{r}_2, \ldots, \vec{r}_N) Q \psi(\vec{r}_1, \vec{r}_2, \ldots, \vec{r}_N) \, d\vec{r}_1 \, d\vec{r}_2 \, \ldots \, d\vec{r}_N}, \qquad (2.33)$$

where we in the second identity have used that each electron gives the same contribution to the expectation value because of their indistinguishability.

A special case is the electron density, $\rho(\vec{r})$. To calculate this, we pick out a value for $\vec{r}$ and ask whether electron 1, 2, $\cdots$, N can be found at this point. This corresponds to

$$\hat{q}(\vec{r}_i) = \delta(\vec{r} - \vec{r}_i), \qquad (2.34)$$

where $\delta(\vec{r} - \vec{r}_i)$ is Dirac's δ function that is briefly described in Section 16.2.

2.6 Problems with answers

1. **Problem:** Consider a particle (mass m) in one dimension that moves in a time-independent potential $V(x)$. The ground state wave function is $\psi(x) = A \sin(kx+\phi)$ for $a \le x \le b$, and $\psi(x) = 0$ for $x \le a$ or $x \ge b$. Let the ground state energy be E_0. From this, determine A, k, ϕ, and $V(x)$.

   Answer: The time-independent Schrödinger equation for this system is

   $$-\frac{\hbar^2}{2m} \frac{d^2}{dx^2} \psi(x) + V(x)\psi(x) = E\psi(x). \qquad (2.35)$$

   For the range $a \le x \le b$, we have then

   $$\frac{\hbar^2}{2m} k^2 \psi(x) + V(x)\psi(x) = E_0 \psi(x) \qquad (2.36)$$

   since the ground state energy equals E_0. From this,

   $$V(x) = E_0 - \frac{\hbar^2 k^2}{2m}, \qquad (2.37)$$

   that actually is a constant.
   Because the wave function is to be continuous, specifically at $x = a$ and $x = b$, we must have

   $$A \sin(ka + \phi) = 0 = A \sin(kb + \phi) \qquad (2.38)$$

or

$$k = \frac{n\pi}{b-a}$$

$$\phi = m\pi - ka \tag{2.39}$$

with n and m being integers. Different values of m lead to wave functions that differ at most in the sign, so that we can set

$$m = 0 \tag{2.40}$$

without restriction.

Finally, the wave function must be normalized:

$$1 = \int_a^b \left[A \sin(kx + \phi) \right]^2 dx. \tag{2.41}$$

Using the integrals in Chapter 17 and the results of equation (2.39), and also assuming that A is real and positive,

$$1 = A^2 \int_a^b \sin^2(kx + \phi)\, dx = A^2 \int_{a+\phi/k}^{b+\phi/k} \sin^2(ky)\, dy$$

$$= A^2 \left[-\frac{1}{4k} \sin(2ky) + \frac{y}{2} \right]_{a+\phi/k}^{b+\phi/k}$$

$$= A^2 \left[-\frac{1}{4k} \sin(2kb + 2\phi) + \frac{1}{4k} \sin(2ka + 2\phi) + \frac{b-a}{2} \right]$$

$$= A^2 \left[-\frac{1}{4k} \sin(2k(b-a)) + \frac{b-a}{2} \right]$$

$$= A^2 \left[-\frac{1}{4k} \sin(2n\pi) + \frac{b-a}{2} \right] = A^2 \left[0 + \frac{b-a}{2} \right] = A^2 \frac{b-a}{2} \tag{2.42}$$

giving

$$A = \sqrt{\frac{2}{b-a}}. \tag{2.43}$$

2. **Problem:** Is $\tilde{\psi}(x,t) = A \cdot \sin(\frac{x\pi}{L} - v \cdot t)$ (A and v are constants) a solution to the time-dependent Schrödinger equation for a particle in the box $0 \le x \le L$? Justify the answer.

Answer: For a particle in the box, $V(x) = 0$ for $0 \le x \le L$, and otherwise ∞, as discussed in more detail in Chapter 4. Inside the box, the time-dependent Schrödinger equation is then

$$-\frac{\hbar^2}{2m} \frac{d^2}{dx^2} \tilde{\psi}(x,t) = i\hbar \frac{\partial}{\partial t} \tilde{\psi}(x,t). \tag{2.44}$$

When we in this insert

$$\tilde{\psi}(x,t) = A \cdot \sin\left(\frac{x\pi}{L} - v \cdot t\right), \tag{2.45}$$

we get

$$\frac{\hbar^2\pi^2}{2mL^2}A \cdot \sin\left(\frac{x\pi}{L} - v \cdot t\right) = -i\hbar A \cdot v \cos\left(\frac{x\pi}{L} - v \cdot t\right), \tag{2.46}$$

which cannot be satisfied for all $0 \le x \le L$. So the answer is: No.

3. **Problem:** The wave functions $\Psi(x,y,z,t)$ are always time-dependent. Can an expectation value of an observable $\hat{C}$ be independent of time? If "yes," then in which cases? Justify the answer.

    Answer: The time-dependent expectation value is

$$\langle C \rangle(t) = \frac{\int\int\int \Psi^*(x,y,z,t)\hat{C}\Psi(x,y,z,t)\,dx\,dy\,dz}{\int\int\int \Psi^*(x,y,z,t)\Psi(x,y,z,t)\,dx\,dy\,dz}. \tag{2.47}$$

One possibility that this is independent of time corresponds to

$$\hat{C}\Psi(x,y,z,t) = c\Psi(x,y,z,t) \tag{2.48}$$

(where c is a constant). Then

$$\langle C \rangle(t) = c. \tag{2.49}$$

Another possibility is found when $\hat{C}$ is independent of time, and Ψ is a stationary wave function,

$$\Psi(x,y,z,t) = \psi(x,y,z)\exp\left(-\frac{iE}{\hbar}t\right). \tag{2.50}$$

By inserting this into equation (2.47), we immediately find that $\langle C \rangle$ is then independent of time.

So the answer is: Yes.

2.7 Problems

1. Consider a particle (mass m) in two dimensions, which moves in a potential $V(x,y)$. The ground state wave function is $\psi(x,y) = A\sin(k_x x + \phi_x)\sin(k_y y + \phi_y)$ for $a_x \le x \le b_x$ and $a_y \le y \le b_y$, and $\psi(x,y) = 0$ for (x,y) outside this range. Furthermore, let the ground state energy be E_0. From this, determine A, k_x, ϕ_x, k_y, ϕ_y, and $V(x,y)$.

2. Consider a particle (mass m) in one dimension that moves in the potential $V(x)$. The ground state wave function is $\psi(x) = A\exp[-\alpha(x-a)^2]$. Let the ground state energy be E_0. Determine A and $V(x)$.

3. Is $\psi(x,y,t) = A \cdot \sin(\frac{x\pi}{L_x} - v_x \cdot t)\sin(\frac{y\pi}{L_y} - v_y \cdot t)$ (A, v_x and v_y are constants) a solution to the time-dependent Schrödinger equation for a particle in the box $0 \le x \le L_x$, $0 \le y \le L_y$? Justify the answer. For this system, $V(x,y) = 0$ for $0 \le x \le L_x$, $0 \le y \le L_y$, and otherwise $V(x,y) = \infty$. Such systems are described in Chapter 4 in more detail.

4. How do we go from the time-dependent Schrödinger equation to the time-independent Schrödinger equation? Explain the role of the energy.

3 Operators and quantum theory

3.1 Operators

In the previous chapter, we saw that operators play a central role in quantum theory. For this reason, we shall discuss the fundamentals of operators in general as well as their application in quantum theory, focusing on the aspects relevant to our purposes.

Functions are, in a certain way, boxes that manipulate numbers and from which numbers return. Examples of functions are

$$g_1(s) = 2s^2$$
$$g_2(x, y) = xy^2$$
$$g_3(a, b, c) = \begin{pmatrix} a + b \\ a - c \end{pmatrix}. \tag{3.1}$$

According to classical physics, all possible physico-chemical quantities can be expressed with the help of functions. This is no longer the case in quantum theory, and instead we will have to use operators.

An operator can also be viewed as a box in a certain way. But for operators, functions are inserted, and functions are returned. We will label operators by a hat. Examples of operators are

$$\hat{A}_1 f_1(x) = \frac{df_1(x)}{dx}$$
$$\hat{A}_2 f_2(x) = \cos[f_2(x)] + \pi + 4[f_2(x)]^2$$
$$\hat{A}_3 f_3(x) = \int_{-\infty}^{x} f_3(y) \, dy,$$
$$\hat{A}_4 f_4(x, y) = \frac{\partial f_4(x, y)}{\partial x} + f_4(0, y^2),$$
$$\hat{A}_5 f_5(x) = \begin{pmatrix} \cos^2[f_5(x)] \\ \exp[f_5(x) + 2] \end{pmatrix},$$
$$\hat{A}_6 f_6(x) = \begin{cases} 4x & \text{for } f_6(x) = 1 \\ 0 & \text{for } f_6(x) = 0, \end{cases}$$
$$\hat{A}_7 f_7(x, y) = f_7(y, x). \tag{3.2}$$

We will use some of these operators to discuss general properties of operators.

First, it is important that the operator and the functions on which the operator acts match. This means, e. g., that the operator $\hat{A}_1$ can only operate on those functions, which depend on exactly one variable. Furthermore, an expression like $\hat{A}_4[x^2 + yz]$ makes no sense, while $\hat{A}_4[x^2 + y]$ does. In addition, the operator $\hat{A}_6$ can operate only on those functions which can assume only the two values 0 and 1.

https://doi.org/10.1515/9783110742206-003

Second, it is possible to have more operators operating one after the other, if the intermediate results are correct, i. e., if the result which is generated from the application of the first operator belongs to the validity range of the second operator.

In general, the product of two operators is defined as

$$\hat{A}\hat{B}f = \hat{A}[\hat{B}f].\tag{3.3}$$

Then, e. g.,

$$\hat{A}_1\hat{A}_2(x^2) = \hat{A}_1[\hat{A}_2(x^2)] = \hat{A}_1[\cos(x^2) + \pi + 4x^4]$$
$$= \frac{d}{dx}[\cos(x^2) + \pi + 4x^4] = -2x\sin(x^2) + 16x^3,\tag{3.4}$$

while an expression $\hat{A}_4\hat{A}_1f(x)$ makes no sense. The expression $\hat{A}_6\hat{A}_2(x^2)$ only makes sense as long as x satisfies that $\hat{A}_2(x^2)$ is either 0 or 1.

Third, the order of the operators is very important. By replacing the order of the operators in equation (3.4), we get

$$\hat{A}_2\hat{A}_1(x^2) = \hat{A}_2[\hat{A}_1(x^2)] = \hat{A}_2[2x] = \cos(2x) + \pi + 16x^2,\tag{3.5}$$

i. e., a result that is different from that of equation (3.4).

However, it may happen that two operators $\hat{A}$ and $\hat{B}$ for all the functions f, on which they can operate, satisfy

$$\hat{A}\hat{B}f = \hat{B}\hat{A}f.\tag{3.6}$$

In this case, we say that $\hat{A}$ and $\hat{B}$ **commute**. Equation (3.6) can also be expressed with the help of the so-called commutator. The commutator is defined as

$$[\hat{A},\hat{B}] \equiv \hat{A}\hat{B} - \hat{B}\hat{A}\tag{3.7}$$

and for commutating operators we have

$$[\hat{A},\hat{B}] = 0.\tag{3.8}$$

Fourth, we mention that the operator $\hat{A}_7$ is an example of a permutation operator. Such operators are important in quantum theory. A permutation operator exchanges two arguments so that, e. g.,

$$\hat{A}_7(x^2 + y) = y^2 + x.\tag{3.9}$$

Most often, we will consider operators operating on wave functions for one or more electrons. For each electron, we have three spatial coordinates and one spin coordinate. In many cases, the position coordinates can take any value between $-\infty$ and $+\infty$, while the spin variable can only take the two values $-\frac{1}{2}$ and $+\frac{1}{2}$.

Finally, we mention again that we have already introduced the Hamilton operator

$$\hat{H} = -\frac{\hbar^2}{2m}\nabla^2 + V(\vec{r},t)\tag{3.10}$$

as the quantum-mechanical operator for energy.

3.2 Expectation value

In the previous chapter, we have already seen that in quantum theory we can rarely give exact values for measured variables, but only for the expectation values, which corresponds to the average value which would be determined by repeated, identical measurements.

To calculate these expectation values, we need integrals over all possible values of the variables or coordinates of the system. We will use the simpler notation for these:

$$\int \ldots d\vec{x}. \tag{3.11}$$

$\vec{x}$ is a combined variable containing all variables. Thus, for a single electron, $\vec{x}$ represents all three position coordinates and the single spin coordinate. The integral sign in equation (3.11) will then be understood as the integration over the three position coordinates and a summation over the two values of the spin coordinate.

For a two-electron system $\vec{x}$ then represents the three-position coordinates for the first electron, the three position coordinates for the second electron, a spin coordinate for the first electron, and finally a spin coordinate for the second electron. In some cases, it is useful to specify this explicitly, and then we can write the integral in equation (3.11) by the two equivalent expressions, e. g.,

$$\int \ldots d\vec{x} = \int \int \ldots d\vec{x}_1 \, d\vec{x}_2, \tag{3.12}$$

where the indices 1 and 2 distinguish the two electrons.

Integrals of the type

$$\int f_1^*(\vec{x}) f_2(\vec{x}) \, d\vec{x} \equiv \langle f_1 | f_2 \rangle \tag{3.13}$$

are called overlap matrix elements, where we have used the so-called bra-ket notation of Dirac (after Paul Andre Maurice Dirac) for these. If the integral in equation (3.13) is equal to 0, it is said that the two functions f_1 and f_2 are **orthogonal**.

We will also use integrals of the type

$$\int f_1^*(\vec{x}) \hat{A} f_2(\vec{x}) \, d\vec{x} = \int f_1^*(\vec{x}) [\hat{A} f_2(\vec{x})] \, d\vec{x} \equiv \langle f_1 | \hat{A} | f_2 \rangle. \tag{3.14}$$

These are called matrix elements for the operator $\hat{A}$, and where we again have used the bra-ket notation of Dirac for these. For us, the bra-ket notation is a convenient, short-hand notation for integrals of the types of equations (3.13) and (3.14).

The expression in equation (3.14) is evaluated as follows. First, we let the operator $\hat{A}$ act on the function f_2. Then the result of this is subsequently multiplied by the function f_1^*, and finally the product is integrated over the complete space of all variables.

The operators may possess two further properties. First, an operator $\hat{A}$ is hermitian if and only if

$$\langle f_1|\hat{A}|f_2\rangle = \langle f_2|\hat{A}|f_1\rangle^* \tag{3.15}$$

is valid for **all** pairs of functions f_1 and f_2. For any (even non-hermitian) operator $\hat{A}$, the hermitian conjugate operator $\hat{A}^\dagger$ can be introduced so that

$$\langle f_1|\hat{A}^\dagger|f_2\rangle = \langle f_2|\hat{A}|f_1\rangle^*, \tag{3.16}$$

is valid for all functions f_1, f_2. Then for a hermitian operator

$$\hat{A}^\dagger = \hat{A}. \tag{3.17}$$

Second, $\hat{A}$ is a linear operator if for all constants c_1 and c_2 and any two functions f_1 and f_2

$$\hat{A}(c_1 f_1 + c_2 f_2) = c_1 \hat{A} f_1 + c_2 \hat{A} f_2 \tag{3.18}$$

is satisfied.

We emphasize that almost all physically and chemically relevant operators are hermitian, and most of those are linear, too. The Hamilton operator is both.

3.3 An example

We illustrate the concepts of the last chapter through a simple example.

We consider such functions of one variable $f(x)$, which vanish outside the range $a \le x \le b$,

$$f(x) = 0 \quad \text{for } x \le a \quad \text{or} \quad x \ge b. \tag{3.19}$$

We consider the operator $\hat{A}_1$ of equation (3.2). Then

$$\hat{A}_1(c_1 f_1 + c_2 f_2) = \frac{d}{dx}(c_1 f_1 + c_2 f_2) = c_1 \frac{d}{dx} f_1 + c_2 \frac{d}{dx} f_2 = c_1 \hat{A}_1 f_1 + c_2 \hat{A}_1 f_2, \tag{3.20}$$

i. e., the operator $\hat{A}_1$ is obviously linear.

On the other hand,

$$\langle f_1|\hat{A}_1|f_2\rangle = \int_a^b f_1^*(x) \frac{d}{dx} f_2(x)\, dx = [f_1^*(x) f_2(x)]_a^b - \int_a^b \left[\frac{d}{dx} f_1^*(x)\right] f_2(x)\, dx$$

$$= -\int_a^b f_2(x) \frac{d}{dx} f_1^*(x)\, dx = -\left[\int_a^b f_2^*(x) \frac{d}{dx} f_1(x)\, dx\right]^*$$

$$= -\langle f_2|\hat{A}_1|f_1\rangle^*. \tag{3.21}$$

Due to the minus sign in the last equation, the operator $\hat{A}_1$ is not hermitian. In the derivation of equation (3.21), we have in the second identity used the standard expression for an integral of a product of one function and the derivative of another,

$$\int f \cdot g' \, dx = f \cdot g - \int g \cdot f' \, dx, \tag{3.22}$$

and in the third identity used equation (3.19). Equation (3.21) also shows that

$$\hat{A}_1^\dagger = -\hat{A}_1 = -\frac{d}{dx}. \tag{3.23}$$

3.4 Eigenvalues and eigenfunctions

Solving the time-independent Schrödinger equation

$$\hat{H}\Psi = E\Psi \tag{3.24}$$

involves determining an eigenvalue E and an eigenfunction Ψ. An eigenfunction f to an operator $\hat{A}$ is a function that obeys that $\hat{A}f = af$ with a being a constant, i. e., the so-called eigenvalue. Equation (3.24) is accordingly a special case of the general problem of finding eigenvalues and eigenfunctions for a given (hermitian and linear) operator $\hat{A}$. We shall here, therefore, consider the general problem of solving

$$\hat{A}f_i = a_i f_i \tag{3.25}$$

where a_i is an eigenvalue and f_i the corresponding eigenfunction. i is used in labeling different solutions to equation (3.25).

Let us consider two examples. Here, the details of the examples are not important, but we can through those explain a few aspects. In the first example, if $\hat{A}$ is the Hamilton operator for a harmonic oscillator in one dimension, the potential equals $V(x) = \frac{1}{2}kx^2$, and the system is discussed in more detail in Chapter 6. Here, we only need to know that the eigenvalues are equal to $(n + \frac{1}{2})\hbar\omega$ with n equal to a nonnegative integer. Further, ω is a constant which depends on the mass and force constant (k) of the oscillator. As we shall see in Chapter 6, the eigenfunctions become Gauss functions multiplied by so-called Hermite polynomials. In this case, the eigenvalues and eigenfunctions are labeled by the integer n, and they are numerable.

For the electron of a hydrogen atom, we have eigenfunctions characterized by the three integers n, l, m, describing the position-space dependence of the wave function, plus a single half-integer, m_s, describing the spin-dependence. The hydrogen atom is discussed in detail in Chapter 9, but most of the details are irrelevant to the discussion here. These eigenfunctions are all bounded, which means that they decay exponentially far away from the nucleus, and they are also numerable, so they can be counted. Their eigenvalues are negative as a consequence of the functions being bounded. There are, however, also a set of so-called continuum states with positive energies. These states are not bounded and the eigenvalues are not discrete but form a continuum. Also these are characterized by three numbers describing the position-space dependence and one that describes the spin-dependence.

In general, therefore, it shall merely be stressed that there exist many (often infinitely many) solutions to equation (3.25), and that the eigenvalues may be either discrete or form a (finite or infinite) continuum.

3.5 Hermitian operators

Since all operators of our interest are hermitian and linear, we shall here discuss the properties of the eigenfunctions and values for such operators.

We consider two eigenfunctions for a hermitian operator

$$\hat{A}f_n = a_n f_n$$
$$\hat{A}f_m = a_m f_m, \tag{3.26}$$

and consider

$$\langle f_m|\hat{A}|f_n\rangle = \langle f_m|\hat{A}f_n\rangle = \langle f_m|a_n f_n\rangle = a_n\langle f_m|f_n\rangle. \tag{3.27}$$

Simultaneously,

$$\langle f_m|\hat{A}|f_n\rangle = \langle f_n|\hat{A}|f_m\rangle^* = \langle f_n|\hat{A}f_m\rangle^*$$
$$= \langle f_n|a_m f_m\rangle^* = a_m^*\langle f_n|f_m\rangle^* = a_m^*\langle f_m|f_n\rangle. \tag{3.28}$$

Combining equations (3.27) and (3.28) yields now

$$(a_n - a_m^*)\langle f_m|f_n\rangle = 0. \tag{3.29}$$

For $n = m$, we obtain accordingly that a_n is real. This result does make a lot of sense: Any physical observable is described quantum-theoretically by a hermitian operator, and the fact that the eigenvalues of these are real means that measuring such an observable leads to a real (in contrast to complex) result (see also Section 3.9).

Equation (3.29) shows subsequently (since the eigenvalues are real) that eigenfunctions belonging to different eigenvalues are **orthonormal**, i. e.,

$$\langle f_m|f_n\rangle = 0 \quad \text{for } a_m \neq a_n. \tag{3.30}$$

There may exist more different eigenfunctions belonging to the same eigenvalue. A well-known example is the electronic eigenfunctions for the hydrogen atom. Here, e. g., the six 2p and two 2s functions (if we take the spin into account) all have the same energy. From those, we can produce eight arbitrary independent linear combinations that then also have the same energy. This is possible because we assume that the operator is linear. Actually, it can be shown that it is always possible to form a new set of any finite set of functions so that the new functions are orthogonal. There are different ways of obtaining a such set of orthogonal functions, including the Schmidt orthogonalization procedure. Here, however, it is not important to discuss how it is done, but only to emphasize that it can be done.

Subsequently, through this procedure we may accordingly replace equation (3.30) by

$$\langle f_m | f_n \rangle = 0 \quad \text{for } m \neq n \tag{3.31}$$

Finally, if any function f_n obeys equation (3.26), so does a constant times f_n this eigenvalue function (again because the operator is assumed being linear),

$$\hat{A}(c_n f_n) = a_n(c_n f_n). \tag{3.32}$$

Since, moreover, $\langle f_n | f_n \rangle$ is a real and positive number, we can choose

$$c_n = \left[\langle f_n | f_n \rangle \right]^{-1/2}. \tag{3.33}$$

This gives us a new set of (scaled) functions that satisfy

$$\langle f_m | f_n \rangle = \delta_{n,m}. \tag{3.34}$$

$\delta_{i,j}$ is the so-called Kronecker δ,

$$\delta_{i,j} = \begin{cases} 1 & i = j \\ 0 & i \neq j. \end{cases} \tag{3.35}$$

Thus, the "new" functions are normalized and also orthogonal. In total, the eigenfunctions are said to be **orthonormal**.

The fact that the eigenvalues a_n are real means that they can be placed along a real axis. Furthermore, for almost all cases of our interest there is always a smallest eigenvalue, so that we can sort the eigenvalues according to increasing size. Denoting the smallest one a_0, we have then

$$a_0 \leq a_1 \leq a_2 \leq \cdots \leq a_{n-1} \leq a_n \leq a_{n+1} \leq \cdots. \tag{3.36}$$

Without proof, we shall give one further important property of the eigenfunctions: They form a complete set. This means that any function g can be expanded in the set $\{f_n\}$,

$$g = \sum_n c_n f_n, \tag{3.37}$$

where the c_n are some constants. Actually, multiplying equation (3.37) by f_m^* and integrating over all variables, we obtain, using the bra-ket notation,

$$\langle f_m | g \rangle = \sum_n c_n \langle f_m | f_n \rangle. \qquad (3.38)$$

Finally, the orthonormality of the functions f_n [Gl. (3.34)] gives

$$c_m = \langle f_m | g \rangle, \qquad (3.39)$$

valid for any m.

It is often said that the functions $\{f_n\}$ span a so-called Hilbert space. For a Hilbert space, one defines an inner product for two elements of the Hilbert space, g_1 and g_2, which in our case is $\langle g_1 | g_2 \rangle$. Furthermore, any other function can be expanded according to the functions $\{f_n\}$: The functions $\{f_n\}$ form a basis for the Hilbert space.

This is hardly different from what we know, e. g., for position vectors in the three-dimensional space. Three arbitrary vectors, $\vec{a}$, $\vec{b}$, and $\vec{c}$, which are linearly independent, form a basis so that any other vector can be written as

$$\vec{r} = x\vec{a} + y\vec{b} + z\vec{c}. \qquad (3.40)$$

Here, $\vec{a}$, $\vec{b}$, and $\vec{c}$ correspond to the functions $\{f_n\}$, and x, y, and z correspond to the coefficients $\{c_n\}$. $\vec{a}$, $\vec{b}$, and $\vec{c}$ do not have to be, but can be orthogonal and normalized. As an inner product, we can in this case use the "normal' scalar product $\vec{r}_1 \cdot \vec{r}_2$.

3.6 Commuting operators

That two (or more) operators commute can be used with advantage. It will be useful to recognize that the concept of commuting operators often is used without explicitly being discussed. One example is the case where one of the two commuting operators is the Hamilton operator of the electrons of a given system and the other is a symmetry operator. The symmetry operator is used in attributing any eigenfunction of the Hamilton operator not only an energy but also a label describing how the function transforms according to the symmetry operator.

A particularly simple example is that the symmetry operator is a mirror operation or an inversion. In that case, the labels g and u are often used in describing whether the function is symmetric (g=gerade, even) or antisymmetric (u=ungerade, odd) with respect to this symmetry operation. Alternatively formulated, g and u quantifies the eigenvalue o of

$$\hat{O}\Psi = o\Psi. \qquad (3.41)$$

Here, $\hat{O}$ is the symmetry operator, and in our case where $\hat{O}$ is a mirror operation or an inversion, o can only take the values +1 and −1. In other cases, where $\hat{O}$ is the symmetry operator, the value o will be different, although we for the symmetry operations always have

$$|o| = 1. \tag{3.42}$$

The fact that the eigenvalues may be nonreal, follows from the fact that the symmetry operators are often non-hermitian.

Another example where commuting operators are used is that of hydrogen-atom-like eigenfunctions, which we will discuss in detail later. As we shall see, the quantum number n describes the Hamilton operator eigenvalue, whereas the quantum numbers l and m describe those for the operators of the square of the angular momentum and of its z component, respectively. These three operators commute, which is the reason for why we can construct functions that simultaneously are eigenfunctions for all three operators, as we now shall demonstrate.

We shall discuss the case that two operators commute. The generalization to several commuting operators will not be discussed. We consider two hermitian, linear operators $\hat{A}$ and $\hat{B}$, that obey

$$[\hat{A}, \hat{B}] = 0. \tag{3.43}$$

The important point is that it is possible to obtain a complete set of functions that simultaneously are eigenfunctions to both operators,

$$\hat{A}f_n = a_n f_n$$
$$\hat{B}f_n = b_n f_n. \tag{3.44}$$

We will prove this only in the case that the eigenfunctions are nondegenerate. Then we apply the operator $\hat{B}$ on the first identity in equation (3.44) and use that $\hat{B}$ is linear:

$$\hat{B}\hat{A}f_n = a_n \hat{B}f_n, \tag{3.45}$$

but because of equation (3.43) we also have

$$\hat{B}\hat{A}f_n = \hat{A}\hat{B}f_n, \tag{3.46}$$

so what

$$\hat{A}(\hat{B}f_n) = a_n(\hat{B}f_n), \tag{3.47}$$

which shows that $(\hat{B}f_n)$ is proportional to f_n, because $\hat{A}$ is linear, and because there is only one eigenfunction to $\hat{A}$ for the eigenvalue a_n to $\hat{A}$:

$$(\hat{B}f_n) = \hat{B}f_n = b_n f_n. \tag{3.48}$$

This is what we wanted to prove. For the degenerate case, the proof will be more complex, but still valid.

3.7 The postulates of quantum theory

Quantum theory is built on three (or four) postulates, which will be briefly discussed here. We begin by formulating the postulates and subsequently discuss them:

1. For a system with n particles, the state of the system can be completely described by a wave function

$$\psi(\vec{r}_1, \vec{r}_2, \ldots, \vec{r}_n, t) \tag{3.49}$$

Here, $\vec{r}_i$ are the position coordinates of the ith particle and t is the time.

2. All experimental observables can be expressed using operators. These are related to position and momentum operators of the particles. For these,

$$[\hat{q}_k, \hat{q}_l] = 0$$
$$[\hat{p}_k, \hat{p}_l] = 0$$
$$[\hat{q}_k, \hat{p}_l] = i\hbar\delta_{k,l}. \tag{3.50}$$

$\hat{q}_k$ is the operator for the kth position coordinate (i. e., the x, y, or z coordinate of one of the particles of the system), and $\hat{p}_k$ is the associated momentum operator.

3. When repeating a measurement of an experimental observable A for identical systems in the same state, one finds an expectation value on average, which can also be calculated,

$$\langle A \rangle = \frac{\langle \psi | \hat{A} | \psi \rangle}{\langle \psi | \psi \rangle}. \tag{3.51}$$

Here, $\hat{A}$ is the operator for the experimental observable and ψ is the wave function for the state of the system.

4. The wave function is interpreted in such a way that the quantity

$$\psi^*(\vec{r}_1, \vec{r}_2, \ldots, \vec{r}_n, t)\psi(\vec{r}_1, \vec{r}_2, \ldots, \vec{r}_n, t)d\vec{r}_1 d\vec{r}_2 \ldots d\vec{r}_n$$
$$= |\psi(\vec{r}_1, \vec{r}_2, \ldots, \vec{r}_n, t)|^2 d\vec{r}_1 d\vec{r}_2 \ldots d\vec{r}_n \tag{3.52}$$

is the probability that, at time t, particle 1 is in the volume element $d\vec{r}_1$ around $\vec{r}_1$, particle 2 is in the volume element $d\vec{r}_2$ around $\vec{r}_2$, and so on, as long as $\langle \psi | \psi \rangle = 1$.

For the individual points, the following comments should be made:

1. This formulation corresponds to the so-called position representation (see below). Furthermore, we have not considered any spin variables here, although they should occur as well. Finally, the wave function can be more or less arbitrary. Above all, it does not have to be the solution to the time-dependent or time-independent Schrödinger equation or another eigenvalue equation.

2. This postulate does not determine what the operators look like. For example, it is possible to use the operators for the x coordinates of the position and the momentum of the ith particle according to

$$\hat{x}_i = x_i$$
$$\hat{p}_{xi} = \frac{\hbar}{i} \frac{\partial}{\partial x_i}.$$

(3.53)

Compared to the classical physics, the position coordinate is not changed, which is why we call this the position representation. Another possibility, which satisfies equation (3.50), is

$$\hat{x}_i = -\frac{\hbar}{i} \frac{\partial}{\partial p_{xi}}$$
$$\hat{p}_{xi} = p_{xi}$$

(3.54)

This corresponds to the momentum representation. Which one ultimately is used, is unimportant in principle and, therefore, you always use what is most convenient for the problem at hand. Also other representations are possible. In this manuscript, we will almost exclusively use the position representation.

3. The fact that one has to determine the operator for the experimental observable is often referred to as Schrödinger's method. But exactly how to construct $\hat{A}$ is not always easy and is often ambiguous. For example, consider a particle in one dimension, and imagine that we measure the product of position and momentum coordinate, $x \cdot p$. Then there are several possible operators. This includes

$$\widehat{xp} = \begin{cases} \hat{x} \cdot \hat{p} \\ \hat{p} \cdot \hat{x} \\ \frac{1}{2}(\hat{x} \cdot \hat{p} + \hat{p} \cdot \hat{x}) \\ \dots \end{cases}$$

(3.55)

For a classic quantity like $x^2 p$, there will be even more options:

$$\widehat{x^2 p} = \begin{cases} \hat{x}^2 \cdot \hat{p} \\ \hat{p} \cdot \hat{x}^2 \\ \frac{1}{2}(\hat{x}^2 \cdot \hat{p} + \hat{p} \cdot \hat{x}^2) \\ \frac{1}{3}(\hat{x}^2 \cdot \hat{p} + \hat{x} \cdot \hat{p} \cdot \hat{x} + \hat{p} \cdot \hat{x}^2) \\ \frac{1}{4}(\hat{x}^2 \cdot \hat{p} + 2\hat{x} \cdot \hat{p} \cdot \hat{x} + \hat{p} \cdot \hat{x}^2) \\ \dots \end{cases}$$

(3.56)

Because the commutator

$$[\hat{x}, \hat{p}] = \hat{x}\hat{p} - \hat{p}\hat{x} \neq 0,$$

(3.57)

the different expressions in equation (3.55) and in equation (3.56) are not identical. We will see an example later (Section 8.6) that it can be important what one chooses.

The wave function ψ in equation (3.51) does not have to be one of the eigenfunctions of $\hat{A}$. In general, ψ can be expanded in the eigenfunctions to $\hat{A}$,

$$\psi = \sum_i c_i f_i \tag{3.58}$$

where f_i is the ith normalized eigenfunction to $\hat{A}$,

$$\hat{A} f_i = a_i f_i$$
$$\langle f_i | f_i \rangle = 1. \tag{3.59}$$

This is a consequence of the fact that the eigenfunctions to $\hat{A}$ form a complete set of functions.

If ψ is normalized,

$$\langle \psi | \psi \rangle = 1, \tag{3.60}$$

then the expectation value becomes

$$\langle A \rangle = \sum_i |c_i|^2 a_i. \tag{3.61}$$

If ψ is one of the eigenfunctions to $\hat{A}$, f_k, one finds

$$\langle A \rangle = a_k, \tag{3.62}$$

so you measure exactly the corresponding eigenvalue.

4. This postulate is not really a true postulate, but corresponds to Born's interpretation of the wave function, which we discussed in Section 2.3. Furthermore, we have assumed that the wave function is normalized, i. e., equation (3.60) is satisfied.

3.8 Position and momentum representation

Here, we will explain the relationship between the position and the momentum representations mainly through a simple example. The results are generally valid (with appropriate modifications).

We consider a particle that moves in one dimension in the potential:

$$V(x) = c \cdot x^4 \tag{3.63}$$

The Hamilton operator is the operator for the energy, i. e., for

$$H = \frac{p^2}{2m} + V(x).$$ (3.64)

The corresponding Hamilton operator can be constructed both in the position representation and in the momentum representation. In the first case, we use equation (3.53) and get

$$\hat{H} = -\frac{\hbar^2}{2m}\frac{d^2}{dx^2} + c \cdot x^4.$$ (3.65)

In the second case, we use equation (3.54) and then get

$$\hat{H} = \frac{p^2}{2m} + c\hbar^4 \frac{d^4}{dp^4}.$$ (3.66)

In both cases, we can set up the time independent Schrödinger equation. In the position representation, this is

$$\left[-\frac{\hbar^2}{2m}\frac{d^2}{dx^2} + c \cdot x^4\right]\psi(x) = E\psi(x),$$ (3.67)

while it in the momentum representation becomes

$$\left[\frac{p^2}{2m} + c\hbar^4 \frac{d^4}{dp^4}\right]\phi(p) = E\phi(p).$$ (3.68)

The two equations are clearly different, but they still must have (and do have) the same energy eigenvalues, E. Furthermore, we know that position and momentum coordinates are not independent of each other, so it may be possible to use $\psi(x)$ to obtain all information about the momentum properties of the particle and, equivalently with the help of $\phi(p)$ we can also obtain all the information about the spatial behavior of the particle.

In fact, it can be shown that the two wave functions $\psi(x)$ and $\phi(p)$ are linked together

$$\phi(p) = \frac{1}{\sqrt{2\pi\hbar}}\int \psi(x)e^{-\frac{ipx}{\hbar}}\,dx$$

$$\psi(x) = \frac{1}{\sqrt{2\pi\hbar}}\int \phi(p)e^{\frac{ipx}{\hbar}}\,dp$$ (3.69)

i. e., through a Fourier transformation.

This also means that it is not necessary to solve both equations (3.67) and (3.68) to get the complete information about the behavior of the particle in both spaces.

Finally, we emphasize that equation (3.69) is generally valid for particles moving in one dimension. For higher dimensions, the equation must be modified accordingly.

3.9 Observables

In Section 2.4, we mentioned the uncertainty relations of Heisenberg,

$$\Delta x \cdot \Delta p \geq \frac{\hbar}{2}. \tag{3.70}$$

Here, x and p are the associated position and momentum coordinates. This relation can be determined using the commutator relations equation (3.50). In the general case for two observables, A and B,

$$\Delta A \cdot \Delta B \geq \frac{1}{2}|\langle [\hat{A}, \hat{B}] \rangle|. \tag{3.71}$$

$\langle [\hat{A}, \hat{B}] \rangle$ is the expectation value of the commutator between the operators $\hat{A}$ and $\hat{B}$. For $\hat{A} = \hat{x}$ and $\hat{B} = \hat{p}$, $[\hat{A}, \hat{B}] = i\hbar$, and equation (3.70) follows immediately.

Similarly, it can be shown that

$$\Delta E \cdot \Delta t \geq \frac{\hbar}{2}, \tag{3.72}$$

whereby an estimate of the uncertainties in energy and time is obtained. This result has important consequences for spectroscopy. We consider a system (e. g., a molecule) that is in an excited state. This state has a very short life time τ and will accordingly fall back to the ground state within a time of the order of τ and thereby emit the energy difference E in the form of light. In an emission spectrum, we would be able to recognize this by a sharp line at the energy E. But because the excited state is very short-lived, this line will be broadened and have a width ΔE, that obeys

$$\Delta E \cdot \tau \geq \frac{\hbar}{2}. \tag{3.73}$$

Otherwise, the uncertainty relation (3.72) of Heisenberg would be violated: We would have known very well what energy the system had at a very specific time. The broadening, ΔE, is called lifetime broadening or natural broadening. We emphasize that this broadening is fundamental and cannot be removed through any experimental design.

Another result that we will not derive is

$$\frac{d\langle A \rangle}{dt} = \frac{i}{\hbar} \langle [\hat{H}, \hat{A}] \rangle, \tag{3.74}$$

that allows for the calculation of the temporal evolution of the average of an observable.

From equation (3.74), one can get

$$\frac{d\langle x \rangle}{dt} = \frac{\langle p_x \rangle}{m}. \tag{3.75}$$

That is, as an average, classical mechanics is also valid in quantum theory.

If we measure some observable and find that the uncertainty of that observable vanishes, then the system must be in an eigenstate of the associated operator. We will prove this, also because we thereby can illustrate the use of different quantum theoretical concepts.

The operator whose expectation value we measure is assumed to be hermitian and linear and is denoted by $\hat{A}$. We have

$$\hat{A}\psi_i = a_i\psi_i. \tag{3.76}$$

Without limitation, we can assume that the eigenfunctions of this operator satisfy

$$\langle\psi_i|\psi_j\rangle = \delta_{i,j}. \tag{3.77}$$

In addition, we know that, since $\hat{A}$ is hermitian, the eigenvalues $\{a_i\}$ are all real.

The state of the system we are studying is described by the wave function ϕ. Because the eigenfunctions in equation (3.76) form a complete set of functions, ϕ can then be expanded in those

$$\phi = \sum_i c_i\psi_i. \tag{3.78}$$

Moreover, ϕ is assumed to be normalized, which means that

$$1 = \langle\phi|\phi\rangle = \left\langle \sum_i c_i\psi_i \middle| \sum_j c_j\psi_j \right\rangle = \sum_{i,j} c_i^* c_j \langle\psi_i|\psi_j\rangle$$
$$= \sum_{i,j} c_i^* c_j \delta_{i,j} = \sum_i |c_i|^2. \tag{3.79}$$

We also have (where we use that $\hat{A}$ is linear)

$$\langle\phi|\hat{A}|\phi\rangle = \left\langle \sum_i c_i\psi_i \middle| \hat{A} \middle| \sum_j c_j\psi_j \right\rangle = \left\langle \sum_i c_i\psi_i \middle| \sum_j c_j\hat{A}\psi_j \right\rangle = \left\langle \sum_i c_i\psi_i \middle| \sum_j c_j a_j\psi_j \right\rangle$$
$$= \sum_{i,j} c_i^* c_j a_j \langle\psi_i|\psi_j\rangle = \sum_{i,j} c_i^* c_j a_j \delta_{i,j} = \sum_i |c_i|^2 a_i \tag{3.80}$$

and

$$\langle\phi|\hat{A}^2|\phi\rangle = \left\langle \sum_i c_i\psi_i \middle| \hat{A}^2 \middle| \sum_j c_j\psi_j \right\rangle = \left\langle \sum_i c_i\psi_i \middle| \hat{A} \middle| \sum_j c_j\hat{A}\psi_j \right\rangle$$
$$= \left\langle \sum_i c_i\psi_i \middle| \hat{A} \middle| \sum_j c_j a_j\psi_j \right\rangle = \left\langle \sum_i c_i\psi_i \middle| \sum_j c_j a_j\hat{A}\psi_j \right\rangle$$
$$= \left\langle \sum_i c_i\psi_i \middle| \sum_j c_j a_j^2\psi_j \right\rangle = \sum_{i,j} c_i^* c_j a_j^2 \langle\psi_i|\psi_j\rangle$$
$$= \sum_{i,j} c_i^* c_j a_j^2 \delta_{i,j} = \sum_i |c_i|^2 a_i^2. \tag{3.81}$$

If the uncertainty vanishes, we have

$$0 = \langle\phi|\hat{A}^2|\phi\rangle - \langle\phi|\hat{A}|\phi\rangle^2 = \sum_i |c_i|^2 a_i^2 - \left[\sum_i |c_i|^2 a_i\right]^2$$

$$= 1 \cdot \sum_i |c_i|^2 a_i^2 - \sum_j |c_j|^2 a_j \sum_i |c_i|^2 a_i$$

$$= \sum_j |c_j|^2 \sum_i |c_i|^2 a_i^2 - \sum_j |c_j|^2 a_j \sum_i |c_i|^2 a_i$$

$$= \sum_{i,j} |c_j|^2 |c_i|^2 (a_i^2 - a_i a_j). \tag{3.82}$$

We now use that the terms for $i = j$ are equal to zero, and that we can then restrict ourselves to terms with $j \geq i$ (if we simultaneously modify the summand):

$$0 = \langle\phi|\hat{A}^2|\phi\rangle - \langle\phi|\hat{A}|\phi\rangle^2 = \sum_{i,j} |c_j|^2 |c_i|^2 (a_i^2 - a_i a_j)$$

$$= \sum_{j\geq i} |c_j|^2 |c_i|^2 (a_i^2 - a_i a_j + a_j^2 - a_j a_i) = \sum_{j\geq i} |c_j|^2 |c_i|^2 (a_i - a_j)^2. \tag{3.83}$$

Because the eigenvalues are real, equation (3.83) can only then be fulfilled if all the coefficients $\{c_i\}$ that are nonzero belong to the same eigenvalue. Accordingly, we have shown that the uncertainty disappears only when the state of the system is an eigenstate of the corresponding operator.

3.10 Problems with answers

1. **Problem:** For which functions does the set of functions $f_n(x) = \sin(\frac{n x \pi}{L})$, $n = 1, 2, 3, \ldots$ and $0 \leq x \leq L$ form a complete set of functions? Justify the answer.
   Answer: The functions are defined in the interval $0 \leq x \leq L$ and are all equal to 0 for $x = 0$ and for $x = L$. Furthermore, $\{f_n\}$ forms the complete set of eigenfunctions for $\hat{H} = -\frac{\hbar^2}{2m}\frac{d^2}{dx^2}$. Therefore, they form a complete set for functions defined in the interval $0 \leq x \leq L$ that equal 0 for $x = 0$ and $x = L$.

2. **Problem:** Consider a particle (mass m) in one dimension that moves in a potential $V(x)$. Let the ground state eigenfunction be $\psi(x) = A\exp[-\alpha(x - a)^2]$. Determine the expectation value for x.
   Answer: Because the function $\psi(x)$ is real and symmetric about $x = a$, the expected value becomes equal to

$$\langle x \rangle = \frac{\int \psi(x) x \psi(x)\, dx}{\int \psi(x)\psi(x)\, dx} = \frac{\int \psi(x)(x-a)\psi(x)\, dx}{\int \psi(x)\psi(x)\, dx} + \frac{\int \psi(x)a\psi(x)\, dx}{\int \psi(x)\psi(x)\, dx} = 0 + a = a. \tag{3.84}$$

Here, we have used that $x = (x - a) + a$ and subsequently that the first term leads to an integrand in the denominator that is antisymmetric about $x = a$, whereas the first term gives a symmetric integrand. Thus, the first term vanishes and only the second term needs to be evaluated.

Alternatively, one may directly use that $\psi(x)$ is symmetric about $x = a$, so that the expectation value has to equal a.

3. **Problem:** Which of the following functions are eigenfunctions for the operator d/dx:
    (a) e^{-ikx},
    (b) $\cos kx$,
    (c) k,
    (d) kx and
    (e) e^{-kx^2}?

    k denotes a constant. Give the eigenvalue in each case.

    Answer:
    (a) $\frac{d}{dx}[e^{-ikx}] = -ik \cdot e^{-ikx}$. Is eigenfunction with eigenvalue $-ik$.
    (b) $\frac{d}{dx}[\cos(kx)] = -k\sin(kx)$. Is only for $k = 0$ an eigenfunction and then the eigenvalue is 0.
    (c) $\frac{d}{dx}[k] = 0$. Is an eigenfunction with the eigenvalue 0.
    (d) $\frac{d}{dx}[kx] = k$. Is only for $k = 0$ an eigenfunction and then every value of the eigenvalue is possible. But this case is not relevant to us: we will not consider wave functions that are identically zero everywhere.
    (e) $\frac{d}{dx}[e^{-kx^2}] = -2kxe^{-kx^2}$. Is only for $k = 0$ an eigenfunction and then the eigenvalue is 0.

4. **Problem:** For an operator $\hat{A}$ and a function ψ, we have $\hat{A}\psi = a \cdot \psi$ with a equal to a constant. Determine $\Delta A = [\int \psi^*(\vec{x})\hat{A}^2\psi(\vec{x})\,d\vec{x} - (\int \psi^*(\vec{x})\hat{A}\psi(\vec{x})\,d\vec{x})^2]^{1/2}$. Consider the two cases: (1) ψ is normalized and (2) ψ is not normalized.

    Answer: In bra-ket notation, we have:

    $$\Delta A = [\langle\psi|\hat{A}^2|\psi\rangle - (\langle\psi|\hat{A}|\psi\rangle)^2]^{1/2} = [a^2\langle\psi|\psi\rangle - (a\langle\psi|\psi\rangle)^2]^{1/2}$$
    $$= a[\langle\psi|\psi\rangle - (\langle\psi|\psi\rangle)^2]^{1/2}. \tag{3.85}$$

    For $\langle\psi|\psi\rangle = 1$ is $\Delta A = 0$, while for other values of $\langle\psi|\psi\rangle$ ΔA any other value (even imaginary ones) can be assumed. Bearing in mind that ΔA as defined in equation (3.85) makes little sense for nonnormalized wave functions.

5. **Problem:** Consider a particle (mass m) moving in the one-dimensional potential $\hat{V} = V(x)$. Let the kinetic energy operator of the particle be $\hat{T}$. Determine the commutator $[\hat{V}, \hat{T}]$.

    Answer: We have

    $$\hat{T} = -\frac{\hbar^2}{2m}\frac{d^2}{dx^2}. \tag{3.86}$$

From this,

$$[\hat{V}, \hat{T}]f(x) = V(x)\frac{-\hbar^2}{2m}\frac{d^2}{dx^2}f(x) - \frac{-\hbar^2}{2m}\frac{d^2}{dx^2}[V(x)f(x)]$$

$$= V(x)\frac{-\hbar^2}{2m}\frac{d^2}{dx^2}f(x) - \frac{-\hbar^2}{2m}\frac{d}{dx}\left[\frac{dV(x)}{dx}f(x) + V(x)\frac{df(x)}{dx}\right]$$

$$= V(x)\frac{-\hbar^2}{2m}\frac{d^2f(x)}{dx^2}$$

$$- \frac{-\hbar^2}{2m}\left[\frac{d^2V(x)}{dx^2}f(x) + 2\frac{dV(x)}{dx}\frac{df(x)}{dx} + V(x)\frac{d^2f(x)}{dx^2}\right]$$

$$= \frac{\hbar^2}{2m}\left[\frac{d^2V(x)}{dx^2}f(x) + 2\frac{dV(x)}{dx}\frac{df(x)}{dx}\right]$$

$$= \frac{\hbar^2}{2m}\left[\frac{d^2V(x)}{dx^2} + 2\frac{dV(x)}{dx}\frac{d}{dx}\right]f(x). \tag{3.87}$$

Thus, the commutator becomes

$$[\hat{V}, \hat{T}] = \frac{\hbar^2}{2m}\left[\frac{d^2V(x)}{dx^2} + 2\frac{dV(x)}{dx}\frac{d}{dx}\right]. \tag{3.88}$$

6. **Problem:** Rewrite $\int[c_1f_1(x) + c_2f_2(x)]^*\hat{A}[c_1f_1(x) - c_2f_2(x)]\,dx$. $\hat{A}$ is a linear and hermitian operator, c_1 and c_2 are constants, and f_1 and f_2 are different orthonormal eigenfunctions to $\hat{A}$ for the eigenvalues a_1 and a_2, respectively.
   Answer: Using the bra-ket notation, we have:

$$\langle c_1f_1 + c_2f_2|\hat{A}|c_1f_1 - c_2f_2\rangle$$

$$= c_1^*c_1\langle f_1|\hat{A}|f_1\rangle - c_1^*c_2\langle f_1|\hat{A}|f_2\rangle + c_2^*c_1\langle f_2|\hat{A}|f_1\rangle - c_2^*c_2\langle f_2|\hat{A}|f_2\rangle$$

$$= c_1^*c_1a_1\langle f_1|f_1\rangle - c_1^*c_2a_2\langle f_1|f_2\rangle + c_2^*c_1a_1\langle f_2|f_1\rangle - c_2^*c_2a_2\langle f_2|f_2\rangle$$

$$= |c_1|^2a_1 - |c_2|^2a_2. \tag{3.89}$$

7. **Problem:** Consider two linear operators, $\hat{A}$ and $\hat{B}$, that commute. They have a common set of eigenfunctions, $\hat{A}f_n = a_nf_n$ and $\hat{B}f_m = b_mf_m$. Which eigenfunctions and eigenvalues have $\hat{A} + 2\hat{B}$, as well as $\hat{A}\hat{B}$?
   Answer:

$$[\hat{A} + 2\hat{B}]f_p = \hat{A}f_p + 2\hat{B}f_p = a_pf_p + 2b_pf_p = [a_p + 2b_p]f_p$$

$$\hat{A}\hat{B}f_p = \hat{A}[\hat{B}f_p] = \hat{A}[b_pf_p] = a_pb_pf_p. \tag{3.90}$$

Thus, the functions f_p are also eigenfunctions to the operators $\hat{A} + 2\hat{B}$ and $\hat{A}\hat{B}$ with the eigenvalues $a_p + 2b_p$ and a_pb_p, respectively.

8. **Problem:** Consider the operator $\hat{A}$, which changes a complex function into its complex conjugate function, $\hat{A}f(x) = f^*(x)$. x is real.
   (a) Is $\hat{A}$ a linear operator? Justify the answer.
   (b) Determine $\hat{A}f(x)$ for $f(x) = e^{-2(x-2i)^2+i}$.

Answer:

(a) We consider

$$\hat{A}[c_1 f_1 + c_2 f_2] = c_1^* f_1^* + c_2^* f_2^* \neq c_1 f_1^* + c_1 f_2^* = c_1 \hat{A} f_1 + c_2 \hat{A} f_2 \qquad (3.91)$$

with c_1 and c_2 being constants and f_1 and f_2 being some functions. The answer is accordingly: no.

(b) We have

$$\hat{A}\left[e^{-2(x-2i)^2+i}\right] = e^{-2(x+2i)^2-i}. \qquad (3.92)$$

3.11 Problems

1. Explain the relationships between the terms "operator," "observable," "eigen-function," and "eigenvalue."
2. Explain the quantum mechanical notion of "orthogonality."
3. Explain the term "complete set of functions."
4. Explain the four postulates of quantum theory.
5. Consider a particle (mass m) in one dimension that moves in a potential $V(x)$. Let the ground state eigenfunction be $\psi(x) = A \exp[-\alpha(x - a)^2]$. Determine the expectation value for x^2.
6. Consider a particle (mass m) in two dimensions, in a potential $V(x,y)$. Let the ground state eigenfunction be $\psi(x) = A \exp[-\alpha_x(x - a_x)^2 - \alpha_y(y - a_y)^2]$. Determine the expectation value for x.
7. How is the expectation value determined, and what does it describe?
8. Define the term operator. Use the position representation to write the expressions for the following operators:
   (a) the operator of the momentum in the z direction,
   (b) the operator of the position coordinate in the y direction,
   (c) the kinetic energy operator,
   (d) the potential energy operator,
   (e) the total energy operator (Hamilton operator),
   (f) the components of the angular momentum operator,

$$\vec{l} = (l_x, l_y, l_z) = (y p_z - z p_y, z p_x - x p_z, x p_y - y p_x) = \vec{r} \times \vec{p}. \qquad (3.93)$$

9. Which of the following functions are eigenfunctions for the d^2/dx^2 operator:
   (a) $a e^{-3x} + b e^{-3ix}$,
   (b) $\sin^2 kx$,
   (c) kx,
   (d) $\cos 5x$ and
   (e) e^{-ax^2}?

Give the eigenvalue in each case.

10. What is meant by the product of two operators? Give an example of two different operators, $\hat{A}$ and $\hat{B}$, which satisfy $\hat{A}\hat{B} = \hat{B}\hat{A}$, and give an example of the case that $\hat{A}\hat{B} \neq \hat{B}\hat{A}$. Determine the following products: (a) $\hat{z}\hat{p}_z$, (b) $\hat{p}_z\hat{z}$, (c) $\hat{y}\hat{p}_x$, (d) $\hat{p}_x\hat{y}$, (e) $\hat{p}_x^2$, (f) $\hat{p}_x\hat{p}_y$, (g) $\hat{x}\hat{p}_x^2$, (h) $\hat{p}_x\hat{x}^2$, and (i) $(d/dx + \hat{x})^3$.

11. Use the bra-ket notation to prove that all eigenvalues of any hermitian operator are real, and that the eigenfunctions of a hermitian operator for different eigenvalues are orthogonal.

12. How is the commutator $[\hat{A}, \hat{B}]$ defined? When are two operators interchangeable and when not? Give an example of interchangeable operators and an example of noninterchangeable operators. Determine the following commutators: (1) $[\hat{y}, \hat{p}_y]$, (2) $[\hat{y}, \hat{x}]$, (3) $[\hat{x}, \hat{T}]$, (4) $[\hat{p}_x, \hat{T}]$, (5) $[\hat{x}, \hat{V}(x)]$, (6) $[\hat{p}_x, \hat{V}(x)]$, (7) $[\hat{x}, \hat{H}]$, and (8) $[\hat{p}_x, \hat{H}]$. Here, $\hat{T}$, $\hat{V}(x)$, and $\hat{H}$ are the operators for the kinetic, the potential, and the total energy, respectively.

13. Consider a particle (mass m) in one dimension, x. Let the wave function of the particle be $\psi(x) = N[\ln x - \sin x]$ with N equal to a constant.
    (a) Determine $\hat{x}\psi(x)$.
    (b) Determine $\hat{x}^2\psi(x)$.
    (c) Determine $\hat{p}_x\psi(x)$.
    (d) Determine $\hat{p}_x^2\psi(x)$.
    (e) What applies to the commutator $[\hat{x}^2, \hat{p}_x^2]$?
    (f) Determine $(\hat{p}_x^2\hat{x}^2 - \hat{x}^2\hat{p}_x^2)\psi(x)$.
    (g) Determine $(\hat{p}_x\hat{x} - \hat{x}\hat{p}_x)^4\psi(x)$.

14. We consider the following three operators that act on the function $f(x)$, $\hat{A}f(x) = \frac{df(x)}{dx}$, $\hat{B}f(x) = \frac{d^2f(x)}{dx^2}$, and $\hat{C}f(x) = [\frac{df(x)}{dx}]^2$. Determine the commutators $[\hat{A}, \hat{B}]$, $[\hat{A}, \hat{C}]$, and $[\hat{B}, \hat{C}]$.

15. Which of the following functions are eigenfunctions for the inversion operator $\hat{I}$ (that replaces x with $-x$ everywhere): (a) $x^3 - kx$, (b) $\cos(kx)$, (c) $\sin(k^2x)$, and (d) $x^2 + 3x - 1$? What is the corresponding eigenvalue?

16. Which of the operators: $\hat{x}$, $\hat{p}_x$, and d/dx are hermitian? Prove the results.

17. Can the d/dx operator describe an observable? Justify the answer.

4 Particle in a box

4.1 The Schrödinger equation and its solutions

The practical way of dealing with the Schrödinger equation and its solutions is best explained through a simple example. The example we will discuss is one of the few systems for which the Schrödinger equation can be solved analytically. There are few others, some of which we will discuss in the following chapters.

Here, we consider a particle (in one dimension) that is confined to a finite range $0 \leq x \leq a$; see Figure 4.1. The potential is correspondingly

$$V(x) = \begin{cases} 0 & 0 \leq x \leq a \\ \infty & \text{otherwise.} \end{cases} \tag{4.1}$$

The time-independent Schrödinger equation for this system is

$$-\frac{\hbar^2}{2m}\frac{d^2}{dx^2}\psi(x) + V(x)\psi(x) = E\psi(x) \tag{4.2}$$

with the energy E equal to a position-independent constant.

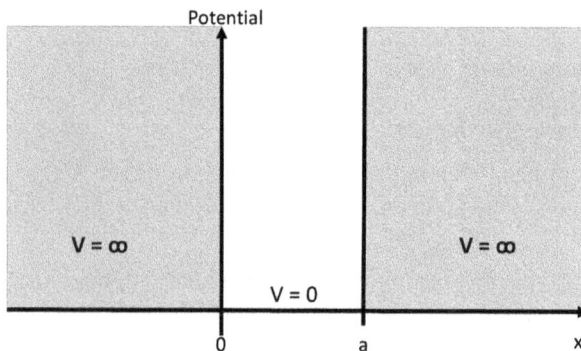

Figure 4.1: The potential of a particle in a box. In the gray area, the potential is infinite, while inside the box the potential is zero.

Equation (4.2) holds for all x, thus also for x outside the box. In order to prevent the energy of the particle from becoming infinite, the wave function must vanish outside the box,

$$\psi(x) = 0, \quad x < 0 \quad \text{or} \quad x > a. \tag{4.3}$$

Because $\psi(x)$ must be continuous, it must also be true that the wave function disappears at the boundaries of the box (otherwise the wave function would be discontinuous),

$$\psi(0) = \psi(a) = 0. \tag{4.4}$$

https://doi.org/10.1515/9783110742206-004

Inside the box, the Schrödinger equation is

$$-\frac{\hbar^2}{2m}\frac{d^2}{dx^2}\psi(x) = E\psi(x). \tag{4.5}$$

This equation has the general solution

$$\psi(x) = A \cdot \sin(kx) + B \cdot \cos(kx) \tag{4.6}$$

with

$$\frac{\hbar^2}{2m}k^2 = E. \tag{4.7}$$

Because $\psi(0) = 0$, B must be 0. The other boundary condition, $\psi(a) = 0$, means that

$$A \cdot \sin(ka) = 0. \tag{4.8}$$

$A = 0$ is not an acceptable solution (then the wave function would be zero everywhere and the particle has disappeared). Instead, we must have

$$\sin(ka) = 0 \tag{4.9}$$

or

$$ka = n\pi \tag{4.10}$$

with $n = 1, 2, 3, \dots$. The wave functions for $n < 0$ are identical to those for $n > 0$ except for the sign, which is irrelevant. For $n = 0$, the wave function becomes 0, so that this value can also be ignored.

We have then

$$\psi(x) = A \cdot \sin\left(\frac{n\pi x}{a}\right) \tag{4.11}$$

and

$$E = \frac{\hbar^2 n^2 \pi^2}{2ma^2}. \tag{4.12}$$

n is a positive integer. Accordingly, the energy is quantized. From this derivation, we also recognize, which is generally valid, that the quantization of the energy originates from the boundary conditions [equation (4.4)].

We can determine the constant A by demanding that the wave function is normalized. Thus, the total probability of finding the particle somewhere should be 1,

$$\int_{-\infty}^{\infty} |\psi(x)|^2 \, dx = |A|^2 \int_0^a \left[\sin\left(\frac{n\pi x}{a}\right)\right]^2 dx = |A|^2 \frac{a}{2} \equiv 1. \tag{4.13}$$

From this,

$$A = \sqrt{\frac{2}{a}}.$$ (4.14)

Here, we have evaluated the integral in equation (4.13) using the formulas in Chapter 17.

Finally, in Figure 4.2 we show the wave functions and their associated densities $[= |\psi(x)|^2]$. Here, we clearly see how the allowed wave functions are those $\sin(kx)$ functions, which are zero for $x = 0$, and which then oscillate exactly so that they become zero also at $x = a$.

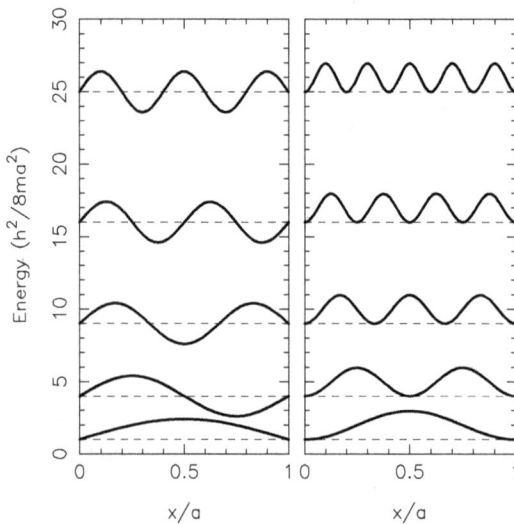

Figure 4.2: The wave functions (left part) and their squares (right part) for the particle in a box. The functions are also shown as a function of energy.

4.2 Time-dependent solutions

The stationary solutions will be those with which we will deal almost exclusively. Nevertheless, it may be interesting to look at the nonstationary solutions, also because we thereby can obtain a comparison between classical physics and quantum physics.

In the upper part of Figure 4.3, we show the temporal evolution of the density of a wave function whose density is originally a Gaussian function localized to the left part of the box. It can be seen that the wave function first widens and then later hits the right wall, whereby a lot of oscillations occur. After reflection at the right wall, the wave function becomes wider and delocalized over the whole box.

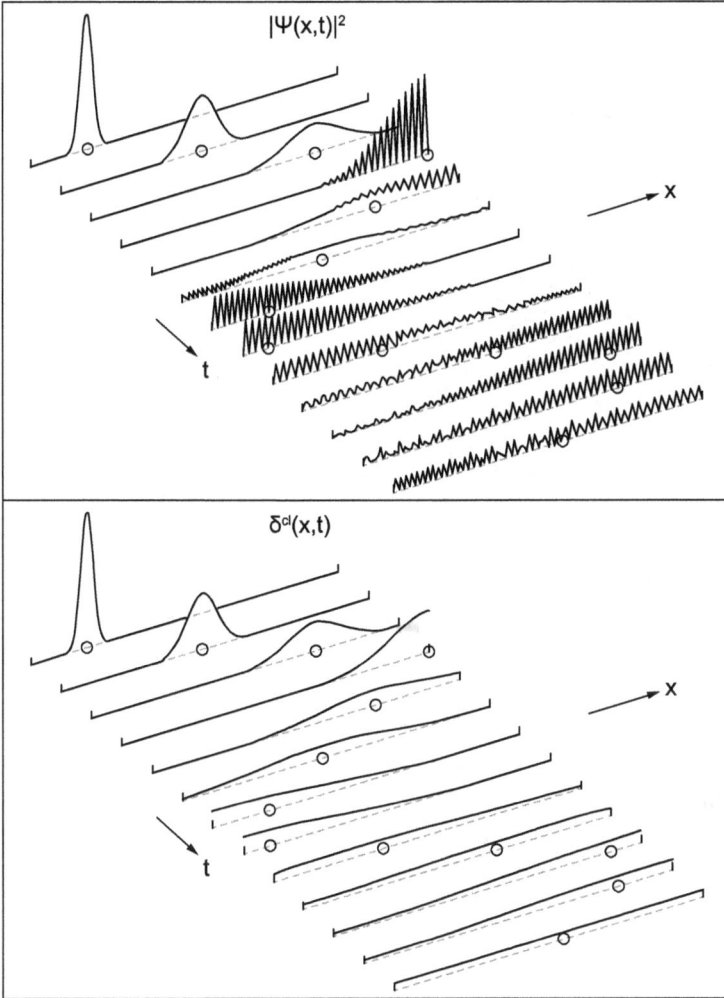

Figure 4.3: The temporal evolution of the density of a wave function of a particle in a one-dimensional box (top) compared to a classical system (bottom). Adapted from S. Brandt and H. D. Dahmen, *The Picture Book of Quantum Mechanics*, Springer-Verlag, 1995.

If we compare this with the behavior of a viscous liquid that is inside a box, we see a similar behavior (see the lower picture in Figure 4.3), although the oscillations do not occur here.

But looking at even longer periods of time, amazing things happen in the quantum world. As Figure 4.4 shows, the wave function will converge later, forming the mirror image of the original wave function in the right half-part of the box. And even later, the original wave function reappears. This behavior is a clear difference to classical behavior.

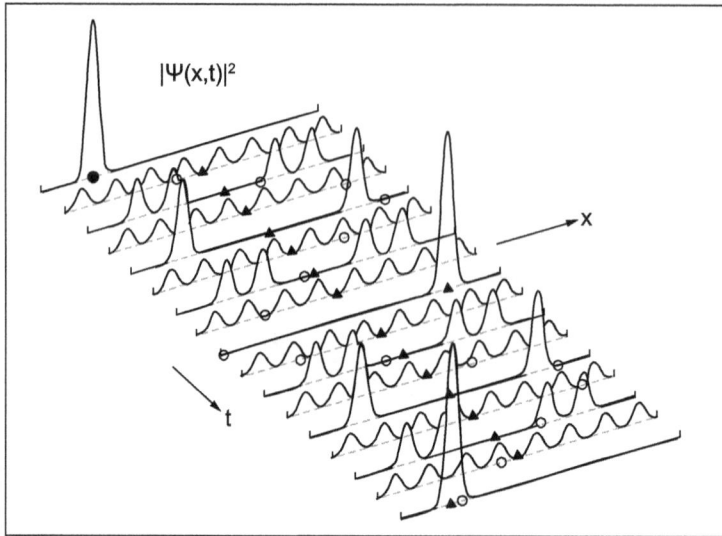

Figure 4.4: As in the upper part of Figure 4.3, but over a period 60 times as long as that of Figure 4.3. Adapted from S. Brandt and H. D. Dahmen, *The Picture Book of Quantum Mechanics*, Springer-Verlag, 1995.

4.3 Expectation values

We will now use the stationary wave functions for the particle in the one-dimensional box to calculate different expectation values and from those also calculate the related uncertainties. First, we get (by taking advantage of the fact that the wave functions are real)

$$\langle x \rangle = \int_0^a \psi(x) x \psi(x)\, dx = \frac{2}{a} \int_0^a x \sin^2\left(\frac{n\pi x}{a}\right) dx$$

$$= \frac{2}{a}\left[\frac{x^2}{4} - \frac{x}{4\alpha}\sin(2\alpha x) - \frac{1}{8\alpha^2}\cos(2\alpha x)\right]_0^a = \frac{a}{2}, \tag{4.15}$$

where we introduced for simplicity

$$\alpha = \frac{n\pi}{a} \tag{4.16}$$

and used the integrals in Chapter 17. Equation (4.15) shows that on average the particle can be found in the middle of the box—a result that makes sense if you look at Figure 4.3.

Subsequently,

$$\langle x^2 \rangle = \int_0^a \psi(x) x^2 \psi(x) \, dx = \frac{2}{a} \int_0^a x^2 \sin^2\left(\frac{n\pi x}{a}\right) dx$$

$$= \frac{2}{a}\left[\frac{x^3}{6} - \frac{x}{4\alpha^2} \cos(2\alpha x) - \left(\frac{x^2}{4\alpha} - \frac{1}{8\alpha^3}\right) \sin(2\alpha x)\right]_0^a$$

$$= \frac{2}{a}\left(\frac{a^3}{6} - \frac{a}{4\alpha^2}\right) = \frac{2}{a}\left(\frac{a^3}{6} - \frac{a^3}{4n^2\pi^2}\right) = a^2\left(\frac{1}{3} - \frac{1}{2n^2\pi^2}\right). \tag{4.17}$$

From equations (4.15) and (4.17), we get the uncertainty of the position coordinate,

$$\Delta x = [\langle x^2 \rangle - \langle x \rangle^2]^{1/2} = a\sqrt{\frac{1}{12} - \frac{1}{2n^2\pi^2}}. \tag{4.18}$$

For the momentum coordinate, we find

$$\langle p \rangle = \int_0^a \psi(x)\frac{\hbar}{i}\frac{d}{dx}\psi(x) \, dx = \int_0^a \psi(x)\left[\frac{\hbar}{i}\frac{d}{dx}\psi(x)\right] dx$$

$$= \frac{\hbar}{i}\frac{2}{a}\int_0^a \sin(\alpha x)\left[\frac{d}{dx}\sin(\alpha x)\right] dx = \frac{\hbar}{i}\frac{2}{a}\alpha\int_0^a \sin(\alpha x)\cos(\alpha x) \, dx$$

$$= \frac{\hbar}{i}\frac{2}{a}\alpha\left[\frac{-1}{4\alpha}\cos(2\alpha x)\right]_0^a = 0 \tag{4.19}$$

and

$$\langle p^2 \rangle = \int_0^a \psi(x)\left[\frac{\hbar}{i}\frac{d}{dx}\right]^2\psi(x) \, dx = \int_0^a \psi(x)\left[\frac{\hbar^2}{i^2}\frac{d^2}{dx^2}\psi(x)\right] dx$$

$$= -\hbar^2\frac{2}{a}\int_0^a \sin(\alpha x)\left[\frac{d^2}{dx^2}\sin(\alpha x)\right] dx = \hbar^2\frac{2}{a}\alpha^2\int_0^a \sin^2(\alpha x) \, dx$$

$$= \hbar^2\alpha^2 = \frac{n^2\pi^2\hbar^2}{a^2}. \tag{4.20}$$

Equation (4.19) makes sense because it is equally likely that the particle will move to the left or to the right, resulting in an average value for the momentum equal to 0.
From equations (4.19) and (4.20), we get then

$$\Delta p = [\langle p^2 \rangle - \langle p \rangle^2]^{1/2} = \frac{n\pi\hbar}{a}. \tag{4.21}$$

And then finally

$$\Delta x \cdot \Delta p = a\sqrt{\frac{1}{12} - \frac{1}{2n^2\pi^2}} \cdot \frac{n\pi\hbar}{a} = \left(\sqrt{\frac{n^2\pi^2}{12} - \frac{1}{2}}\right)\hbar \geq 0.568\hbar, \tag{4.22}$$

where in the last identity we have inserted that value of n, 1, which leads to the smallest value of $\Delta x \cdot \Delta p$. Thus, Heisenberg's uncertainty principle is fulfilled.

The equations also show that the more the particle is delocalized (i. e., the larger a becomes), the narrower is the distribution of momentum values, i. e., delocalization in position space is coupled to localization in momentum space and vice versa.

4.4 Complete set of functions

The (normalized) eigenfunctions for a particle in a box,

$$\psi(x) = \psi_n(x) = \sqrt{\frac{2}{a}} \sin\left(\frac{n\pi x}{a}\right), \quad n = 1, 2, \ldots, \tag{4.23}$$

are inside the interval

$$0 \le x \le a \tag{4.24}$$

defined and fulfill in addition

$$\psi(0) = \psi(a) = 0. \tag{4.25}$$

They form a complete set of functions, so that all functions that are defined in the interval of equation (4.24), and satisfy equation (4.25) can be expanded according to this set of functions,

$$f(x) = \sum_n c_n \psi_n(x). \tag{4.26}$$

An example of such a function is the parabola

$$f(x) = \begin{cases} k \cdot x \cdot (a - x) & 0 \le x \le a \\ 0 & \text{otherwise,} \end{cases} \tag{4.27}$$

with k equal to some constant.

The expansion coefficients c_n are given by

$$c_n = \langle \psi_n | f \rangle = \int_0^a \psi_n(x) f(x)\, dx. \tag{4.28}$$

We will calculate these and thereby use that

$$\alpha = \frac{n\pi}{a}$$
$$\sin(\alpha \cdot 0) = 0$$
$$\sin(\alpha \cdot a) = 0$$

$$\cos(\alpha \cdot 0) = 1$$

$$\cos(\alpha \cdot a) = (-1)^n \tag{4.29}$$

as well as the integrals in Chapter 17. Then we get

$$c_n = \int_0^a \sqrt{\frac{2}{a}}\sin(\alpha x)k(ax - x^2)\,dx = k\sqrt{\frac{2}{a}}\int_0^a [ax\sin(\alpha x) - x^2\sin(\alpha x)]\,dx$$

$$= k\sqrt{\frac{2}{a}}\left[\frac{a}{\alpha^2}\sin(\alpha x) - \frac{ax}{\alpha}\cos(\alpha x) - \frac{2}{\alpha^3}\cos(\alpha x) - \frac{2x}{\alpha^2}\sin(\alpha x) + \frac{x^2}{\alpha}\cos(\alpha x)\right]_0^a$$

$$= k\sqrt{\frac{2}{a}}\left[-\frac{a^2}{\alpha}(-1)^n - \frac{2}{\alpha^3}(-1)^n + \frac{a^2}{\alpha}(-1)^n + \frac{2}{\alpha^3}\right]$$

$$= k\sqrt{\frac{2}{a}}\frac{2}{\alpha^3}[1 - (-1)^n] = k\sqrt{\frac{2}{a}}\frac{2a^3}{n^3\pi^3}[1 - (-1)^n]. \tag{4.30}$$

The fact that the coefficients vanish for even n is easily explained by the symmetry properties of the functions:

$$f(x) = f(a - x)$$

$$\psi_n(x) = -(-1)^n\psi_n(a - x). \tag{4.31}$$

Thus, for even n, $\psi_n(x)$ is antisymmetric if $x \to a - x$ is changed, while for odd n the function is symmetric, which also holds for $f(x)$.

4.5 Kinetic energy

The wave function

$$\psi(x) = \sqrt{\frac{2}{a}}\sin\left(\frac{n\pi x}{a}\right) \tag{4.32}$$

is non-zero only in the range $0 \le x \le a$. In this range, the potential is zero, which means that the energy of the particle,

$$E = \frac{\hbar^2\pi^2}{2ma^2}n^2, \tag{4.33}$$

comes solely from the kinetic energy. We recognize from equation (4.33) that the kinetic energy decreases as the length of the box widens, though it is not easy to identify the cause of this at first. After all, there is no particular reason why the particle slows down if it can move in a larger area.

But in quantum theory, the kinetic energy can be calculated using the wave function in position space. We have

$$\langle E_{\text{kin}} \rangle = \int_0^a \psi^*(x) \left[-\frac{\hbar^2}{2m} \frac{d^2}{dx^2} \psi(x) \right] dx = -\frac{\hbar^2}{2m} \int_0^a \psi^*(x) \left[\frac{d^2}{dx^2} \psi(x) \right] dx$$

$$= -\frac{\hbar^2}{2m} \left[\psi^*(x) \frac{d}{dx} \psi(x) \right]_0^a + \frac{\hbar^2}{2m} \int_0^a \frac{d}{dx} \psi^*(x) \frac{d}{dx} \psi(x)\, dx$$

$$= \frac{\hbar^2}{2m} \int_0^a \frac{d}{dx} \psi^*(x) \frac{d}{dx} \psi(x)\, dx = \frac{\hbar^2}{2m} \int_0^a \left| \frac{d}{dx} \psi(x) \right|^2 dx, \tag{4.34}$$

where we have used that the wave function disappears at $x = 0$ and $x = a$. Equation (4.34) is generally valid if we consider systems for which the wave function disappears at the boundaries of the region inside which it is defined.

Equation (4.34) shows that the kinetic energy becomes large when the wave function oscillates strongly and thus has a large derivative. For the particle in a box, this means that as the box gets larger, the wave function becomes wider and, at the same time, generally smaller because of the normalization condition. Hence, the wave function oscillates less and the kinetic energy is reduced. Basically, therefore, the kinetic energy is lowest when the wave function has as few nodes as possible and is distributed over a larger range.

4.6 Momentum representation

The general expression for the wave function in momentum space when calculated from the position-space wave function is given by equation (3.69), which in the present case becomes

$$\phi_n(p) = (2\pi\hbar)^{-1/2} \int e^{-\frac{ipx}{\hbar}} \psi_n(x)\, dx = (2\pi\hbar)^{-1/2} \int_0^a e^{-\frac{ipx}{\hbar}} \sqrt{\frac{2}{a}} \sin\left(\frac{n\pi x}{a} \right) dx$$

$$= (2a\hbar)^{-1/2} \frac{1}{2i} \int_0^a \left[e^{i(\frac{n\pi}{a} - \frac{p}{\hbar})x} - e^{i(-\frac{n\pi}{a} - \frac{p}{\hbar})x} \right] dx$$

$$= (2a\hbar)^{-1/2} \frac{1}{2i} \left[\frac{e^{i(\frac{n\pi}{a} - \frac{p}{\hbar})x}}{i(\frac{n\pi}{a} - \frac{p}{\hbar})} - \frac{e^{i(-\frac{n\pi}{a} - \frac{p}{\hbar})x}}{i(-\frac{n\pi}{a} - \frac{p}{\hbar})} \right]_0^a$$

$$= \frac{1}{2} (2a\hbar)^{-1/2} \left[\frac{1}{\frac{n\pi}{a} - \frac{p}{\hbar}} + \frac{1}{\frac{n\pi}{a} + \frac{p}{\hbar}} - \frac{e^{i(\frac{n\pi}{a} - \frac{p}{\hbar})a}}{\frac{n\pi}{a} - \frac{p}{\hbar}} + \frac{e^{i(-\frac{n\pi}{a} - \frac{p}{\hbar})x}}{-\frac{n\pi}{a} - \frac{p}{\hbar}} \right]$$

$$= \frac{1}{2} (2a\hbar)^{-1/2} [1 - (-1)^n e^{-ipa/\hbar}] \left[\frac{1}{\frac{n\pi}{a} - \frac{p}{\hbar}} + \frac{1}{\frac{n\pi}{a} + \frac{p}{\hbar}} \right]$$

$$= \frac{1}{2} \sqrt{\frac{a}{\pi\hbar}} \left[\frac{1}{n\pi - y} + \frac{1}{n\pi + y} \right] \cdot [1 - (-1)^n e^{-iy}] \tag{4.35}$$

with

$$y = pa/\hbar. \tag{4.36}$$

Here, we used the familiar identity

$$\sin s = \frac{1}{2i}(e^{is} - e^{-is}). \tag{4.37}$$

It is realized that

$$\phi_n(-p) = \phi_n^*(p) \tag{4.38}$$

so it is equally likely that the particle is moving in the positive or in the negative direction. Moreover, $|\phi_n(p)|^2$ has larger contributions from larger p with increasing n. This is consistent with the fact that the total energy consists only of kinetic energy and increases with n.

In Figure 4.5, the densities in the momentum space, $|\phi_n(p)|^2$, for $n = 1$ to $n = 9$ are shown as a function of y of equation (4.36) and without the factor $\frac{a}{4\pi\hbar}$ on the ordinate axis. In this figure, we see that the density has maxima for $y = \pm n\pi$, although the density for other y values is not exactly zero. Therefore, Heissenberg's uncertainty principle is not violated. For $y = \pm n\pi$,

$$\left|\phi_n\left(\pm\frac{n\pi\hbar}{a}\right)\right| = \frac{1}{2n\pi}\sqrt{\frac{a}{\pi\hbar}}. \tag{4.39}$$

We can understand the occurrence of the maxima for $y = \pm n\pi$ by means of the relation of de Broglie,

$$\lambda = \frac{h}{p} = \frac{h}{y\hbar/a} = \frac{2\pi a}{\pm n\pi} = \pm\frac{2a}{n} \tag{4.40}$$

or

$$a = \pm n\frac{\lambda}{2}. \tag{4.41}$$

Thus, the maxima corresponds to the case that an integral number of half-wavelengths fit inside the box. That this is the case is easily seen in Figure 4.2.

4.7 Experimental realizations: conjugated molecules

The particle in a box is a model system that can be handled relatively easily mathematically and at the same time illustrates the quantization of energy. But the particle in a box also provides a surprisingly accurate description of the properties of π electrons in long conjugated systems, e. g., in β-carotene. For a long, conjugated molecule,

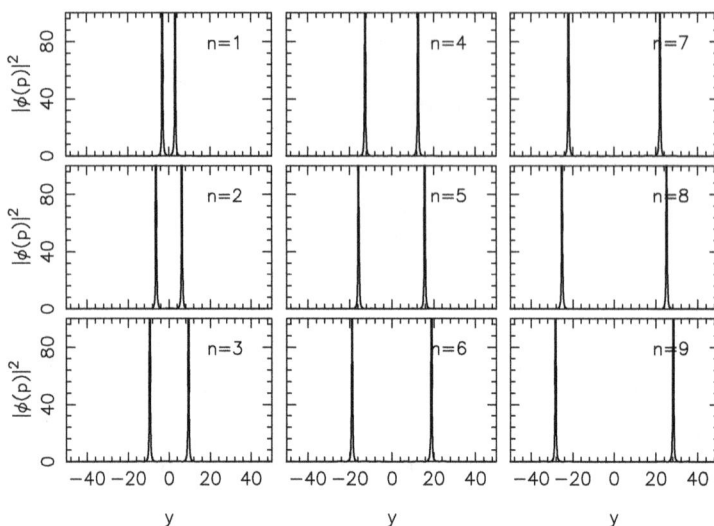

Figure 4.5: Scaled densities in momentum space for a particle in a box for the nine lowest energy states.

delocalized π electrons can be excited by absorption of ultraviolet or visible light. Assuming that the π electrons move freely and independently in a box of length L, the energy of the lowest energetic excitation can be estimated. With $2n\,\pi$ electrons, the energies of the energetically highest occupied molecular orbital and the energetically lowest, unoccupied molecular orbital are equal to

$$\epsilon_{\text{HOMO}} = \frac{\hbar^2\pi^2}{2mL^2}n^2$$

$$\epsilon_{\text{LUMO}} = \frac{\hbar^2\pi^2}{2mL^2}(n+1)^2 \tag{4.42}$$

(HOMO = Highest Occupied Molecular Orbital; LUMO = Lowest Unoccupied Molecular Orbital). This gives the lowest excitation energy

$$\Delta\epsilon = \epsilon_{\text{LUMO}} - \epsilon_{\text{HOMO}} = \frac{\hbar^2\pi^2}{2mL^2}(2n+1). \tag{4.43}$$

If this energy is known, e. g., the length of the molecule can be estimated.

Consider Figure 4.6. The two halves show the energy levels and their occupation of two different molecules that differ in size: the molecule in the right half-part is twice as large and has twice as many valence electrons as that in the left half-part. The energetically lowest excitations of the two molecules are also shown, and it can be seen clearly how these excitation energies become smaller, the larger the molecule becomes.

In an older publication by Hans Kuhn, Helv. Chim. Acta **31**, 1441–1455 (1948), it was shown how the model of the particle in a box can be used for conjugated molecules,

Energy   Energy

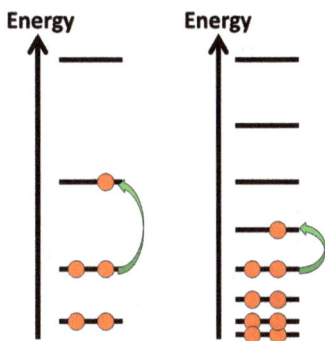

Figure 4.6: The model of the particle in the box as applied for two long molecules of different lengths. The molecule to the left is half as long and has half as many electrons as that to the right. The arrows show the energetically lowest excitations.

also to be able to make quantitative predictions. We will briefly discuss two examples of his work here.

The left part of Figure 4.7 shows the structure of a symmetrical cyanine dye. Here, we will focus on the behavior of the π electrons. Each C-atom of the polymethine chain as well as the two N-atoms on both sides of the chain are bonded to neighboring atoms via three σ-bonds. For these bonds, each C-atom and each N-atom use three valence electrons, leaving one valence electron of each C-atom and a total of three valence electrons of the two N-atoms. The number of π electrons is then

$$N = Z + 1, \tag{4.44}$$

where Z represents the number of atoms which are connected together to form a chain of resonating single and double bonds. Furthermore, (see Figure 4.7)

$$Z = 2j + 5 \tag{4.45}$$

with j shown in Figure 4.7.

Similar to equations (4.42) and (4.43), we find

$$\epsilon_{HOMO} = \frac{\hbar^2 \pi^2}{2mL^2} \left(\frac{N}{2} \right)^2$$

$$\epsilon_{LUMO} = \frac{\hbar^2 \pi^2}{2mL^2} \left(\frac{N}{2} + 1 \right)^2$$

$$\Delta\epsilon = \epsilon_{LUMO} - \epsilon_{HOMO} = \frac{\hbar^2 \pi^2}{2mL^2} (N + 1). \tag{4.46}$$

We will let l be the typical length of a C–C or C–N bond and have then

$$L = N \cdot l \tag{4.47}$$

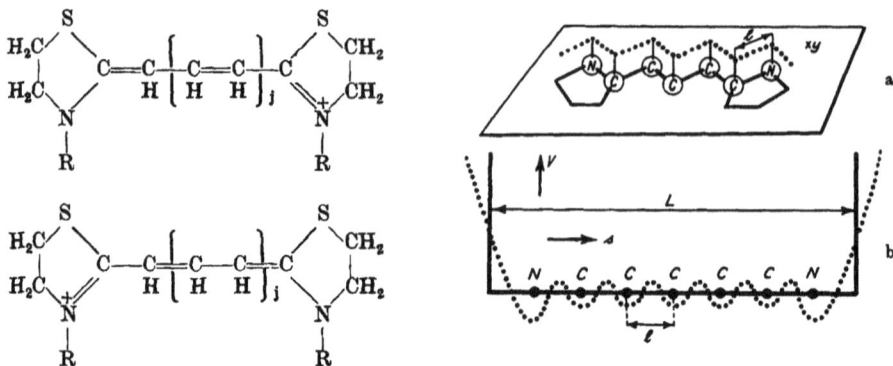

Figure 4.7: The model of the particle in a box applied to a conjugated molecule. Reproduced with permission from John Wiley & Sons from Hans Kuhn: *Elektronengasmodell zur quantitativen Deutung der Lichtabsorption von organischen Farbstoffen I.*, Helv. Chim. Acta **31**, 1441–1455 (1948).

(see Figure 4.7). Moreover, we shall use the relationships between wavelength, frequency, and energy,

$$\lambda = \frac{c}{v} = \frac{hc}{hv} = \frac{hc}{\Delta\epsilon}. \tag{4.48}$$

We then use equations (4.46) and (4.47) and obtain thereby

$$\lambda = \frac{8mc}{h} \frac{l^2 N^2}{N+1}. \tag{4.49}$$

Using a typical value of $l = 1.39$ Å, this gives

$$\lambda = 637 \text{ Å} \cdot \frac{2N^2}{N+1}. \tag{4.50}$$

With equations (4.44) and (4.45), this gives for $j = 1$ a value of $\lambda = 4530$ Å, while an experimental value is $\lambda = 4450$ Å, i. e., a surprisingly good agreement when taking the simplicity of the model into account.

When passing to the systems of Figure 4.8, only equation (4.45) needs to be modified. In this case,

$$Z = 2j' + 9. \tag{4.51}$$

Through this modification, we obtain the wavelengths given in Table 4.1, which also lists experimental results. Again, one recognizes a surprisingly good agreement, especially considering how simple the model of Hans Kuhn is.

4.8 Experimental realizations: chains of metal atoms

In a much more recent work, *Realization of a particle-in-a-box: electron in atomic Pd chain*, published in J. Phys. Chem. B **109**, 20657–20660 (2005), Niklas Nilius, Thomas

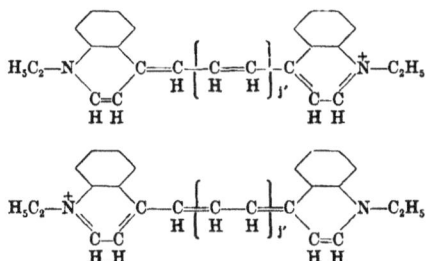

Figure 4.8: The model of the particle in a box as applied to a conjugated molecule. Reproduced with permission from John Wiley & Sons from Hans Kuhn: *Elektronengasmodell zur quantitativen Deutung der Lichtabsorption von organischen Farbstoffen I.*, Helv. Chim. Acta **31**, 1441–1455 (1948).

Table 4.1: Calculated and experimental values of the wavelengths (in nm) of the first electronic excitation of the molecules in Figure 4.8 for different j'.

Z	N	j'	λ (Theory)	λ (Experiment)
9	10	0	579	590
11	12	1	706	710
13	14	2	834	820
15	16	3	959	930

M. Wallis, and Wilson Ho have presented results for a system that is quite close to the particle in a box.

Using a scanning tunneling microscope (see later, Section 5.3), the authors first generated chains of Pd atoms on a NiAl surface. These chains have up to 20 atoms and are linear (see Figure 4.9). Subsequently, one can use an experimental setup as shown schematically in Figure 4.10 to inject electrons from a metal tip through the chain of Pd atoms into the substrate. If the position of the tip is determined very accurately, then one can obtain information on the probability that an electron will be injected at a specific position along the Pd chain. These probabilities depend not only on the position but also on the energy of the injected electron, i. e., one obtains information on the available orbitals of the chains of Pd atoms as a function of position and energy. It is interesting that the results are quite similar to those obtained from the wave functions of a particle in a box, although there are deviations (the system is not quite one-dimensional, and the potential is not quite constant along the Pd chain).

This is shown in Figure 4.11. This figure shows the contour lines on the right and left sides of the densities of various orbitals, while the center images show the densities along the chain. The voltages given are those used to inject the electrons. The higher the voltage, the higher the energy of the wave function.

It can be seen that at the lowest voltage (1.45 V) the density is very similar to that of the energetically lowest state of the particle in a box. Also the next density (for 1.55 V), the similarity to the corresponding wave function for the particle in a box is recogniz-

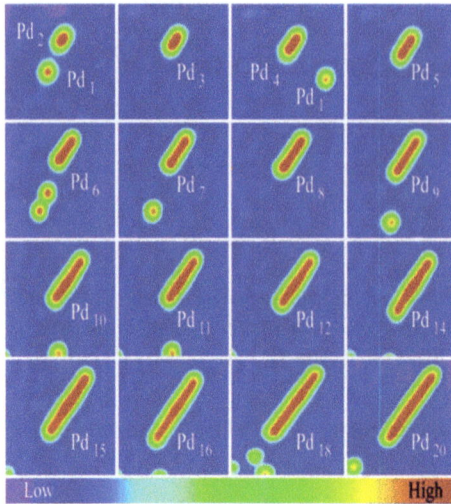

Figure 4.9: The chains of Pd atoms on a NiAl surface. Reproduced with permission from American Chemical Society from N. Nilius, T. M. Wallis, and W. Ho: *Realization of a particle-in-a-box: electron in an atomic Pd chain*, J. Phys. Chem. B **109**, 20657–20660 (2005).

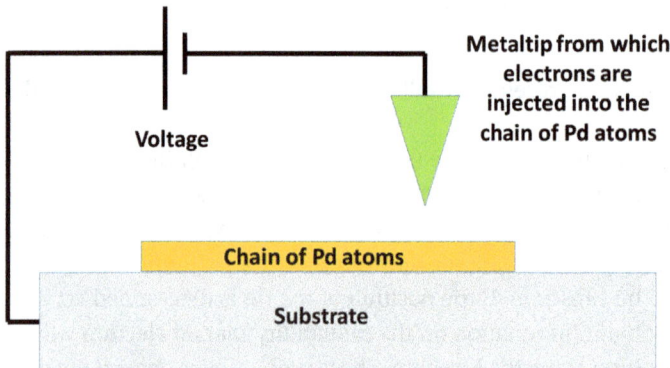

Figure 4.10: Sketch of an experimental setup, showing how electrons can be injected at different positions along the chain of Pd atoms.

able. At even higher voltages, the similarity slowly diminishes, but it is still possible to recognize the basic principles of the wave functions of the particle in a box.

4.9 A model of chemical bonds

The results of this chapter can also be used to gain a first insight into chemical bonds. We will discuss this with the help of the example in Figure 4.12.

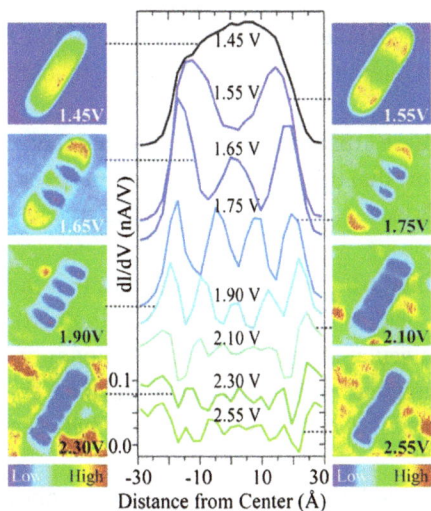

Figure 4.11: Electron densities of different orbitals of a Pd_{20} chain. Reproduced with permission from American Chemical Society from N. Nilius, T. M. Wallis, and W. Ho: *Realization of a particle-in-a-box: electron in an atomic Pd chain*, J. Phys. Chem. B **109**, 20657–20660 (2005).

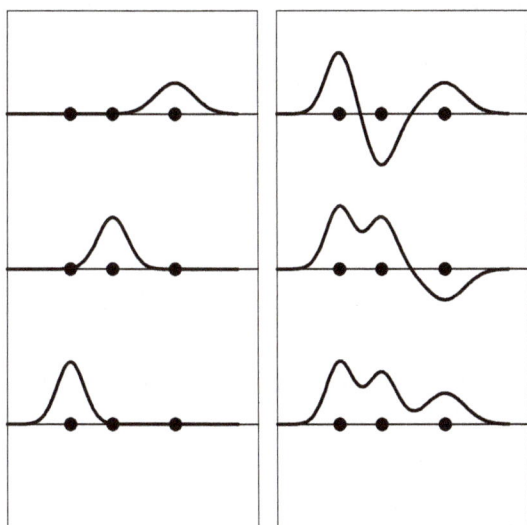

Figure 4.12: Schematic representation of a simple model for chemical bonds of a three-atomic molecule. The left part shows schematically the orbitals localized to the individual atoms, while the right part shows the molecular orbitals that are generated from these atomic orbitals. The three dots mark the positions of the three atomic nuclei.

We consider a linear, three-atomic molecule. For the three isolated, non-interacting atoms, we have atom-centered orbitals, as shown in the left part. Bringing together the three atoms in the molecule, we can create three molecular orbitals out of the three atom-centered orbitals (this will be discussed later in this manuscript in much more details), as shown in the right half of Figure 4.12.

First, we recognize that the electrons in these orbitals have more available space. From the model of the particle in a box, we know that this gives us a lower (kinetic) energy: that is, the delocalization of the electrons leads to a lowering of the total energy— a stabilization of the system.

This applies to all the three molecular orbitals. But in addition, we know that the fewer nodes (zeros) of a wave function, the lower is the kinetic energy of the particle. Accordingly, electrons occupying that molecular orbital, which has no nodes between the atoms, have the lowest kinetic energy. In order to obtain a low total energy, electrons will preferentially occupy molecular orbitals without nodes between the atoms.

Finally, we see that the molecular orbitals that have no nodes between the atoms, also have an increased probability density exactly between the atoms. The electrons located in such regions are attracted simultaneously to two neighboring atomic nuclei and have therefore a particularly low (i. e., more negative) potential energy. Also this leads to a stabilization of the system.

The conclusion is that molecular orbitals, which are highly delocalized and have no nodes between the adjacent atoms, are energetically favored and preferentially occupied by electrons. Such orbitals stabilize the molecule. Conversely, strongly oscillating orbitals with nodes between neighboring atoms are destabilizing the molecule.

4.10 More particles

In the previous two sections, we have considered the electrons in a chain and treated them as if they were particles in a box. As is known, it is not possible to have more than one electron in an orbital. Instead, the particles occupy the levels from below in energy. Then, for N electrons, the total electron density would be the sum of the densities of the N energetically lowest wave functions,

$$\rho(x) = \sum_{n=1}^{N} \rho_n(x) = \sum_{n=1}^{N} \left| \sqrt{\frac{2}{a}} \sin\left(\frac{n\pi x}{a}\right) \right|^2. \tag{4.52}$$

This function, together with the contributions of the individual wave functions, ρ_n, for n and N between 1 and 10 are all shown in Figure 4.13. In this simple example, we did not attempt to take spin variables into account and have, accordingly, nowhere multiplied the individual $\rho_n(x)$ by 2.

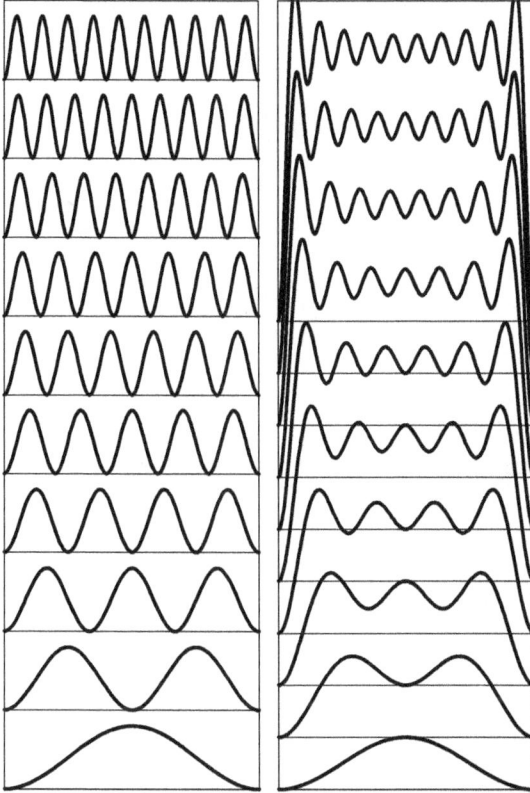

Figure 4.13: The left part shows the contributions $\rho_n(x)$ of the individual wave functions of equation (4.52) for n from 1 (bottom) to 10 (top), while the right part shows the total density for N from 1 (bottom) to 10 (top).

4.11 More dimensions

As another example, we consider a particle in a three-dimensional box. The potential is

$$V(x, y, z) = \begin{cases} 0 & 0 \leq x \leq a,\, 0 \leq y \leq b,\, 0 \leq z \leq c, \\ \infty & \text{otherwise,} \end{cases} \qquad (4.53)$$

and the (time-independent) Schrödinger equation is, similar to the one-dimensional case,

$$\left[-\frac{\hbar^2}{2m}\left(\frac{\partial^2}{\partial x^2} + \frac{\partial^2}{\partial y^2} + \frac{\partial^2}{\partial z^2} \right) + V(x, y, z) \right] \psi(x, y, z) = E \cdot \psi(x, y, z). \qquad (4.54)$$

$\psi(x, y, z)$ must vanish outside the box and, therefore, also be identical to 0 on the boundaries of the box (otherwise the wave function would not be continuous).

We make the Ansatz that $\psi(x, y, z)$ can be written as a product of three functions, each function depending on only one of the three coordinates,

$$\psi(x, y, z) = \psi_x(x) \cdot \psi_y(y) \cdot \psi_z(z). \tag{4.55}$$

We insert this assumption into equation (4.54) and subsequently divide by the product in equation (4.55) and get then

$$\frac{-\frac{\hbar^2}{2m} \frac{\partial^2 \psi_x(x)}{\partial x^2}}{\psi_x(x)} = E - \frac{-\frac{\hbar^2}{2m} \frac{\partial^2 \psi_y(y)}{\partial y^2}}{\psi_y(y)} - \frac{-\frac{\hbar^2}{2m} \frac{\partial^2 \psi_z(z)}{\partial z^2}}{\psi_z(z)}. \tag{4.56}$$

The left-hand side depends only on x, but not on y or z. On the other hand, the right-hand side depends only on y and z, but not on x. So both sides must be independent of all three coordinates and accordingly equal to some constant, E_x. From this, we get two equations

$$-\frac{\hbar^2}{2m} \frac{\partial^2 \psi_x(x)}{\partial x^2} = E_x \psi_x(x) \tag{4.57}$$

and

$$\frac{-\frac{\hbar^2}{2m} \frac{\partial^2 \psi_y(y)}{\partial y^2}}{\psi_y(y)} = E - E_x - \frac{-\frac{\hbar^2}{2m} \frac{\partial^2 \psi_z(z)}{\partial z^2}}{\psi_z(z)}. \tag{4.58}$$

For the latter equation, we apply arguments analogous to those above, i. e., that the left-hand side depends only on y, but not on z, whereas the right-hand side depends only on z, but not on y. So both sides have to be independent of the two coordinates and accordingly equal to some other constant, E_y.

This leads to two equations

$$-\frac{\hbar^2}{2m} \frac{\partial^2 \psi_y(y)}{\partial y^2} = E_y \psi_y(y) \tag{4.59}$$

and

$$-\frac{\hbar^2}{2m} \frac{\partial^2 \psi_z(z)}{\partial z^2} = E_z \psi_z(z) \tag{4.60}$$

as well as

$$E = E_x + E_y + E_z. \tag{4.61}$$

The three equations (4.57), (4.59), and (4.60) are identical to the equation we treated for a particle in a one-dimensional box, so we can directly adopt the solutions from there. Then we get

$$\psi(x,y,z) = \psi_{n_x,n_y,n_z}(x,y,z) = \sqrt{\frac{8}{abc}} \sin\left(\frac{n_x \pi x}{a}\right) \sin\left(\frac{n_y \pi y}{b}\right) \sin\left(\frac{n_z \pi z}{c}\right)$$

$$E = E_{n_x,n_y,n_z} = \frac{\hbar^2 \pi^2}{2m}\left(\frac{n_x^2}{a^2} + \frac{n_y^2}{b^2} + \frac{n_z^2}{c^2}\right). \tag{4.62}$$

We have here explicitly specified that the wave function and the energy depend on three **quantum numbers** n_x, n_y, and n_z. Each of the three quantum numbers is a positive integer.

A special case occurs for

$$a = b = c. \tag{4.63}$$

Then

$$E_{n_x,n_y,n_z} = \frac{\hbar^2 \pi^2}{2ma^2}(n_x^2 + n_y^2 + n_z^2). \tag{4.64}$$

As an example, we can consider the states for $(n_x, n_y, n_z) = (1,8,5)$, $(1,5,8)$, $(5,1,8)$, $(5,8,1)$, $(8,1,5)$, $(8,5,1)$, $(4,7,5)$, $(4,5,7)$, $(7,4,5)$, $(7,5,4)$, $(5,4,7)$, and $(5,7,4)$ that all have the same energy. These states are therefore said to be **energetically degenerate**. Such **degeneracies** can also be found for other cases of (a, b, c).

In the simpler case of a two-dimensional box, we obtain in an analogous way

$$\psi(x,y) = \psi_{n_x,n_y}(x,y) = \sqrt{\frac{4}{ab}} \sin\left(\frac{n_x \pi x}{a}\right) \sin\left(\frac{n_y \pi y}{b}\right)$$

$$E = E_{n_x,n_y} = \frac{\hbar^2 \pi^2}{2m}\left(\frac{n_x^2}{a^2} + \frac{n_y^2}{b^2}\right). \tag{4.65}$$

Some of the lowest-energy wave functions for the case $a = b$ are shown in Figure 4.14. That the wave functions can be written as products is easily recognized.

Creating pictures as in Figure 4.14 is not easy for the inexperienced. Instead, such functions are often represented through contour curves. Thereby one draws the (most often closed) curves in a two-dimensional plane, which one obtains, if one connects all points, that have the same function value. In our case, this means that we connect those (x, y) points that have the same value for $\psi(x,y)$. Such representations for the wavefunctions in Figure 4.14 are shown in Figure 4.15.

From the examples in this chapter, we also recognize a general principle: The number of quantum numbers is identical to the dimensionality of the system. For a single particle in a one-dimensional potential, we have a single quantum number. For a particle in a three-dimensional potential, we have three quantum numbers. Accordingly, for seven particles in three dimensions, we have $7 \cdot 3 = 21$ quantum numbers, though we use only a few of them in most such cases.

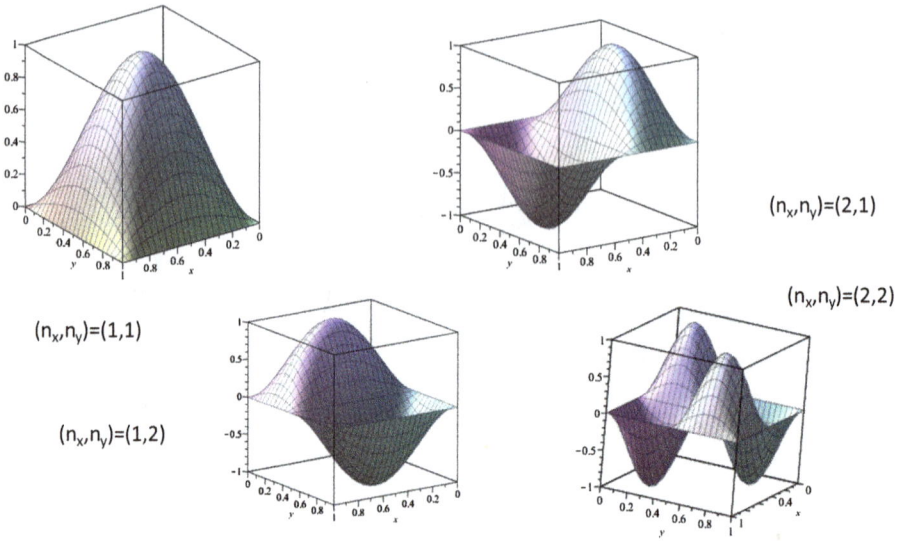

Figure 4.14: Some wave functions for a particle in a two-dimensional square-shaped box.

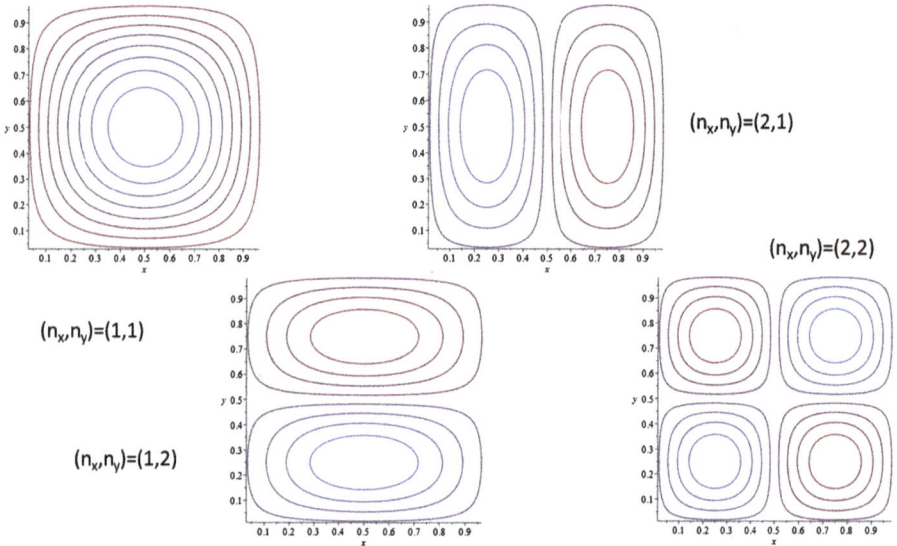

Figure 4.15: Some wave functions of a particle in a two-dimensional square-shaped box represented through contour curves.

4.12 Problems with answers

1. **Problem:** Consider a particle in a three-dimensional rectangular box. The particle is confined in a three-dimensional volume with $0 < x < L_1$, $0 < y < L_2$, and

$0 < z < L_3$. Inside this region, the potential energy is zero, on the walls it rises abruptly to infinity. The mass of the particle is m.

(a) What does the Hamilton operator look like for this problem?

(b) Which boundary conditions must the wave function fulfill?

(c) Show that $\psi(x, y, z) = N \cdot \sin(k_x x) \cdot \sin(k_y y) \cdot \sin(k_z z)$ for (x, y, z) within the box is an eigenfunction to the Hamiltonian operator.

(d) Show that $k_x = \frac{n_x \pi}{L_1}$, $k_y = \frac{n_y \pi}{L_2}$, and $k_z = \frac{n_z \pi}{L_3}$ where n_x, n_y, and n_z are positive integers.

(e) Determine the value of N.

(f) Calculate the expectation value for $\hat{x}$ for $n_x = 1$, $n_y = 2$, and $n_z = 1$.

(g) Calculate the expected value for $\hat{p}_z^2$ for $n_x = 1$, $n_y = 2$, and $n_z = 2$.

(h) Consider the ground state of the system, $n_x = n_y = n_z = 1$ and $L_2 = 10$ nm and $L_1 = 9$ nm. What is the probability that the particle will be (1) between $x = 6.01$ nm and $x = 6.2$ nm, (2) between $x = 6.01$ nm and $x = 6.2$ nm, and simultaneously between $y = 7.1$ nm and $y = 7.9$ nm and (3) between $x = 6.01$ nm and $x = 6.2$ nm and at the same time between $y = 11$ nm and $y = 11.101$ nm?

Answer:

(a) In three dimensions, the Hamiltonian is

$$\hat{H} = -\frac{\hbar^2}{2m}\left[\frac{\partial^2}{\partial x^2} + \frac{\partial^2}{\partial y^2} + \frac{\partial^2}{\partial z^2}\right] + V(x, y, z). \tag{4.66}$$

In our case, we have that

$$V(x, y, z) = \begin{cases} 0 & 0 < x < L_1, \ 0 < y < L_2, \ 0 < z < L_3 \\ \infty & \text{otherwise} \end{cases} \tag{4.67}$$

(b) Only inside the box, the wave function is not identically equal to zero, because otherwise the energy would be infinitely large. In order for the wave function to remain continuous, we must then have

$$\psi(0, y, z) = \psi(L_1, y, z) = \psi(x, 0, z)$$
$$= \psi(x, L_2, z) = \psi(x, y, 0) = \psi(x, y, L_3) = 0. \tag{4.68}$$

(c) Inside the box, we have

$$\hat{H}\psi(x, y, z) = -\frac{\hbar^2}{2m}\left[\frac{\partial^2}{\partial x^2} + \frac{\partial^2}{\partial y^2} + \frac{\partial^2}{\partial z^2}\right]N \sin(k_x x) \sin(k_y y) \sin(k_z z)$$

$$= \frac{\hbar^2}{2m}(k_x^2 + k_y^2 + k_z^2)N \sin(k_x x) \sin(k_y y) \sin(k_z z). \tag{4.69}$$

This proves that the function is an eigenfunction to the Hamiltonian and that the energy eigenvalue is

$$E = \frac{\hbar^2}{2m}(k_x^2 + k_y^2 + k_z^2). \tag{4.70}$$

Furthermore, we realize that half-part of the boundary conditions are automatically fulfilled for this function,

$$\psi(0, y, z) = \psi(x, 0, z) = \psi(x, y, 0) = 0. \tag{4.71}$$

(d) For the other boundary conditions, we must demand that

$$\sin(k_x L_1) = 0$$
$$\sin(k_y L_2) = 0$$
$$\sin(k_z L_3) = 0. \tag{4.72}$$

If $\sin(kL) = 0$, then $kL = n\pi$ (with n being an integer), or $k = n\pi/L$. In our case, we have

$$k_x = \frac{n_x \pi}{L_1}$$
$$k_y = \frac{n_y \pi}{L_2}$$
$$k_z = \frac{n_z \pi}{L_3}. \tag{4.73}$$

If $n_x = 0$, $n_y = 0$, or $n_z = 0$, then $\psi(x, y, z) = 0$, which is meaningless. Furthermore, if we change $n_x \rightarrow -n_x$ and/or $n_y \rightarrow -n_y$ and/or $n_z \rightarrow -n_z$, the wave function hardly changes: $\psi(x, y, z) \rightarrow \pm\psi(x, y, z)$, so that negative values of n_x, n_y, and/or n_z describe the same wave functions as those with positive n_x, n_y, n_z. Therefore, we can restrict ourselves to $n_x > 0$, $n_y > 0$, and $n_z > 0$.

(e) N is determined from the normalization condition

$$1 = \int_0^{L_3} \int_0^{L_2} \int_0^{L_1} |\psi(x, y, z)|^2 \, dx \, dy \, dz$$

$$= N^2 \int_0^{L_1} \sin^2(n_x \pi x / L_1) \, dx \int_0^{L_2} \sin^2(n_y \pi y / L_2) \, dy \int_0^{L_3} \sin^2(n_z \pi z / L_3) \, dz$$

$$= N^2 \frac{L_1}{2} \frac{L_2}{2} \frac{L_3}{2} \tag{4.74}$$

with the help of the formulas in Chapter 17. We then get that

$$N = \sqrt{\frac{8}{L_1 L_2 L_3}} \tag{4.75}$$

assuming that N is positive and real.

(f) Regardless of the values of the quantum numbers,

$$\langle x \rangle = \int_0^{L_3} \int_0^{L_2} \int_0^{L_1} x |\psi(x,y,z)|^2 \, dx \, dy \, dz$$

$$= N^2 \int_0^{L_1} x \sin^2(n_x \pi x/L_1) \, dx \int_0^{L_2} \sin^2(n_y \pi y/L_2) \, dy \int_0^{L_3} \sin^2(n_z \pi z/L_3) \, dz$$

$$= \frac{8}{L_1 L_2 L_3} \frac{L_1^2}{4} \frac{L_2}{2} \frac{L_3}{2} = \frac{L_1}{2}, \tag{4.76}$$

where we have used the formulas in Chapter 17.

(g) In this case,

$$\langle p_z^2 \rangle = \int_0^{L_3} \int_0^{L_2} \int_0^{L_1} \psi^*(x,y,z) \hat{p}_z^2 \psi(x,y,z) \, dx \, dy \, dz$$

$$= N^2 \int_0^{L_1} \sin^2(n_x \pi x/L_1) \, dx \int_0^{L_2} \sin^2(n_y \pi y/L_2) \, dy$$

$$\cdot \int_0^{L_3} \sin(n_z \pi z/L_3)(-\hbar^2)\left[\frac{d^2}{dz^2} \sin(n_z \pi z/L_3)\right] dz$$

$$= N^2 \hbar^2 \pi^2 \frac{n_z^2}{L_3^2} \int_0^{L_1} \sin^2(n_x \pi x/L_1) \, dx \int_0^{L_2} \sin^2(n_y \pi y/L_2) \, dy$$

$$\cdot \int_0^{L_3} \sin(n_z \pi z/L_3) \sin(n_z \pi z/L_3) \, dz = \hbar^2 \pi^2 \frac{n_z^2}{L_3^2} \tag{4.77}$$

because of equation (4.75). For $n_z = 2$, we have $\langle p_z^2 \rangle = \frac{4\hbar^2 \pi^2}{L_3^2}$.

(h) In the first case, we set $x_a = 6.01\,\text{nm}$ and $x_e = 6.2\,\text{nm}$. Then the searched probability becomes

$$P_1 = N^2 \int_{x_a}^{x_e} \sin^2(n_x \pi x/L_1) \, dx \int_0^{L_2} \sin^2(n_y \pi y/L_2) \, dy \int_0^{L_3} \sin^2(n_z \pi z/L_3) \, dz$$

$$= N^2 \left[-\frac{L_1}{4 n_x \pi} \sin(2 n_x \pi x/L_1) + \frac{x}{2}\right]_{x_a}^{x_e} \frac{L_2}{2} \frac{L_3}{2}$$

$$= \frac{2}{L_1}\left[\frac{x_e - x_a}{2} - \frac{L_1}{4 n_x \pi} \sin\left(\frac{2 n_x \pi x_e}{L_1}\right) + \frac{L_1}{4 n_x \pi} \sin\left(\frac{2 n_x \pi x_a}{L_1}\right)\right]$$

$$= \frac{x_e - x_a}{L_1} - \frac{1}{2 n_x \pi} \sin\left(\frac{2 n_x \pi x_e}{L_1}\right) + \frac{1}{2 n_x \pi} \sin\left(\frac{2 n_x \pi x_a}{L_1}\right)$$

$$= \frac{6.2 - 6.01}{9} - \frac{1}{2\pi} \sin \frac{12.4\pi}{9} + \frac{1}{2\pi} \sin \frac{12.02\pi}{9}$$

$$= 0.021111 + 0.147566 - 0.138384 = 0.0303. \tag{4.78}$$

Again, we have used the integrals in Chapter 17.

In the second case, we similarly set $x_a = 6.01\,\text{nm}$ and $x_e = 6.2\,\text{nm}$ and $y_a = 7.1\,\text{nm}$ and $y_e = 7.9\,\text{nm}$. As above, we then get

$$P_2 = N^2 \int_{x_a}^{x_e} \sin^2(n_x\pi x/L_1)\,dx \int_{y_a}^{y_e} \sin^2(n_y\pi y/L_2)\,dy \int_{0}^{L_3} \sin^2(n_z\pi z/L_3)\,dz$$

$$= N^2 \left[-\frac{L_1}{4n_x\pi}\sin(2n_x\pi x/L_1) + \frac{x}{2} \right]_{x_a}^{x_e} \left[-\frac{L_2}{4n_y\pi}\sin(2n_y\pi y/L_2) + \frac{y}{2} \right]_{y_a}^{y_e} \frac{L_3}{2}$$

$$= \left[\frac{x_e - x_a}{L_1} - \frac{1}{2n_x\pi}\sin\left(\frac{2n_x\pi x_e}{L_1}\right) + \frac{1}{2n_x\pi}\sin\left(\frac{2n_x\pi x_a}{L_1}\right) \right]$$

$$\cdot \left[\frac{y_e - y_a}{L_2} - \frac{1}{2n_y\pi}\sin\left(\frac{2n_y\pi y_e}{L_2}\right) + \frac{1}{2n_y\pi}\sin\left(\frac{2n_y\pi y_a}{L_2}\right) \right]$$

$$= \left[\frac{6.2 - 6.01}{9} - \frac{1}{2\pi}\sin\frac{12.4\pi}{9} + \frac{1}{2\pi}\sin\frac{12.02\pi}{9} \right]$$

$$\cdot \left[\frac{7.9 - 7.1}{10} - \frac{1}{2\pi}\sin\frac{15.8\pi}{10} + \frac{1}{2\pi}\sin\frac{14.2\pi}{10} \right]$$

$$= 0.030293 \cdot 0.08 = 0.00242. \tag{4.79}$$

In the third case, we use that the range $11\,\text{nm} < y < 11.101\,\text{nm}$ lies outside the box, so that in this region $\psi(x, y, z) = 0$. From this, we get immediately

$$P_3 = 0. \tag{4.80}$$

4.13 Problems

1. Consider a particle (mass m) in a one-dimensional box between $x = -L$ and $x = 3L$.
   (a) What does the Hamiltonian look like?
   (b) Which boundary conditions must the wave functions fulfill?
   (c) Show that $\psi(x) = N \cdot \sin[\alpha(x+\beta)]$ is an eigenfunction to the Hamilton operator.
   (d) Determine α and β for the ground state.
   (e) Determine N.
   (f) Calculate the expectation value for $\hat{x}$ for this function.
   (g) Calculate the expected value for $\hat{p}_x$ for this function.
   (h) Calculate the expectation value for the operator $\hat{B} = (\hat{x}\hat{p}_x - \hat{p}_x\hat{x})^4$ for this function.
   (i) What is the probability that the particle will be found between (i) $x = -3L$ and $x = L$, (ii) $x = 0$ and $x = L$, and (iii) $x = -L$ and $x = L$?

2. Consider a particle in a 2-dimensional box, $a \leq x \leq 2a$, $2a \leq y \leq 4a$. Determine the eigenfunctions and eigenvalues for the Hamilton operator of this system. Is there any energetic degeneracies (justify the answer)?

3. Explain the connection between "particles in a box" and the π electrons of a conjugated molecule, including optical absorption and spatial extent of the molecule.

4. Show that the wave function $\sqrt{\frac{2}{L}} \sin(\frac{x\pi}{L})$, $0 \leq x \leq L$, satisfies the uncertainty principle of Heissenberg.

5. Explain the relationship between kinetic energy and wave function in position space.

6. The wave function for a particle in a one-dimensional box of length L is equal to $\Psi_n = N \sin(\frac{n\pi x}{L})$ with $n = 1, 2, 3\ldots$ and $N = $ a constant. Normalize the function in the interval $0 \leq x \leq L$. Verify that the wave functions for different n are orthogonal.

7. Depict the three first energy levels and corresponding wave functions and probability densities for a particle in a one-dimensional box, $0 \leq x \leq L$.

5 More or less free particles

5.1 Free particle in one dimension

There are few systems for which one can solve the Schrödinger equation analytically. Some of these will be briefly discussed here, also because they have relevance to physical/chemical phenomena.

In this chapter, we first consider a free particle in one dimension. For this, the potential is constant everywhere and can be set equal to 0 without restriction. The stationary Schrödinger equation is then

$$-\frac{\hbar^2}{2m}\frac{d^2\psi(x)}{dx^2} = E \cdot \psi(x), \tag{5.1}$$

where m is the mass of the particle. The general solution of this equation can be written as

$$\psi(x) = Ae^{ikx} + Be^{-ikx}, \tag{5.2}$$

with

$$k = \sqrt{\frac{2mE}{\hbar^2}}. \tag{5.3}$$

In principle, the wave function should be normalized, i. e.,

$$\lim_{L\to\infty}(|A|^2 + |B|^2)L = 1, \tag{5.4}$$

whereby we have introduced L as the length of the region inside which the particle is located and which ultimately should become infinitely large. Here, however, we will not worry about the normalization, as it is not relevant for our arguments.

We consider

$$\hat{p}e^{\pm ikx} = \frac{\hbar}{i}\frac{d}{dx}e^{\pm ikx} = \pm\frac{\hbar}{i}ike^{\pm ikx} = \pm\hbar ke^{\pm ikx}, \tag{5.5}$$

which shows that each of the two parts in equation (5.2) describes an eigenfunction to the momentum operator. The first function, e^{ikx}, corresponds to a particle that moves to the right with the momentum $p = \hbar k$, while the other function, e^{-ikx}, is a function that moves to the left with the momentum $p = -\hbar k$. Each of these functions is called a **plane wave**, and if either $A = 0$ or $B = 0$, the function in equation (5.2) is a function with only one single value for the momentum. Therefore, $\Delta p = 0$ for such a function.

This can also be recognized by the fact that

$$\langle p \rangle = \int \psi^*(x)\hat{p}\psi(x)\,dx = \int \psi^*(x)(\pm\hbar k)\psi(x)\,dx = \pm\hbar k$$

https://doi.org/10.1515/9783110742206-005

$$\langle p^2 \rangle = \int \psi^*(x)\hat{p}^2\psi(x)\, dx = \int \psi^*(x)\hat{p}[\hat{p}\psi(x)]\, dx$$

$$= \int \psi^*(x)\hat{p}[\pm\hbar k\psi(x)]\, dx = (\hbar k)^2$$

$$\Delta p = [\langle p^2 \rangle - \langle p \rangle^2]^{1/2} = 0. \tag{5.6}$$

This applies only if one of the two constants A or B is 0, because then the wave function is also an eigenfunction to the momentum operator. Otherwise $\Delta p > 0$.

That in this single case $\Delta p = 0$ is not in conflict with Heisenberg's uncertainty principle. For $A = 0$ or $B = 0$, the wave function in equation (5.2) is completely delocalized, so that $\Delta x \to \infty$.

If a wave function is formed by a linear combination of several plane waves, a function can be obtained which is increasingly localized in position space, that is, Δx becomes smaller. This is illustrated in Figure 5.1. But at the same time, the number of plane waves increases, so that Δp increases. This illustrates Heisenberg's uncertainty principle in a different way.

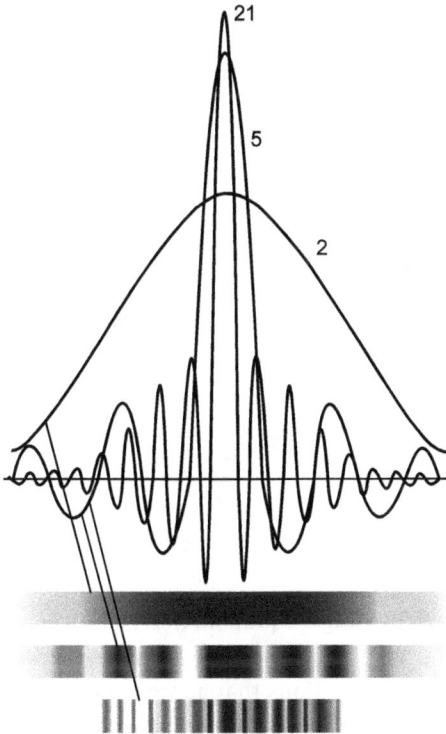

Figure 5.1: Wave functions composed of an increasing number of plane waves. The numbers indicate the number of $(+k, -k)$ pairs used to generate the various functions. Adapted from the book Peter W. Atkins, *Kurzlehrbuch Physikalische Chemie*, Wiley-VCH, 2001.

Generally, one can write the linear combination of several plane waves as

$$\psi(x) = \sum_k c_k e^{ikx} \tag{5.7}$$

We assume that the wave function is normalized such that an equation like equation (5.4) is satisfied,

$$\sum_k |c_k|^2 = \frac{1}{L}. \tag{5.8}$$

For the function in equation (5.7), we get then that

$$\langle \psi | \hat{p} | \psi \rangle = \sum_k L(\hbar k |c_k|^2)$$

$$\langle \psi | \hat{p}^2 | \psi \rangle = \sum_k L(\hbar^2 k^2 |c_k|^2). \tag{5.9}$$

Thereby

$$\Delta p = \left\{ \sum_k (\hbar^2 k^2 L |c_k|^2) - \left[\sum_k (\hbar k L |c_k|^2) \right]^2 \right\}^{1/2}. \tag{5.10}$$

Because of equation (5.8) we have

$$0 \leq L|c_k|^2 \leq 1, \tag{5.11}$$

where $L|c_k|^2 = 1$ holds only if all but one c_k are equal to zero. Only in that case, Δp can vanish.

5.2 Steps

Also for the potential

$$V(x) = \begin{cases} 0 & x < 0 \\ V_0 & x > 0 \end{cases} \tag{5.12}$$

we can determine stationary solutions to the Schrödinger equation. Here, however, as an illustration that also has relevance for the next system we shall consider, we will instead examine the temporal evolution of wave functions that approach and pass such a step. In doing so, we will gain deeper insights into the quantum world.

Accordingly, we consider a wave function with an expectation value of the energy equal to E, and that approaches the step. In Figure 5.2, examples for $V_0 > 0$ are shown; once for $E > V_0$ and once for $E < V_0$.

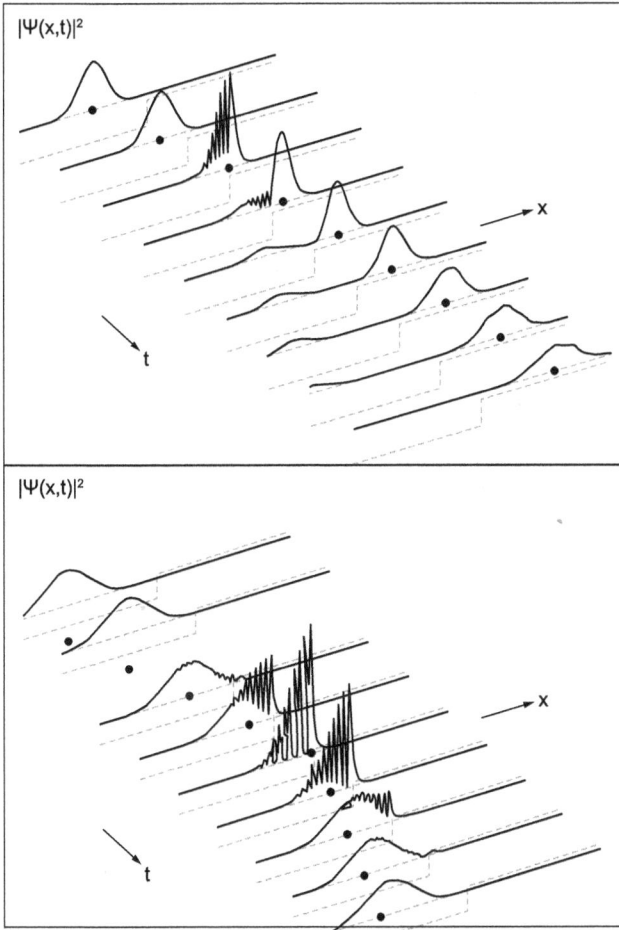

Figure 5.2: The temporal evolution of wave functions that approach a step. In the upper part, the energy of the particle is higher than that of the step; in the lower one smaller. Adapted from S. Brandt and H. D. Dahmen, *The Picture Book of Quantum Mechanics*, Springer-Verlag, 1995.

In the first case, it can be seen how the wave function approaches the step and that many oscillations develop near the step, similar to what we saw in Section 4.2 for a particle in a box. A little later, the wave function has split into two parts, and it can be seen that the function to the right of the step propagates more slowly than the reflected function to the left of the step. This is because some of the energy to the right of the step is no longer kinetic energy, but has turned into potential energy and, therefore, only a smaller part of the energy remains as kinetic energy.

Although it may appear differently, it should be emphasized that this case does **not** imply splitting the individual particles into two parts. Instead, it means that when a very large number of particles approach the step, one part is reflected, and the other continues to move past the step. This case also shows a deviation from the classical behavior: even if $E > V_0$, there is a finite probability that a particle will be reflected.

In the second case in Figure 5.2, it can be seen that for $E < V_0$ there is also a finite probability that a particle penetrates the classically forbidden range $x > 0$ (i. e., the region where the energy of the particle is lower than the potential energy), albeit only in a smaller part of it.

In Figure 5.3, we show the case $V_0 < 0$. Again, you can see the oscillations that occur when the particle is near the step.

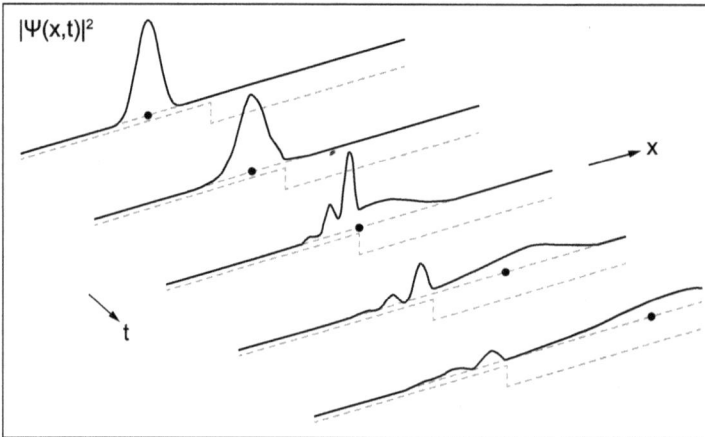

Figure 5.3: The temporal evolution of a wave function that approaches a step. Adapted from S. Brandt and H. D. Dahmen, *The Picture Book of Quantum Mechanics*, Springer-Verlag, 1995.

5.3 Tunnel effect

The examples in the last chapter indicate that for a particle with the energy E, which approaches a potential barrier (height V_0) of finite width L, there exists a nonzero probability to tunnel through this barrier, even if $E < V_0$. Equivalently, there is also a nonzero probability that the particle will be reflected even if $E > V_0$. This effect is the so-called **tunnel effect**.

As a model system, we consider a particularly simple potential,

$$V(x) = \begin{cases} 0 & x < 0 \\ V_0 & 0 < x < L \\ 0 & x > L. \end{cases} \tag{5.13}$$

The situation is shown schematically in Figure 5.4, whereby a plane wave hits the barrier at $x = 0$, after which it partially continues (albeit weakened) and partly is reflected.

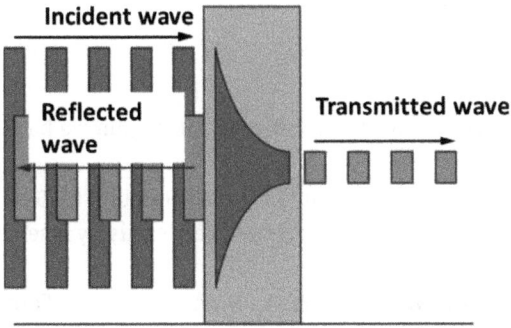

Figure 5.4: The tunnel effect. Adapted from the book of Peter W. Atkins, *Physikalische Chemie*, Wiley-VCH, 2001.

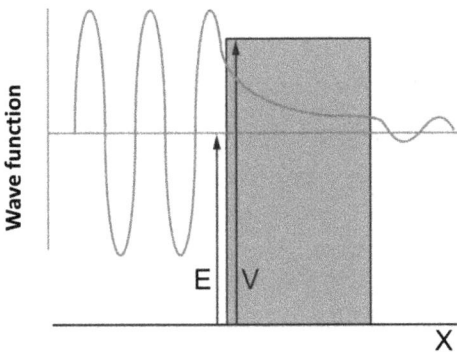

Figure 5.5: Schematic representation of a wavefunction for the tunnel effect. Adapted from the book of Peter W. Atkins, *Physikalische Chemie*, Wiley-VCH, 2001.

On the other side of the barrier, at $x = L$, the wave function propagates as a plane wave but with a smaller amplitude.

In order to solve the time-independent Schrödinger equation for this case, it is convenient to consider the three regions of equation (5.13) separately. In each of those, the potential is constant and the solutions can thereby be written similar to what we did for the free particle. All in all this gives (see also Figure 5.5)

$$\psi(x) = \begin{cases} A_1 e^{ikx} + B_1 e^{-ikx} \equiv \psi_1(x) & x < 0 \\ A_2 e^{ixx} + B_2 e^{-ixx} \equiv \psi_2(x) & 0 < x < L \\ A_3 e^{ikx} \equiv \psi_3(x) & x > L. \end{cases} \tag{5.14}$$

Here, we have introduced

$$k = \sqrt{\frac{2mE}{\hbar^2}}$$

$$\kappa = \sqrt{\frac{2m(E - V_0)}{\hbar^2}}. \tag{5.15}$$

Notice that there is no term of the type $B_3 e^{-ikx}$ for $x > L$, since in this region we have only a transmitted but no reflected wave.

For $E < V_0$, κ is imaginary. That $\psi(x)$ from equation (5.14) with equation (5.15) is a solution to the stationary Schrödinger equation, can be proved very easily by inserting.

For a given energy E, we have five unknown constants, A_1, B_1, A_2, B_2, and A_3. Four of these can be expressed in terms of the fifth, taking advantage of the fact that the wave function in equation (5.14) must be continuous and differentiable everywhere, especially at $x = 0$ and $x = L$. With the expressions in equation (5.14) this means

$$\psi_1(0) = \psi_2(0)$$
$$\psi_1'(0) = \psi_2'(0)$$
$$\psi_2(L) = \psi_3(L)$$
$$\psi_2'(L) = \psi_3'(L) \tag{5.16}$$

(the primes mark the differentation with respect to x) or

$$A_1 + B_1 = A_2 + B_2$$
$$k(A_1 - B_1) = \kappa(A_2 - B_2)$$
$$A_2 e^{i\kappa L} + B_2 e^{-i\kappa L} = A_3 e^{ikL}$$
$$\kappa(A_2 e^{i\kappa L} - B_2 e^{-i\kappa L}) = kA_3 e^{ikL}. \tag{5.17}$$

We will not carry this calculation through to the end, although it is relatively simple but somewhat tedious. Instead, we realize that the quantity of interest is the proportion of a plane wave at $x < 0$ that is transmitted through the barrier. This part of the wave function is quantified through $\frac{A_3}{A_1}$. The probability that a particle tunnels through the barrier is the so-called transmission coefficient,

$$T = \left|\frac{A_3}{A_1}\right|^2, \tag{5.18}$$

while the reflected component is represented by the reflection coefficient,

$$R = 1 - T = 1 - \left|\frac{A_3}{A_1}\right|^2 = \left|\frac{B_1}{A_1}\right|^2. \tag{5.19}$$

Here, we use that the term $B_1 e^{-ikx}$ describes the reflected wave of the particle.

The result is (without derivation)

$$T = \left\{ 1 + \frac{V_0^2}{4E(V_0 - E)} \sinh^2 \left[\frac{2mL^2(V_0 - E)}{\hbar^2} \right]^{1/2} \right\}^{-1}$$

$$= \left\{ 1 + \frac{1}{4x(1 - x)} \sinh^2 \left(\sqrt{\alpha(1 - x)} \right) \right\}^{-1} \tag{5.20}$$

with

$$x = \frac{E}{V_0}$$

$$\alpha = \frac{2mL^2}{\hbar^2 V_0}. \tag{5.21}$$

Furthermore,

$$\sinh(s) = \frac{1}{2}(e^s - e^{-s}) \tag{5.22}$$

is the hyperbolic sine function.

This equation shows the following:

- Also for $E < V_0$ there is a finite probability that a particle will tunnel through the barrier.
- For $E > V_0$ there is a finite probability that a particle will be reflected..
- For $E \to \infty$, $T \to 1$.
- The probabilities depend on the mass of the particle: the heavier the particle, the closer T approaches the values of classical physics, i. e., $T = 1$ for $E > V_0$ and $T = 0$ for $E < V_0$. This is shown in Figures 5.6 and 5.7.

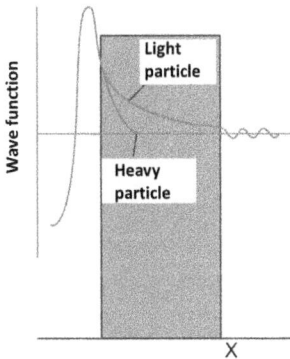

Figure 5.6: The difference between light and heavy particles in the tunnel effect. Adapted from the book Peter W. Atkins, *Physikalische Chemie*, Wiley-VCH, 2001.

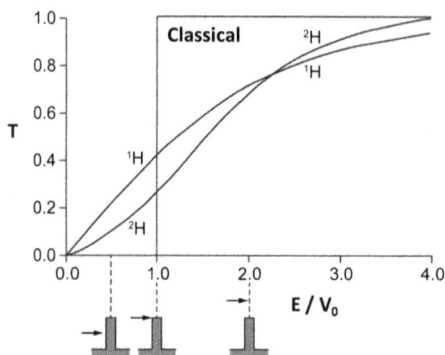

Figure 5.7: The difference between light and heavy particles in the tunnel effect. Adapted from the book Peter W. Atkins, *Physikalische Chemie*, Wiley-VCH, 1990.

- For $E > V_0$, $T = 1$, when

$$E - V_0 = \frac{\hbar^2 \pi^2}{2mL^2} n^2, \quad n = 1, 2, 3, \ldots, \tag{5.23}$$

i. e., if the energy of the particle relative to the barrier height equals the energy that a particle has in a box of the same length. This can be interpreted as a so-called resonance effect.

All these five points are illustrated in Figure 5.8.

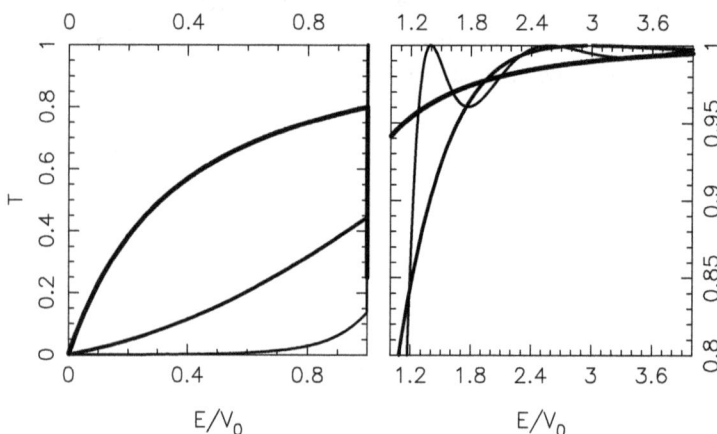

Figure 5.8: The tunneling probability as a function of the energy of the particle for three different values of α from equation (5.21): $\alpha = 1$, 5 and 25 for the thicker, middle, and thinner curve.

As a further illustration of the tunnel effect, Figure 5.9 shows the temporal evolution of the density of a wave function that approaches a potential barrier. The three cases

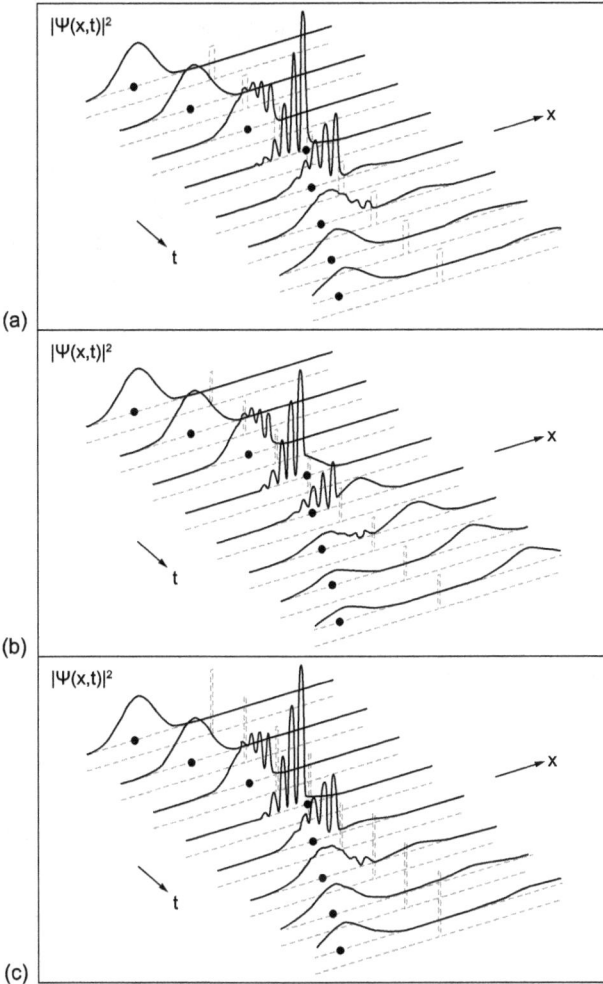

Figure 5.9: Temporal evolution of a wave function that hits a potential barrier. From (a) to (b), the width of the barrier is halved, while the height from (b) to (c) is doubled. Adapted from S. Brandt and H. D. Dahmen, *The Picture Book of Quantum Mechanics*, Springer-Verlag, 1995.

differ in the width and height of the potential barrier. It can be seen how the density is separated into two parts: one part that is transmitted and a second part that is reflected. The narrower and lower the barrier, the larger the transmitted part. As mentioned above, it should be emphasized here that no particle is split into two parts. Instead, for a very large number of particles that hit the barrier, one part is reflected and the other part is transmitted.

The tunnel effect is important for different physical and chemical phenomena. In this chapter, we have studied a very simple example, while for "true" systems, the potential as a function of position often looks very differently. Accordingly, the quantitative formulas for the probability that a particle can tunnel through will be different, but the qualitative statement remains: there is a finite probability that a particle will tunnel through a potential barrier.

As one example, in Figure 5.10 we show the potential that a particle (e. g., an α or β particle) feels inside an atomic nucleus. In some cases, even though the energy of the particle is lower than the barrier height, the particle may tunnel through the potential barrier and leave the nucleus. Then the nucleus has decayed and was **radioactive**. The energy of the particle as well as the width, height, and shape of the barrier depend on the atomic nucleus and, therefore, different nuclei are differently radioactive.

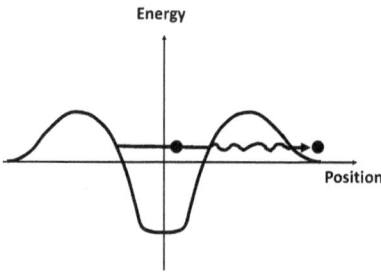

Figure 5.10: Schematic representation of a radioactive decay that can be explained through the tunnel effect.

Another example is that of chemical reactions, which we will discuss briefly in Section 5.5. The system, consisting of the interacting molecules, must often overcome an energy barrier in order to react to products. In some cases, this process can be explained through the tunnel effect.

5.4 Scanning tunneling microscope

A modern application of the tunnel effect is scanning tunneling microscopy, an experiment we have already briefly met in Section 4.8. In this experiment, a metal tip is moved along a surface of some substrate. There is a (very) small gap between tip and surface. By applying a voltage between tip and surface, electrons can tunnel through this gap. This leads to a current flow that can be measured. If the distance between the metal tip and the surface suddenly becomes smaller, for instance due to additional atoms on the surface (see Figure 5.11), the tunneling probability increases and therefore also the current. When keeping the current constant, one must then move the tip further away from the surface. Measuring this movement (see Figure 5.12) one obtains an image of the atomic structure of the surface. Figure 5.13 shows an example of the outcome of such an experiment. Here, the experiment demonstrates the presence of Cs atoms on a GaAs surface.

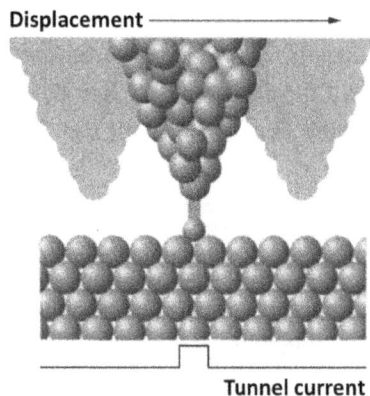

Displacement

Tunnel current

Figure 5.11: The principle of the scanning tunneling microscope. A metal tip is placed over a surface at a very short distance and the tunnel current is measured as a function of the position of the tip. The figure shows three such positions. Adapted from the book of Peter W. Atkins, Kurzlehrbuch Physikalische Chemie, Wiley-VCH, 2001.

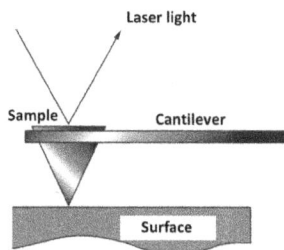

Laser light

Sample   Cantilever

Surface

Figure 5.12: The principle of the scanning tunneling microscope. If the tunnel current is kept constant, the lateral position of the metal tip above the surface is determined by means of the laser beam. If the cantilever is, for instance, moved upward to keep the tunnel current constant, the laser radiation is reflected in another direction, which can be measured. A picture of the topology of the surface is thereby obtained. Adapted from the book of Peter W. Atkins, *Kurzlehrbuch Physikalische Chemie*, Wiley-VCH, 2001.

5.5 Chemical reactions

There are indications that the tunnel effect also plays a role in chemical reactions. In a chemical reaction or transformation (see Figure 5.14), the reactants must overcome an energy barrier that is often so high that it is extremely unlikely that the thermal energy will be sufficient to overcome it. Instead, one can imagine that the reactants tunnel along the reaction coordinate through the energy barrier, as exemplified in Figure 5.14 for some reaction. Here, one should not overstrain the interpretation: in the end, the reaction coordinate is not a true position coordinate. However, since the probability of an energy barrier being passed depends on the mass of the system, one can experimentally test whether this interpretation is realistic. By replacing some atoms with

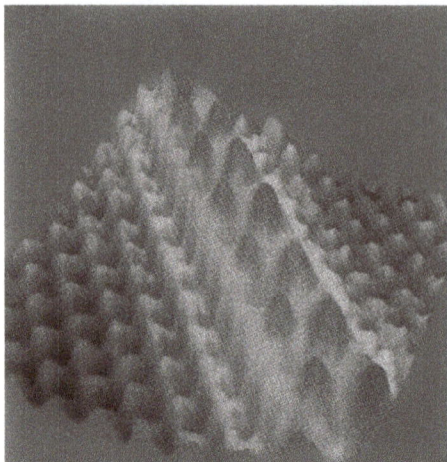

Figure 5.13: Example of the results of scanning tunneling microscopy experiments. Shown are Cs atoms, which have been deposited on a GaAs surface. Adapted from the book of Peter W. Atkins, *Kurzlehrbuch Physikalische Chemie*, Wiley-VCH, 2001.

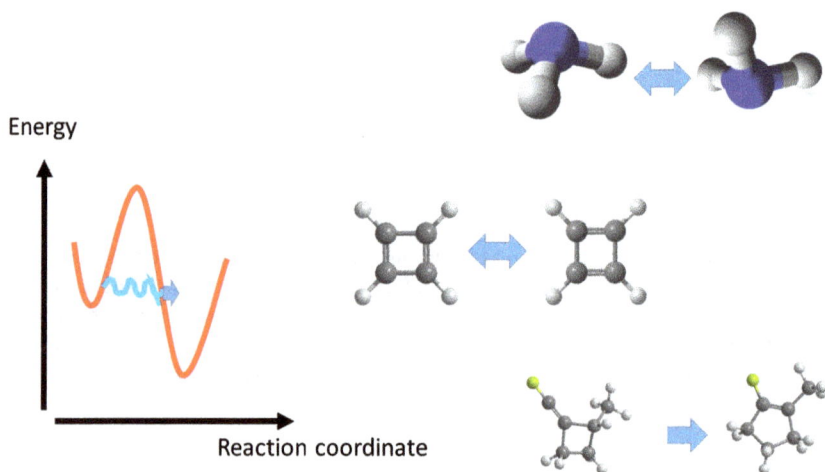

Figure 5.14: The tunnel effects in chemical reactions. On the left, the variation of the energy along the reaction coordinate is sketched, while on the right, examples of chemical reactions are shown, where the tunneling effect is important. These include the so-called umbrella movement of the NH_3 molecule (above), the change in C-C bond lengths in cyclobutadiene (center), and a chemical transformation (bottom).

corresponding isotopes, the mass can be changed. Subsequently, one can investigate to what extent the reaction rate changes, as one would expect, if the tunnel effect were relevant. And indeed, it has been possible to demonstrate the relevance of the tunneling effect to chemical reactions.

5.6 Problems with answers

1. **Problem:** Consider the time-dependent Schrödinger equation for a free particle in one dimension x. Show that $\psi(x,t) = e^{i(kx-\omega t)}$ is a solution, and establish a relationship between k and ω.

   Answer: Into the time-dependent Schrödinger equation for the free particle in one dimension,

   $$-\frac{\hbar^2}{2m}\frac{\partial^2}{\partial x^2}\psi(x,t) = i\hbar\frac{\partial}{\partial t}\psi(x,t),$$ (5.24)

   we insert

   $$\psi(x,t) = e^{i(kx-\omega t)}$$ (5.25)

   and then get

   $$-\frac{\hbar^2}{2m}(-k^2)e^{i(kx-\omega t)} = i\hbar(-i\omega)e^{i(kx-\omega t)},$$ (5.26)

   what is actually fulfilled, if

   $$\frac{\hbar}{2m}k^2 = \omega.$$ (5.27)

2. **Problem:** Determine a normalized eigenfunction for a particle (mass m) that is moving in the potential $V(x)$. The energy of the particle is $E < V_0$, and $V(x)$ is given as

   $$V(x) = \begin{cases} 0 & 0 \le x \le L \\ V_0 & L \le x \\ \infty & x < 0, \end{cases}$$ (5.28)

   with $V_0 > 0$ (see Figure 5.15).

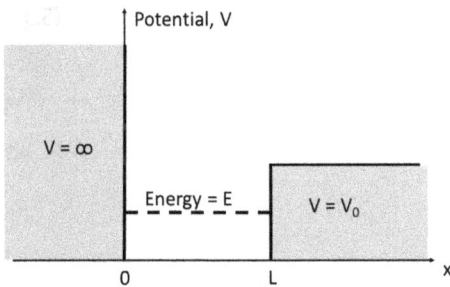

Figure 5.15: The potential in problem 2 in Section 5.6, as well as the energy of the particle.

Answer: The time-independent Schrödinger equation is

$$-\frac{\hbar^2}{2m}\frac{d^2}{dx^2}\psi(x) + V(x)\psi(x) = E\psi(x).$$ (5.29)

For $x < 0$, $V(x)$ is infinitely large. On the other hand, E is a constant and therefore independent of position. In order to avoid that E becomes infinitely large, we must have

$$\psi(x) = 0 \quad \text{for } x < 0.$$ (5.30)

Then also

$$\psi(0) = 0,$$ (5.31)

because $\psi(x)$ should be continuous.

For $0 \le x \le L$, we have

$$\psi(x) = A\sin(kx) + B\cos(kx)$$ (5.32)

with

$$\frac{\hbar^2}{2m}k^2 = E.$$ (5.33)

Because of equation (5.31),

$$B = 0.$$ (5.34)

For $x > L$ the Schrödinger equation is

$$-\frac{\hbar^2}{2m}\frac{d^2}{dx^2}\psi(x) = (E - V_0)\psi(x).$$ (5.35)

Because $E < V_0$, it is most convenient to write the general solution as

$$\psi(x) = Ce^{-\kappa x} + De^{\kappa x}$$ (5.36)

Here,

$$\frac{\hbar^2}{2m}\kappa^2 = V_0 - E,$$ (5.37)

where we choose the positive solution:

$$0 < \kappa = \sqrt{\frac{2m(V_0 - E)}{\hbar^2}}.$$ (5.38)

$\psi(x)$ must be normalized so that

$$1 = \int_{-\infty}^{\infty} |\psi(x)|^2\, dx = \int_{-\infty}^{0} |\psi(x)|^2\, dx + \int_{0}^{L} |\psi(x)|^2\, dx + \int_{L}^{\infty} |\psi(x)|^2\, dx$$ (5.39)

exists. In equation (5.39), the first integral is equal to zero because of equation (5.30). In order to assure that the last integral is finite, we must have

$$D = 0. \tag{5.40}$$

Thus, in total,

$$\psi(x) = \begin{cases} 0 & x < 0 \\ A\sin(kx) & 0 \le x \le L \\ Ce^{-\kappa x} & L \le x. \end{cases} \tag{5.41}$$

Since L and V_0 are fixed, we shall see that the energy E can not be arbitrary. As we have seen in other cases, the quantization of energy comes from the boundary conditions. Specifically, this means that $\psi(x)$ should be continuous at $x = L$,

$$A\sin(kL) = Ce^{-\kappa L}. \tag{5.42}$$

$\frac{d}{dx}\psi(x)$ should also be continuous at $x = L$,

$$Ak\cos(kL) = -\kappa Ce^{-\kappa L} = -\kappa A\sin(kL) \tag{5.43}$$

by applying equation (5.42). From equation (5.43), we get

$$\tan(kL) = -\frac{k}{\kappa} \tag{5.44}$$

or

$$\tan\left(\sqrt{\frac{2mE}{\hbar^2}}L\right) = -\frac{\sqrt{\frac{2mE}{\hbar^2}}}{\sqrt{\frac{2m(V_0-E)}{\hbar^2}}} = -\sqrt{\frac{E}{V_0-E}}, \tag{5.45}$$

this (so-called transcendental) equation cannot be solved analytically. It has, however, solutions only for discrete values of E.

We assume that we have found the energies with the help of equation (5.45). Without limitation, we can assume that the wave function is real. For the normalization of the wave function, we need

$$\int\limits_{-\infty}^{0} |\psi(x)|^2\,dx = 0$$

$$\int\limits_{0}^{L} |\psi(x)|^2\,dx = \int\limits_{0}^{L} A^2\sin^2(kx)\,dx = A^2\left[-\frac{1}{4k}\sin(2kx) + \frac{x}{2}\right]_0^L$$

$$= A^2\left[\frac{L}{2} - \frac{1}{4k}\sin(2kL)\right]$$

$$\int\limits_{L}^{\infty} |\psi(x)|^2\,dx = \int\limits_{L}^{\infty} C^2 e^{-2\kappa x}\,dx = C^2 \left[\frac{-1}{2\kappa} e^{-2\kappa x}\right]_{L}^{\infty}$$

$$= C^2 \frac{1}{2\kappa} e^{-2\kappa L} = A^2 \frac{1}{2\kappa} \sin^2(kL), \tag{5.46}$$

using equation (5.42). We have used integrals from Chapter 17 and in the last equation also equation (5.42). The normalization condition in equation (5.39) then becomes

$$1 = A^2 \left[\frac{L}{2} - \frac{1}{4k} \sin(2kL) + \frac{1}{2\kappa} \sin^2(kL)\right] \tag{5.47}$$

what results in

$$A = \left[\frac{L}{2} - \frac{1}{4k} \sin(2kL) + \frac{1}{2\kappa} \sin^2(kL)\right]^{-1/2}, \tag{5.48}$$

if we (arbitrarily) choose the positive value. Finally, C is determined from equation (5.42):

$$C = A \sin(kL)e^{\kappa L}. \tag{5.49}$$

5.7 Problems

1. Specify the general expression for a wave function of a free particle with energy E in three dimensions? Which special cases correspond to a particle that propagates in the positive or negative z direction?
2. Explain the principle of the scanning tunneling microscope.
3. Explain the term "tunnel effect."
4. Sketch the wave function of a particle moving in one dimension in a potential $V(x)$. $V(x) = V_0$ for $a \leq x \leq b$ and otherwise 0. Let the energy of the particle be $E < V_0$ with $V_0 > 0$.
5. Explain briefly the relationship between tunnel effect and chemical reactions.
6. Sketch the transmission coefficient T as a function of the energy of the particle E for a tunnel effect where the particle tunnels through a potential barrier with the height V_0. Both $E < V_0$ and $E > V_0$ should be considered.
7. Discuss briefly the resonance behavior of the transmission coefficient as a function of the energy of the particle.

6 Vibrations

6.1 Energy of molecular systems

In Section 1.8, we saw how the measurement of spectra of, e. g., molecular systems provide information on the energy levels of the molecules. Absorption and emission spectra have—in principle—maxima at energies corresponding to the energy differences of the molecules. Thus, such spectra contains information about the molecules that can be used to their characterization.

Unfortunately, it is not easy to gain clear insights into the structure and composition of the molecules from the information on the energy levels of the molecules. Often you need several, different sets of spectroscopic information to obtain a reasonably clear knowledge of the molecules (and only if they are not very large). Furthermore, the spectroscopic information can be useful only if one also has ideas of the origin of the energy levels of a molecule. In this manuscript, we will discuss some of the basics for this question.

For a molecule of M atoms, we have a total of $3M$ structural degrees of freedom. Because we can assign a time-dependent movement to each of them, there are $3M$ different types of such displacements and, moreover, the energy of each is quantized. If we choose as three coordinates the center of mass, the associated movements describe translational displacements of the whole molecule; the molecule moves as a rigid body like a particle in a box with a mass equal to the total mass of the molecule. If the molecule is nonlinear, we can choose 3 additional coordinates so that the associated motions describe rotations of the whole molecule around the so-called major axes. For a linear molecule, the number is only 2. The remaining $3M - 6$ (or $3M - 5$ for linear molecules) coordinates describe internal movements, i. e., vibrations within the molecule.

Ultimately, the electrons will also have quantized energies. Thereby, we see that the energy of a molecule can be written as a sum of the energies of the individual contributions,

$$E = E_{\text{trans}} + E_{\text{rot}} + E_{\text{vib}} + E_{\text{el}}, \tag{6.1}$$

i. e., as the sum of a translational, rotational, vibrational, and electronic energy.

Each individual energy type is quantized and depends on the characteristics of the system. By measuring the energies of the individual types of energy, we therefore obtain information about the molecule. The information obtained from the different types of energy is not identical, but rather complementary, so that measuring different types of energy is beneficial in characterizing a molecular system.

In addition, the typical energies of the individual types of energy have very different scales. Typically, the energy splitting of the levels of translational energy is very small; that of the rotational movement larger; that of vibrational energy even larger;

https://doi.org/10.1515/9783110742206-006

and the largest is found for the electronic energy. This also means that different experimental methods have to be used to measure the different types of energy. Finally, it should also be mentioned that the energies of different types are seldom independent, though this will hardly be of relevance in the present manuscript.

In Chapters 4 and 5, we discussed the quantum mechanical treatment of an (almost) free particle. We can use this description for the translational energy and then recognize that the information that you receive from such spectra is mainly the total mass of the molecule. In Chapter 7, we will discuss the rotational energy, and later in this manuscript we will discuss many details of the electronic energy. We obtain thereby also a more detailed understanding of chemical bonding (including, e. g., orbital theory). But first, in this chapter, we will discuss the vibrational energy.

6.2 Vibrations of a molecule

In this chapter, we will briefly explain two models that are used to describe vibrations in molecules. For nonlinear molecules with M atoms, there are $3M - 6$ different vibrational modes, while for linear molecules there are $3M - 5$ different vibrational modes. For each mode of vibration, the movement of the nth atom can be approximately described through

$$\vec{R}_{nk}(t) = \vec{R}_{n0} + \vec{u}_{nk} \cos(\omega_k t). \tag{6.2}$$

Here, $\vec{R}_{n0}$ is the equilibrium position of the nth atom, and k describes the vibrational mode. Furthermore, $\vec{u}_{nk}$ is the displacement of the nth atom for the kth vibrational mode. ω_k is the vibrational frequency for the kth vibrational mode that is common for all atoms.

It can be seen that for the calculation of the energy associated with this vibration, only knowledge of the variation of the energy during the simultaneous movement of all atoms is needed. The quantum-mechanical calculation of this energy is therefore carried through by solving the Schrödinger equation

$$-\frac{\hbar^2}{2\mu} \frac{d^2}{ds^2} \psi(s) + V(s)\psi(s) = E'\psi(s). \tag{6.3}$$

Here, μ is the mass that vibrates (and is rarely easy to determine), and $V(s)$ describes the variation of the energy for the vibrational mode of interest. We shall here discuss two models that differ in the form of $V(s)$. For reasons that will become clear later, we have put the energy on the right-hand side equal to E' instead of E. We emphasize that one does not actually know the vibrational modes and that one has to determine them by a somewhat more complicated procedure (see Section 15.3), but for our purposes it is sufficient to know that the vibrational energies can be determined by solving a one-dimensional Schrödinger equation as in equation (6.3).

6.3 Harmonic oscillator

In the simplest approximation, $V(s)$ is approximated by the potential of the **harmonic oscillator**. Then it is assumed that the energy of the molecule depends quadratically on the deviation from the equilibrium position,

$$V(s) = V_0 + \frac{1}{2}k(s - s_0)^2. \tag{6.4}$$

Here, s is the coordinate describing the vibration.

For this potential, we have the following time-independent Schrödinger equation:

$$-\frac{\hbar^2}{2\mu}\frac{d^2}{ds^2}\psi(s) + \left[V_0 + \frac{1}{2}k(s - s_0)^2\right]\psi(s) = E'\psi(s). \tag{6.5}$$

μ is the mass that vibrates. It is often difficult to determine this mass, but for a diatomic molecule it corresponds to the so-called reduced mass,

$$\mu = \frac{m_1 \cdot m_2}{m_1 + m_2} \tag{6.6}$$

with m_1 and m_2 equal to the masses of the two individual atoms.

We introduce

$$x = s - s_0$$
$$E = E' - V_0 \tag{6.7}$$

and thus obtain the modified Schrödinger equation

$$-\frac{\hbar^2}{2\mu}\frac{d^2}{dx^2}\psi(x) + \frac{1}{2}kx^2\psi(x) = E\psi(x). \tag{6.8}$$

Solving this equation is not very easy and, therefore, the details of how this is done will not be presented here. Instead, we will briefly mention only a few aspects of the outcomes.

- The energy is quantized:

$$E = E_n = \hbar\sqrt{\frac{k}{\mu}}\left(n + \frac{1}{2}\right), \quad n = 0, 1, 2, 3, \ldots. \tag{6.9}$$

- Even the ground state ($n = 0$) has a nonzero energy. This is the so-called zero-point energy.
- That the zero point energy is not equal to zero can be interpreted as a consequence of Heisenberg's uncertainty principle. If the energy were zero, the particle would not move and at the same time would always be at $x = 0$. Then $\Delta p = \Delta x = 0$, and Heisenberg's uncertainty principle would be violated.

- As equation (6.9) shows, the energies are equidistant—a difference to the energies of the particle in a box.
- The wave functions, shown for some values of n in Figure 6.1, have small contributions in the classically forbidden range where the potential is larger than the energy of the system. This is equivalent to the tunnel effect.

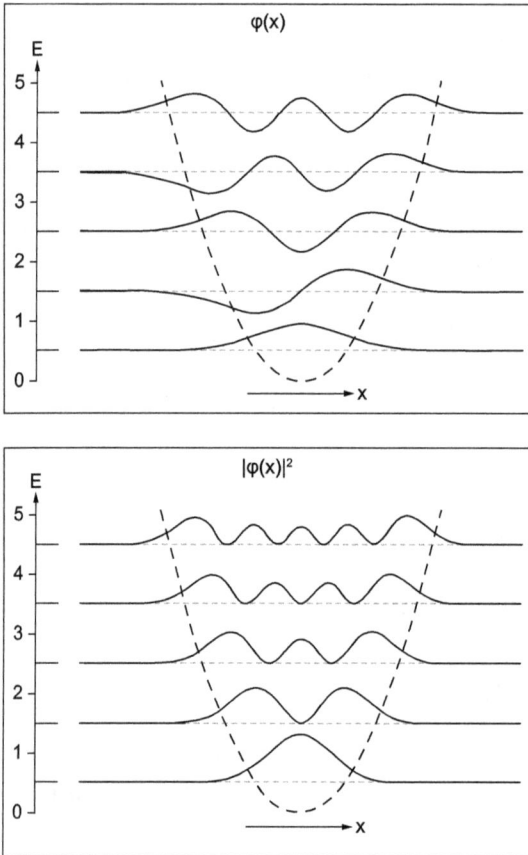

Figure 6.1: The wave functions and their densities for a harmonic oscillator. Adapted from S. Brandt and H. D. Dahmen, *The Picture Book of Quantum Mechanics*, Springer-Verlag, 1995.

- The wave function of the ground state is a Gaussian function,

$$\psi_0(x) = \left(\frac{2\alpha}{\pi}\right)^{1/4} e^{-\alpha x^2}$$

$$\alpha = \frac{\sqrt{\mu k}}{2\hbar}. \tag{6.10}$$

This is the only function for which the "=" in Heisenberg's uncertainty principle holds: $\Delta x \cdot \Delta p = \frac{\hbar}{2}$. Therefore, the Gaussian function is also called "minimum uncertainty function."

–   In the general case, the (normalized) solutions can be written as

$$\psi_n(x) = N_n \cdot H_n(y) \cdot e^{-y^2/2}$$
$$N_n = (\beta\pi^{1/2}2^n n!)^{-1/2}$$
$$\beta = \sqrt{\frac{\hbar}{\sqrt{\mu k}}}$$
$$y = x/\beta. \tag{6.11}$$

Here, $H_n(y)$ is the nth Hermite polynomial (after Charles Hermite). The first of these polynomials are given by

$$H_0(y) = 1$$
$$H_1(y) = 2y$$
$$H_2(y) = 4y^2 - 2$$
$$H_3(y) = 8y^3 - 12y$$
$$H_4(y) = 16y^4 - 48y^2 + 12$$
$$H_5(y) = 32y^5 - 160y^3 + 120y$$
$$H_6(y) = 64y^6 - 480y^4 + 720y^2 - 120. \tag{6.12}$$

For larger n, a useful recursion formula is

$$H_{n+1}(y) = 2yH_n(y) - 2nH_{n-1}(y). \tag{6.13}$$

–   As n becomes larger, the wave function is distributed equally over the entire area of the classically allowed range. This is illustrated in Figure 6.2 and is an example of the so-called Bohr correspondence principle.

The correspondence principle, which is generally valid, states that for very large values of the quantum number(s), the probability density is approaching the classical density.

Finally, in Figure 6.3 we show the temporal evolution of Gaussian functions, $Ne^{-\beta(x-x_0)^2}$, which move in a harmonic potential. Only when the width of the function is that of the ground state [i. e., $\beta = \alpha$ from equation (6.10)], the form of the function remains unchanged. Otherwise, it remains a Gauss function but with varying width. In Figure 6.3, the upper panel shows the case that the initial width $(1/\beta)$ is smaller than the width of the ground state, $1/\alpha$, whereas the middle panel shows the case that the initial width is larger than $1/\alpha$. In the lowest panel, the initial width equals $1/\alpha$.

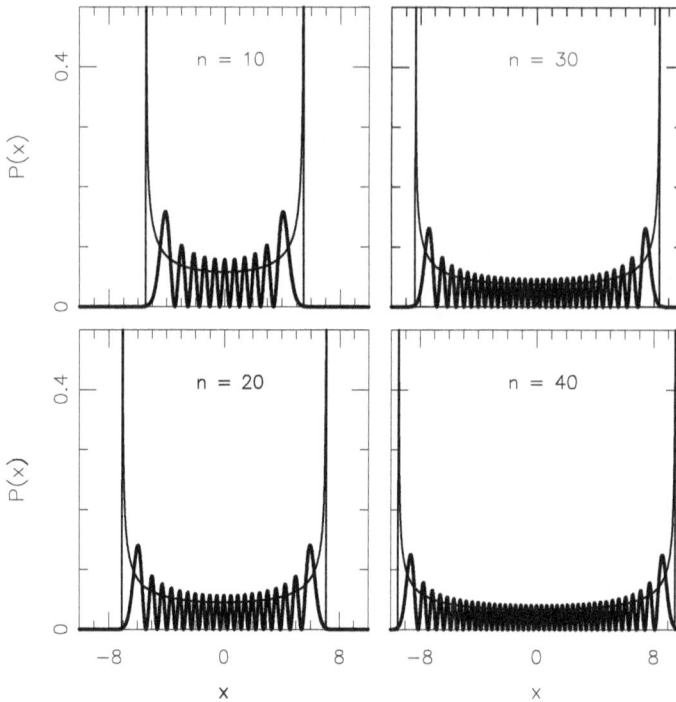

Figure 6.2: The correspondence principle for the harmonic oscillator. The thicker curves show $P_{QM}(x) = |\psi_n(x)|^2$ for the given values of n, while the thinner curves show the classical probability density $P_{Cl}(x)$ for the case that the particle has the same energy as in the quantum case. In both cases $P(x)dx$ is the probability that the particle is found in the range $[x, x + dx]$.

6.4 Momentum representation

The harmonic oscillator has a special symmetry between position and momentum representation. The wave function in momentum space obeys the Schrödinger equation

$$\frac{p^2}{2\mu}\phi(p) - \frac{1}{2}\hbar^2 k\frac{d^2}{dp^2}\phi(p) = E\phi(p). \tag{6.14}$$

By comparison with the equation for the wave function in position space, equation (6.8), we see that the equations are mathematically completely identical, so that if we in equation (6.8) substitute

$$x \to p$$
$$\psi \to \phi$$
$$k \to \frac{1}{\mu}$$
$$\frac{1}{\mu} \to k \tag{6.15}$$

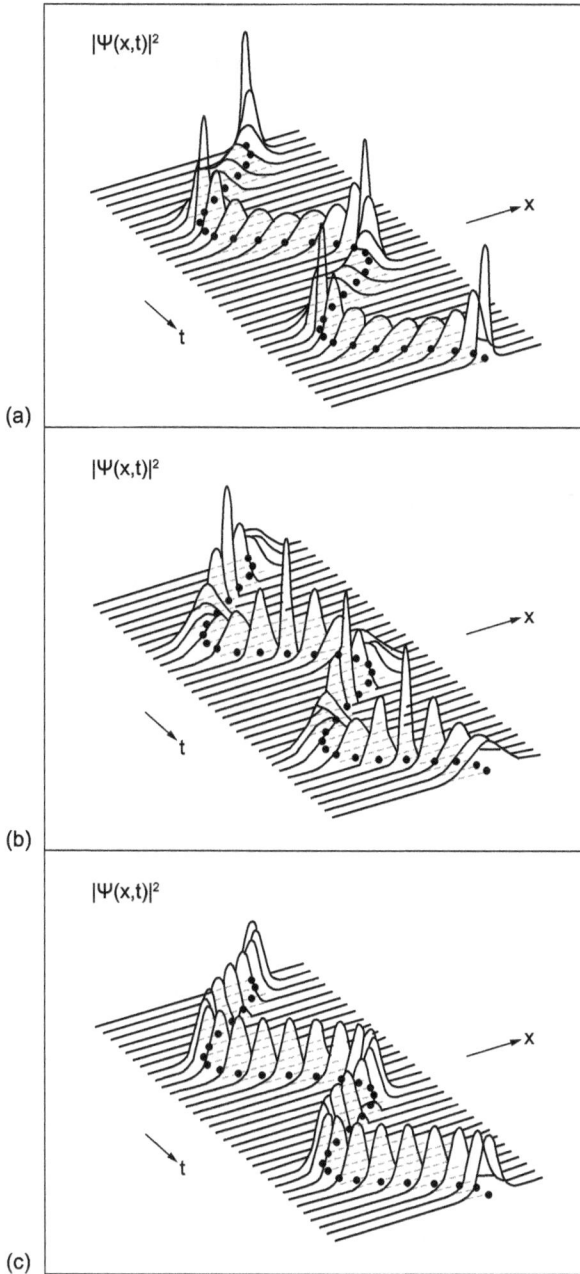

Figure 6.3: Time evolution of Gaussian functions moving in a harmonic potential. For details about the widths of the Gaussians, see the text. Adapted from S. Brandt and H. D. Dahmen, *The Picture Book of Quantum Mechanics*, Springer-Verlag, 1995.

we obtain equation (6.14). Specifically, we thereby get that the energies are unchanged (which must be). Furthermore, the wave functions in spatial and momentum space look identical (except for the substitution above), which is a peculiarity of the harmonic oscillator.

6.5 Morse oscillator

The harmonic potential

$$V(s) = V_0 + \frac{1}{2}k(s - s_0)^2, \tag{6.16}$$

increases both for $s \to \infty$ and for $s \to -\infty$. For us, the harmonic potential is especially important for the application in the description of vibrations in molecules. If we consider a diatomic molecule as a simple example, this means that we assume that the energy of the molecule as a function of the distance between the two atoms approaches infinity, when this distance becomes very large. This is not realistic. Rather, the energy will approach a constant value for larger distances. This can be modelled by replacing the harmonic potential in equation (6.16) with the Morse potential (after Philip McCord Morse),

$$V(s) = V_0 + D\left(e^{-a(s-s_0)} - 1\right)^2. \tag{6.17}$$

In Figure 6.4, the two potentials are compared with each other (where we have set $V_0 = 0$ for the sake of simplicity). The values of the parameters a and D of the Morse potential are chosen so that the curvature for $s = s_0$ is identical to the curvature of the harmonic potential,

$$k = 2Da^2. \tag{6.18}$$

It can be clearly seen that the Morse potential approaches a constant value ($V_0 + D$) as $s \to \infty$.

The Morse potential initially looks mathematically complicated, but has the great advantage that one can solve the associated Schrödinger equation analytically. This is the main reason for considering this model, besides its ability to provide a realistic overall shape for the total energy as a function of bond length. It is hardly possible to derive this model based on physical or chemical considerations.

For the energy one obtains [with k from equation (6.18)]

$$E_n = \hbar\sqrt{\frac{k}{\mu}}\left(n + \frac{1}{2}\right) - \frac{\hbar^2\frac{k}{\mu}}{4D}\left(n + \frac{1}{2}\right)^2. \tag{6.19}$$

As can be seen, the distances between the energies become smaller with increasing n, which can also be seen in Figure 6.4. Furthermore, in contrast to the harmonic potential, there is only a finite number of bound states, i. e., states for which $E_n < D$ and that

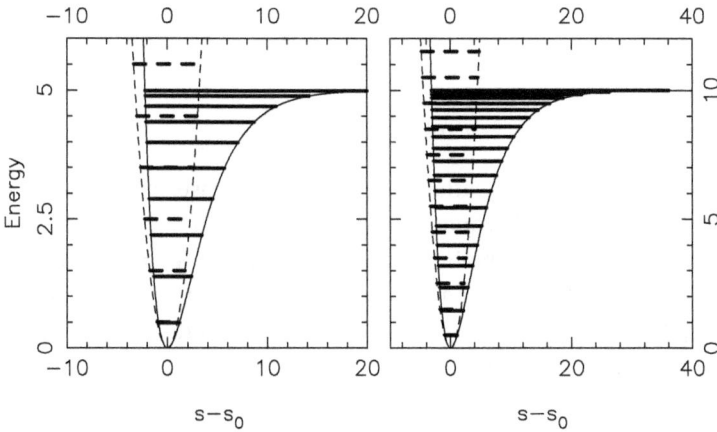

Figure 6.4: A comparison between harmonic and Morse oscillator. In the right panel, D is twice as large as in the left panel, while k has the same value. The potential of the harmonic oscillator and its energy levels are shown by the dashed curves, while the solid curves show the results for the Morse potential.

decay essentially exponentially outside a finite interval in position space. The number of bound states can be estimated in two different ways. At first, we can set

$$E_n = D. \tag{6.20}$$

With

$$g = \hbar \sqrt{\frac{k}{\mu}} \left(n + \frac{1}{2} \right) \tag{6.21}$$

this gives

$$(g - 2D)^2 = 0 \tag{6.22}$$

or

$$n = \frac{2D}{\hbar} \sqrt{\frac{\mu}{k}} - \frac{1}{2}, \tag{6.23}$$

it must be remembered that n must be an integer. Accordingly, the value sought for n equals to the largest integer which is less than the expression in equation (6.23).

Alternatively, the energy difference can be examined,

$$E_{n+1} - E_n = \hbar \sqrt{\frac{k}{\mu}} - \frac{1}{2D} \left(\hbar \sqrt{\frac{k}{\mu}} \right)^2 (n + 1). \tag{6.24}$$

Only values for n, for which this difference is positive, are allowed. Thus, considering

$$E_{n+1} - E_n = 0 \tag{6.25}$$

gives

$$n = \frac{2D}{\hbar} \sqrt{\frac{\mu}{k}} - 1,$$ (6.26)

i. e., a value very close to the value of equation (6.23). Also in this case, only the largest integer smaller than the expression in equation (6.26) is relevant. If two different values are obtained from this, the value obtained by equation (6.23) is the right one.

Thus, with the Morse oscillator we have only a finite number of bound states whose energies obey $E < D$ and that are discrete. However, in addition, we have a continuum of states with energies $E > D$.

6.6 Relevance to experiment

Vibrational spectroscopy is used to measure the energies of transitions between different vibrational energy levels,

$$\Delta E_{nm} = E_n - E_m.$$ (6.27)

Independently of whether one uses the harmonic oscillator or the Morse oscillator for the interpretation, one obtains information on the force constant k and on the "effective" mass μ that vibrates. Of course, these depend on the existing chemical bonds and the masses of the atoms, but unfortunately rarely in a simple way. An exception is that a vibration is strongly localized to a few atoms. In that case, the energy levels are very similar to those found for other molecules with similar groups of atoms. By comparing one can thereby identify such groups. For other vibrations, such assignments are much more difficult, although these vibrational energies may ultimately be important in characterizing the molecule. However, because the vibrational frequencies depend on the masses of the vibrating atoms, isotope substitution offers a possibility to obtain additional information on the vibrations. For example, substituting H by D or ^{12}C by ^{13}C and once again measuring the vibrational spectrum, the frequencies of those vibrations will be changed, in which C and/or H atoms are involved.

6.7 Problems with answers

1. **Problem:** Determine the ground state energy of a particle (mass m) that moves in the potential $V(x) = \frac{k}{2}(x - a)(x + a)$. a is a constant length.
    Answer: We rewrite the potential:

$$V(x) = \frac{k}{2}(x - a)(x + a) = \frac{k}{2}x^2 - \frac{k}{2}a^2.$$ (6.28)

The Schrödinger equation for this system,

$$-\frac{\hbar^2}{2m}\frac{d^2}{dx^2}\psi(x) + \frac{k}{2}x^2\psi(x) - \frac{k}{2}a^2\psi(x) = E\psi(x), \qquad (6.29)$$

is very similar to that of a harmonic oscillator, except that there is an additional constant potential, $-\frac{k}{2}a^2$. Therefore, the energies are identical to those of the harmonic oscillator plus this additional, constant number,

$$E_n = \hbar\sqrt{\frac{k}{m}}\left(n + \frac{1}{2}\right) - \frac{k}{2}a^2, \quad n = 0, 1, 2, \ldots. \qquad (6.30)$$

2. **Problem:** Consider the wave function $\psi(x) = N \cdot e^{-\alpha(x-c)^2}$.
   (a) What is the value of N?
   (b) What is the value of $\int \psi^*(x)\hat{x}\psi(x)\,dx$?
   (c) What is the value of $\int \psi^*(x)\hat{x}^2\psi(x)\,dx$?
   (d) What is the value of $\int \psi^*(x)\hat{p}_x\psi(x)\,dx$?
   (e) What is the value of $\int \psi^*(x)\hat{p}_x^2\psi(x)\,dx$?
   (f) What is the value of $\Delta x \cdot \Delta p_x$?
   (g) What is the probability that the particle is between $x = -\infty$ and $x = c$? Justify the answer.

Answer:
   (a) Without limitation, we can assume that N is real and positive. Then

$$1 = \int_{-\infty}^{\infty} |\psi(x)|^2\,dx = N^2 \int_{-\infty}^{\infty} e^{-2\alpha(x-c)^2}\,dx = N^2 \int_{-\infty}^{\infty} e^{-2\alpha z^2}\,dz$$

$$= 2N^2 \int_{0}^{\infty} e^{-2\alpha z^2}\,dz = N^2\sqrt{\frac{\pi}{2\alpha}}, \qquad (6.31)$$

where we have substituted

$$z = x - c \qquad (6.32)$$

and have used formulas from Chapter 17.
From equation (6.31), we get then

$$N = \left(\frac{2\alpha}{\pi}\right)^{1/4}. \qquad (6.33)$$

   (b)

$$\int \psi^*(x)\hat{x}\psi(x)\,dx = \int_{-\infty}^{\infty} \psi(x)x\psi(x)\,dx = N^2 \int_{-\infty}^{\infty} xe^{-2\alpha(x-c)^2}\,dx$$

$$= N^2 \int_{-\infty}^{\infty} [(x-c) + c]e^{-2\alpha(x-c)^2}\,dx = N^2 \int_{-\infty}^{\infty} [z + c]e^{-2\alpha z^2}\,dz$$

$$= N^2 \int\limits_{-\infty}^{\infty} z e^{-2\alpha z^2}\, dz + N^2 \int\limits_{-\infty}^{\infty} c e^{-2\alpha z^2}\, dz = N^2 \int\limits_{-\infty}^{\infty} c e^{-2\alpha z^2}\, dz$$

$$= c N^2 \int\limits_{-\infty}^{\infty} e^{-2\alpha z^2}\, dz = c \tag{6.34}$$

where we have exploited that z is antisymmetric and $e^{-2\alpha z^2}$ is symmetric with respect to $z \to -z$, so that

$$\int\limits_{-\infty}^{\infty} z e^{-2\alpha z^2}\, dz = 0. \tag{6.35}$$

Furthermore, we have used equation (6.31).

(c) In a similar way, we get

$$\int \psi^*(x)\hat{x}^2\psi(x)\, dx = \int\limits_{-\infty}^{\infty} \psi(x) x^2 \psi(x)\, dx = N^2 \int\limits_{-\infty}^{\infty} x^2 e^{-2\alpha(x-c)^2}\, dx$$

$$= N^2 \int\limits_{-\infty}^{\infty} [(x-c)+c]^2 e^{-2\alpha(x-c)^2}\, dx = N^2 \int\limits_{-\infty}^{\infty} [z+c]^2 e^{-2\alpha z^2}\, dz$$

$$= N^2 \int\limits_{-\infty}^{\infty} (z^2 + 2cz + c^2) e^{-2\alpha z^2}\, dz = N^2 \int\limits_{-\infty}^{\infty} (z^2 + c^2) e^{-2\alpha z^2}\, dz$$

$$= N^2 \frac{1}{4\alpha} \sqrt{\frac{\pi}{2\alpha}} + c^2 N^2 \sqrt{\frac{\pi}{2\alpha}} = \frac{1}{4\alpha} + c^2. \tag{6.36}$$

Here, we have used that

$$\int\limits_{-\infty}^{\infty} s^2 e^{-as^2}\, ds = 2 \int\limits_{0}^{\infty} s^2 e^{-as^2}\, ds, \tag{6.37}$$

and then used the integrals given in Chapter 17.

(d) We find

$$\int \psi^*(x)\hat{p}_x\psi(x)\, dx = \int\limits_{-\infty}^{\infty} \psi^*(x)\left[\frac{\hbar}{i}\frac{\partial}{\partial x}\psi(x)\right] dx$$

$$= N^2 \int\limits_{-\infty}^{\infty} e^{-\alpha(x-c)^2}\left[\frac{\hbar}{i}\frac{\partial}{\partial x} e^{-\alpha(x-c)^2}\right] dx$$

$$= N^2(-i\hbar)(-2\alpha) \int\limits_{-\infty}^{\infty} e^{-\alpha(x-c)^2}(x-c)e^{-\alpha(x-c)^2}\, dx$$

$$= N^2(-i\hbar)(-2\alpha) \int\limits_{-\infty}^{\infty} e^{-\alpha z^2} z e^{-\alpha z^2}\, dz = 0. \tag{6.38}$$

(e) We find

$$\int_{-\infty}^{\infty} \psi^*(x)\hat{p}_x^2\psi(x)\,dx = \int_{-\infty}^{\infty} \psi^*(x)\left[-\hbar^2\frac{\partial^2}{\partial x^2}\psi(x)\right]dx$$

$$= N^2 \int_{-\infty}^{\infty} e^{-\alpha(x-c)^2}\left[-\hbar^2\frac{\partial^2}{\partial x^2}e^{-\alpha(x-c)^2}\right]dx$$

$$= N^2(-\hbar^2)\int_{-\infty}^{\infty} e^{-\alpha(x-c)^2}[4\alpha^2(x-c)^2 - 2\alpha]e^{-\alpha(x-c)^2}\,dx$$

$$= N^2(-\hbar^2)\int_{-\infty}^{\infty} e^{-\alpha z^2}[4\alpha^2 z^2 - 2\alpha]e^{-\alpha z^2}\,dz$$

$$= -N^2\hbar^2\left[4\alpha^2\frac{1}{4\alpha} - 2\alpha\right]\sqrt{\frac{\pi}{2\alpha}} = \hbar^2\alpha. \tag{6.39}$$

(f) We get from the previous results

$$(\Delta x)^2 = \langle\psi|x^2|\psi\rangle - \langle\psi|x|\psi\rangle^2 = \left[\frac{1}{4\alpha} + c^2\right] - c^2 = \frac{1}{4\alpha}$$

$$(\Delta p_x)^2 = \langle\psi|p_x^2|\psi\rangle - \langle\psi|p_x|\psi\rangle^2 = \alpha\hbar^2 - 0^2 = \alpha\hbar^2. \tag{6.40}$$

Thus

$$\Delta x \cdot \Delta p_x = \sqrt{\frac{1}{4\alpha}} \cdot \sqrt{\alpha\hbar^2} = \frac{1}{2}\hbar. \tag{6.41}$$

As already mentioned, the uncertainty relation is exactly fulfilled for this (Gaussian) function ('=' instead of '>').

(g) The sought probability is

$$P = \int_{-\infty}^{c} |\psi(x)|^2\,dx = N^2 \int_{-\infty}^{c} e^{-2\alpha(x-c)^2}\,dx = N^2 \int_{-\infty}^{0} e^{-2\alpha z^2}\,dz$$

$$= N^2\frac{1}{2}\int_{-\infty}^{\infty} e^{-2\alpha z^2}\,dz = \frac{1}{2}. \tag{6.42}$$

3. **Problem:** How many eigenstates lie in the interval $[\frac{5}{6}\hbar\omega, \frac{29}{6}\hbar\omega]$ for a one-dimensional harmonic oscillator (mass m, force constant k, $\omega = \sqrt{\frac{k}{m}}$)?
    Answer: The energies are given as

$$E_n = \hbar\sqrt{\frac{k}{m}}\left(n + \frac{1}{2}\right) \tag{6.43}$$

with n being nonnegative and integral.
Between $\frac{5}{6}\hbar\omega = \frac{5\pi}{3}\hbar\omega \approx 5.24\hbar\omega$ and $\frac{29}{6}\hbar\omega = \frac{29\pi}{3}\hbar\omega \approx 30.37\hbar\omega$ we have then the energy levels for $n = 6, 7, \ldots, 30$, i.e., a total of 25 eigenstates.

4. **Problem:** Consider a particle (mass m) that moves in the potential $V(x,y)$. $V(x,y) = c \cdot (y - d)^2$ for $a \le x \le b$ and $V(x,y) = \infty$ otherwise. Determine the energy eigenvalues of this particle. Prove that for certain values of the parameters, energetical degeneracies can occur.

Answer: The Schrödinger equation,

$$-\frac{\hbar^2}{2m}\left(\frac{\partial^2}{\partial x^2} + \frac{\partial^2}{\partial y^2}\right)\psi(x,y) + V(x,y)\psi(x,y) = E\psi(x,y),\tag{6.44}$$

can be solved most easily through a product Ansatz,

$$\psi(x,y) = \psi_x(x)\psi_y(y)\tag{6.45}$$

Substituting this into the Schrödinger equation, one obtains in the "usual" way

$$-\frac{\hbar^2}{2m}\frac{d^2}{dx^2}\psi_x = E_x\psi_x(x)$$
$$\psi_x(a) = \psi_x(b) = 0$$

$$-\frac{\hbar^2}{2m}\frac{d^2}{dy^2}\psi_y(y) + c \cdot (y - d)^2\psi_y(y) = E_y\psi_y(y)$$

$$E_x + E_y = E.\tag{6.46}$$

The equation for ψ_x is the equation for a particle in a one-dimensional box with length $L = b - a$. From this, we obtain the energies

$$E_x = \frac{\hbar^2\pi^2}{2m(b - a)^2}n_x^2\tag{6.47}$$

with $n_x = 1, 2, 3, \ldots$.

Similarly, we see that the equation for ψ_y equals the equation for a harmonic oscillator with force constant $k = 2c$. From that, we get the energies

$$E_y = \hbar\sqrt{\frac{2c}{m}}\left(n_y + \frac{1}{2}\right) = \hbar\sqrt{\frac{c}{2m}}(2n_y + 1)\tag{6.48}$$

with $n_y = 0, 1, 2, 3, \ldots$.

The total energy is then

$$E = E_{n_x,n_y} = E_x + E_y = \frac{\hbar^2\pi^2}{2m(b - a)^2}n_x^2 + \hbar\sqrt{\frac{c}{2m}}(2n_y + 1).\tag{6.49}$$

Energetic degeneracy for two states characterized by the indices 1 and 2 can occur when

$$\frac{\hbar^2\pi^2}{2m(b - a)^2}(n_{x1}^2 - n_{x2}^2) = \hbar\sqrt{\frac{c}{2m}}2(n_{y2} - n_{y1})\tag{6.50}$$

that can have solution if, for instance,

$$\frac{\hbar^2 \pi^2}{2m(b-a)^2} = \hbar\sqrt{\frac{c}{2m}} \cdot (2N+1). \tag{6.51}$$

$N \geq 0$ is an integer. Setting

$$E_0 = \hbar\sqrt{\frac{c}{2m}} \tag{6.52}$$

we then have

$$E_{n_x,n_y} = E_0\left[(2N+1)n_x^2 + (2n_y+1)\right]. \tag{6.53}$$

By inserting one can easily convince oneself that energetic degeneracies may occur. For example, for $N = 0$ the energy levels $(n_x, n_y) = (1, 13)$, $(3, 9)$, and $(5, 1)$ are energetically degenerate.

6.8 Problems

1. Which functions are used to describe the eigenfunctions of a harmonic oscillator? How do the energies depend on the force constant and on the mass of the oscillator?

2. Determine the constants A, α, and a, so that $\psi(x) = A \cdot e^{-\alpha(x-a)^2}$ is a normalized eigenfunction to the Schrödinger equation for a particle (mass m) in the potential $V(x) = c \cdot (x-1)^2$.

3. Explain the term "Zero point energy."

4. Determine the ground state energy of a particle (mass $3m$) that moves in one dimension in the potential $V(z) = k(2z-3)(2z+3)$.

5. Consider a particle in a one-dimensional box with length L and a particle moving in a one-dimensional harmonic oscillator with force constant k. In each case, the mass of the particle equals m. What relationship between L and k must apply so that the ground state energy of both systems is equal. The lowest potential energy of the particle in the box and the harmonic oscillator are both zero.

6. Describe how vibrational spectra can be calculated theoretically.

7. Consider two bodies with the masses $3m$ and $5m$, and connected by a spring (force constant $2k$). Determine the energy levels of this system expressed with the help of m, k, and $\hbar$.

8. Explain the correspondence principle.

9. Describe briefly the Morse oscillator and its energy levels.

10. Consider a particle (mass m) that moves in the potential $V(x,y) = c \cdot (y-d)^2 + 4c \cdot (x-a)^2$ (a, c, and d are constants). Determine the energy eigenvalues for this system. Which energy levels are energetic degenerate?

11. Sketch the first three energy levels, wave functions, and the corresponding probability densities of a one-dimensional harmonic oscillator.
12. Discuss briefly why isotope substitution can be helpful in characterizing a molecule.

7 Rotations

7.1 Angular momentum and moment of inertia

Before discussing the quantum theoretical treatment of rotations, we will briefly describe two fundamental concepts of physics. They should actually be well known(!), but it may be useful to repeat their definitions here.

The first quantity is that of the angular momentum. We first consider a particle moving on a circular path; see Figure 7.1. The angular momentum $\vec{l}$ is a vector that is defined as

$$\vec{l} = \vec{r} \times \vec{p}. \tag{7.1}$$

Here, $\vec{r}$ is the vector that points from the center of the circle to the particle, and $\vec{p}$ is the momentum of the particle. All three vectors are shown in Figure 7.1. It is easy to see that although $\vec{r}$ and $\vec{p}$ change directions along the circle, this is not the case for $\vec{l}$. In fact, according to classical physics, $\vec{l}$ is a constant of motion, which means that $\vec{l}$ can only be changed under the influence of external forces. Other, perhaps better known, constants of motion are energy and momentum. The fact that $\vec{l}$ is a constant of motion also explains why you can ride a bicycle and that, in principle, this becomes easier the faster you drive.

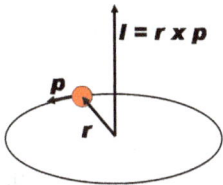

Figure 7.1: The angular momentum of a particle moving on a circular path.

Some movements are more complex and cannot be described by a simple circular path. Figure 7.2 shows an example of such a movement. Each small animal undergoes a superposition of two circular movements: on the one hand, the whole system with all the animals rotates around a common axis, and on the other hand, each cup with one animal rotates around its axis. Also, for this composite motion, there is an angular momentum vector, namely the sum of the vectors of the individual motions.

This is shown in Figure 7.3 for an example. Here, the three angular momentum vectors for three circular motions, which are not even in the same plane, are vectorially added together to form a total angular momentum vector.

There is a simple reason why we are interested in angular momenta. Electrons or atomic nuclei, which make any closed, more or less circular movements, ultimately form nothing but small circuits. Such can be influenced by magnetic fields, which is

https://doi.org/10.1515/9783110742206-007

Figure 7.2: A more complex movement composed of several circular movements.

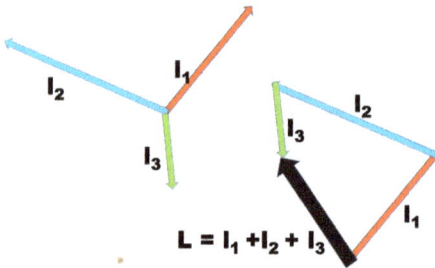

Figure 7.3: The total angular momentum, $\vec{L}$, for a motion composed of three separate rotations with angular momenta $\vec{l}_1, \vec{l}_2$, and $\vec{l}_3$, respectively.

what is exploited in magnetic resonance spectroscopy (NMR, ESR, EPR, for example). Because, as always, the energy is quantized (i. e., can only assume certain values), a small circuit in a magnetic field can only have a finite number of energies, whose energy differences then can be measured by means of spectroscopy. These energy levels depend not only on the applied magnetic field but also on a property of the circuits— their angular momenta. We will discuss this relationship in more detail in Section 7.5.

The second quantity that we will briefly reintroduce is the moment of inertia. Again we consider the system of Figure 7.1. The particle, with mass m, rotates about an axis (containing the angular momentum vector) at a distance r to this. The moment of inertia around this axis is defined as

$$I = mr^2. \tag{7.2}$$

If we have several particles that can have different masses and distances to the axis, the total moment of inertia is given by

$$I = \sum_i m_i r_i^2. \tag{7.3}$$

Here, m_i is the mass of the ith particle and r_i its distance to the axis. We emphasize that the moment of inertia makes sense only if you define the associated axis.

Equation (7.3) can be generalized so that also larger bodies can be treated. We let $\rho(\vec{s})$ be the density of the body at the point $\vec{s}$ and then have in general the analogy to above,

$$I = \int \rho(\vec{s}) r^2(\vec{s}) \, d\vec{s}. \tag{7.4}$$

$r(\vec{s})$ is the distance from the point $\vec{s}$ to the axis.

7.2 2D rotor

Before we present the correct quantum mechanical treatment of the two-dimensional rotor, we discuss a simplified treatment based mainly on classical arguments.

We consider a system as in Figure 7.1, a particle moving at a constant angular velocity on a circular path. Our arguments, however, are not limited to this example, but the result will be generally valid if we consider a body with a given moment of inertia rotating about an axis with constant angular velocity. But let us start with the system of Figure 7.1.

The total energy of the particle is equal to the kinetic energy because we assume that no forces are acting on the system. So

$$E = E_{\text{kin}} = \frac{p^2}{2m}. \tag{7.5}$$

At the same time, the magnitude of the angular momentum equals

$$l = pr. \tag{7.6}$$

We have set the mass and momentum of the particle equal to m and p, respectively, while r is the radius of the circular orbit.

By combining equations (7.5) and (7.6), we get

$$E = \frac{l^2}{2mr^2} = \frac{l^2}{2I}. \tag{7.7}$$

Here is

$$I = mr^2 \tag{7.8}$$

the moment of inertia of the system.

At this point, we introduce the de Broglie relation, which provides a link between the wave and particle nature of quantum mechanical objects,

$$p = \frac{h}{\lambda}. \tag{7.9}$$

λ is the wavelength of the system if the system is considered as a quantum mechanical wave. This wave is accordingly located on the circular path. Requiring that the wave does not annihilate itself (in other words, requiring that there is a constructive interference), a whole number of waves must fit on the circular path. Accordingly, the length of this path, $2\pi r$, must be equal to an integral multiple of λ,

$$2\pi r = n\lambda, \tag{7.10}$$

with $n = 1, 2, 3, \ldots$.

From equations (7.6), (7.9), and (7.10), we obtain then

$$l = pr = \frac{hr}{\lambda} = \frac{hr}{2\pi r/n} = n \cdot \frac{h}{2\pi} = n \cdot \hbar, \tag{7.11}$$

and according to equation (7.7) the energy becomes then

$$E = \frac{\hbar^2}{2I} n^2. \tag{7.12}$$

This means that both the energy and the magnitude of the angular momentum are quantized.

We will now show that this result is also obtained by a "correct" quantum theoretical treatment. For this, we assume that the particle is moving in the (x, y) plane and that it moves around the origin of the coordinate system. The two-dimensional Schrödinger equation then becomes

$$-\frac{\hbar^2}{2m}\left(\frac{\partial^2}{\partial x^2} + \frac{\partial^2}{\partial y^2}\right)\psi = E \cdot \psi. \tag{7.13}$$

Actually, we only have one motional degree of freedom because

$$x^2 + y^2 = r^2 \tag{7.14}$$

is constant. Therefore, equation (7.14) is less useful. Instead, we introduce polar coordinates in two dimensions,

$$x = r \cdot \cos\phi$$
$$y = r \cdot \sin\phi. \tag{7.15}$$

If the above expressions are used in equation (7.14), we need expressions like

$$\frac{\partial}{\partial x} = \frac{\partial}{\partial r} \cdot \frac{\partial r}{\partial x} + \frac{\partial}{\partial \phi} \cdot \frac{\partial \phi}{\partial x} \tag{7.16}$$

and

$$\frac{\partial^2}{\partial x^2} = \frac{\partial^2}{\partial r^2} \cdot \left(\frac{\partial r}{\partial x}\right)^2 + \frac{\partial}{\partial r} \cdot \frac{\partial^2 r}{\partial x^2} + \frac{\partial^2}{\partial \phi^2} \cdot \left(\frac{\partial \phi}{\partial x}\right)^2 + \frac{\partial}{\partial \phi} \cdot \frac{\partial^2 \phi}{\partial x^2} \tag{7.17}$$

and the similar expressions for the derivatives with respect to y. If you do not make mistakes (or you just look up in a mathematical collection of formulas where you can easily find those formulas), you get

$$\frac{\partial^2}{\partial x^2} + \frac{\partial^2}{\partial y^2} = \frac{\partial^2}{\partial r^2} + \frac{1}{r}\frac{\partial}{\partial r} + \frac{1}{r^2}\frac{\partial^2}{\partial \phi^2}. \tag{7.18}$$

In our case, the whole becomes particularly simple: because r is constant, all differentiations with respect to r can be ignored (r can not be varied), so that the Schrödinger equation (7.13) can be written as

$$-\frac{\hbar^2}{2m}\frac{1}{r^2}\frac{\partial^2}{\partial \phi^2}\psi(\phi) = E\psi(\phi) \tag{7.19}$$

Here, we have also taken advantage of the fact that the wave function depends only on the angle ϕ.

Equation (7.19) can be rewritten as

$$-\frac{\hbar^2}{2I}\frac{\partial^2}{\partial \phi^2}\psi(\phi) = E\psi(\phi). \tag{7.20}$$

This equation has the solutions

$$\psi(\phi) = A \cdot e^{im_l\phi} \tag{7.21}$$

with A equal to a constant and

$$m_l = \pm\sqrt{\frac{2IE}{\hbar^2}}. \tag{7.22}$$

We must have (see Figure 7.4) that

$$\psi(\phi + 2\pi) = \psi(\phi), \tag{7.23}$$

(this is called cyclic boundary conditions), from which we get that m_l must be integral. Then equation (7.23) becomes

$$E = \frac{\hbar^2 m_l^2}{2I}, \quad m_l = 0, \pm1, \pm2, \ldots, \tag{7.24}$$

i. e., just as in equation (7.12).

The angular momentum vector in this case lies along the z axis, so that its x and y components vanish, and the length of the angular momentum vector becomes equal to the absolute value of its z component. The quantum mechanical operator of its z component is

$$\hat{l}_z = \hat{x} \cdot \hat{p}_y - \hat{y} \cdot \hat{p}_x = \frac{\hbar}{i}\left(x\frac{\partial}{\partial y} - y\frac{\partial}{\partial x}\right) = \frac{\hbar}{i}\frac{\partial}{\partial \phi} \tag{7.25}$$

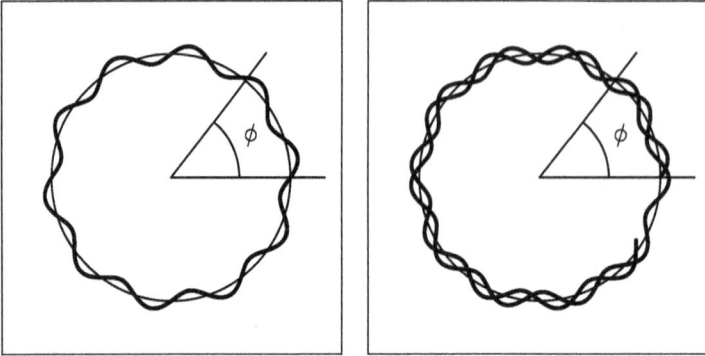

Figure 7.4: The cyclic boundary conditions for the two-dimensional rotor. The picture shows the imaginary part of $e^{im\phi}$, i.e. $\sin(im\phi)$, for (left) m integral and (right) m nonintegral. The thicker curve shows the function value relative to 0 (shown as the thinner curve) as a function of the angle ϕ. It can be seen how only in the left panel the function is mapped onto itself, when ϕ is increased by 2π.

by expressing it in terms of polar coordinates. For the functions of equation (7.21), we obtain immediately,

$$\hat{l}_z \psi(\phi) = \frac{\hbar}{i} \frac{\partial}{\partial \phi} (A \cdot e^{im_l \phi}) = \hbar m_l A \cdot e^{im_l \phi} = \hbar m_l \psi(\phi), \tag{7.26}$$

i. e., the wave function is also an eigenfunction to $\hat{l}_z$ with the eigenvalue $m_l \hbar$.

Finally, we determine the normalization constant A of the wave function. Because the wave function is to be normalized over its entire domain (i. e., ϕ from 0 to 2π), we have

$$\int_0^{2\pi} |\psi(\phi)|^2 \, d\phi = |A|^2 \cdot 2\pi \equiv 1, \tag{7.27}$$

where we choose

$$A = \frac{1}{\sqrt{2\pi}}. \tag{7.28}$$

In Figure 7.5, the real and imaginary parts of the wave functions are shown for some of the smallest, nonnegative values of m_l. This figure also shows the angular momentum vectors.

7.3 3D rotor

For the two-dimensional rotor, we considered a particle that could move in a two-dimensional plane, with the restriction that the distance to a certain point, which

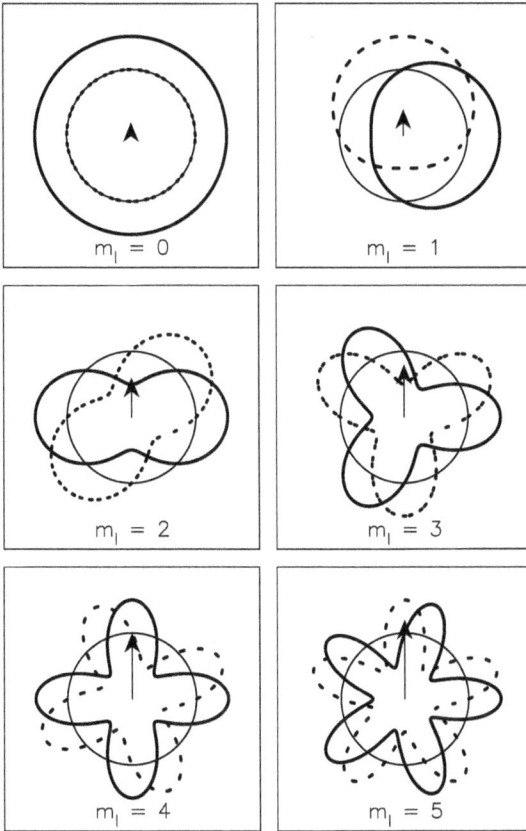

Figure 7.5: The six energetic lowest states for the two-dimensional rotor. The thick solid curves represent the real parts of the wave functions, and the thick, dashed curves represent the imaginary parts. The thin curves represent 0, and the arrows in the center show the angular momentum vectors, which are supposed to be perpendicular to the plane of the images.

we later chose as the origin of the coordinate system, is constant. For the three-dimensional rotor, we proceed in a similar way. This time, the particle can move in a three-dimensional space with the restriction that the distance to a certain point, which in turn we will choose as the origin of the coordinate system, must be constant. In this case, we will present only the quantum theoretical treatment.

As with the two-dimensional rotor, it is again convenient to use a suitable coordinate system. This time, we will use the polar coordinates in three dimensions, also called spherical coordinates. Their definition is shown in Figure 7.6.

We have

$$x = r \sin \theta \cos \phi$$
$$y = r \sin \theta \sin \phi$$
$$z = r \cos \theta. \tag{7.29}$$

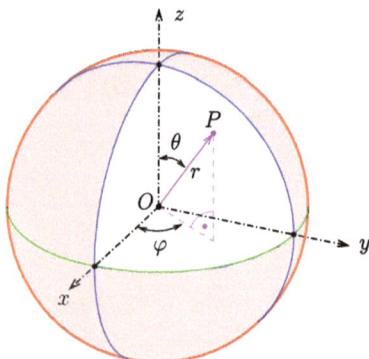

Figure 7.6: The definition of polar coordinates in three dimensions, also called spherical coordinates.

Also, for these coordinates, we need the expression of the Laplace operator [see equation (2.4)]

$$\nabla^2 = \frac{\partial^2}{\partial x^2} + \frac{\partial^2}{\partial y^2} + \frac{\partial^2}{\partial z^2}$$

$$= \frac{\partial^2}{\partial r^2} + \frac{2}{r}\frac{\partial}{\partial r} + \frac{1}{r^2 \sin^2\theta}\frac{\partial^2}{\partial\phi^2} + \frac{1}{r^2 \sin\theta}\frac{\partial}{\partial\theta}\sin\theta\frac{\partial}{\partial\theta}$$

$$\equiv \frac{\partial^2}{\partial r^2} + \frac{2}{r}\frac{\partial}{\partial r} + \frac{1}{r^2}\hat{\Lambda}^2. \tag{7.30}$$

We have here introduced the operator

$$\hat{\Lambda}^2 = \frac{1}{\sin^2\theta}\frac{\partial^2}{\partial\phi^2} + \frac{1}{\sin\theta}\frac{\partial}{\partial\theta}\sin\theta\frac{\partial}{\partial\theta} \tag{7.31}$$

that contains all differentiations with respect to the two angles ϕ and θ.

For the three-dimensional rotor, r is constant, so that in the Schrödinger equation all terms with derivatives with respect to r can be ignored. Then the Schrödinger equation becomes

$$-\frac{\hbar^2}{2mr^2}\hat{\Lambda}^2\psi(\theta,\phi) = E\psi(\theta,\phi), \tag{7.32}$$

where we have explicitly stated that the wave function ψ depends on the two angles θ and ϕ, but not on r. Equation (7.32) can also be written as

$$\hat{\Lambda}^2\psi(\theta,\phi) = -\frac{2IE}{\hbar^2}\psi(\theta,\phi), \tag{7.33}$$

where I is the moment of inertia about the origin of the coordinate system (in contrast to the previous section, not about an axis but about a point, because the particle is moving about a single point).

In Sections 2.2 and 4.11, we have seen that it can be helpful to use a product Ansatz in solving differential equations that depend on multiple variables. Thus, we set

$$\psi(\theta, \phi) = \Theta(\theta) \cdot \Phi(\phi). \tag{7.34}$$

By inserting this into equation (7.33), we get

$$\hat{\Lambda}^2 \psi = \frac{1}{\sin^2 \theta} \Theta \Phi'' + \Phi \frac{1}{\sin \theta} \frac{\partial}{\partial \theta} [\sin \theta \Theta'] = -\frac{2IE}{\hbar^2} \Theta \Phi. \tag{7.35}$$

By dividing by the product $\Theta \Phi$, it follows

$$\frac{\Phi''}{\Phi} + \frac{1}{\Theta} \sin \theta \frac{\partial}{\partial \theta} [\sin \theta \Theta'] = -\frac{2IE}{\hbar^2} \sin^2 \theta \tag{7.36}$$

or

$$\frac{\Phi''}{\Phi} = -\frac{1}{\Theta} \sin \theta \frac{\partial}{\partial \theta} [\sin \theta \Theta'] - \frac{2IE}{\hbar^2} \sin^2 \theta = -m_l^2. \tag{7.37}$$

In the last identity, we have exploited that the left-hand side depends only on ϕ, while the expression in the middle depends only on θ. This means that they can only be identical if they are a constant. We call this constant $-m_l^2$, for reasons that will become clear soon.

From equation (7.37), we obtain two equations: one for Θ and one for Φ. The latter is

$$\frac{\Phi''}{\Phi} = -m_l^2. \tag{7.38}$$

This has the solution

$$\Phi(\phi) = A e^{im_l \phi}, \tag{7.39}$$

whereby m_l can be positive, zero, or negative.

As for the two-dimensional rotor, cyclic boundary conditions also apply here:

$$\Phi(\phi + 2\pi) = \Phi(\phi) \tag{7.40}$$

where

$$\Phi(\phi) = \frac{1}{\sqrt{2\pi}} e^{im_l \phi}. \tag{7.41}$$

m_l is accordingly equal to an integer. Furthermore, we have chosen A in equation (7.39) such that $\Phi(\phi)$ is normalized,

$$\int_0^{2\pi} |\Phi(\phi)|^2 \, d\phi = |A|^2 \cdot 2\pi \equiv 1. \tag{7.42}$$

The other equation, which is given in equation (7.37), is the equation for $\Theta(\theta)$. This equation is

$$\sin\theta\frac{\partial}{\partial\theta}[\sin\theta\Theta'] + \left[\frac{2IE}{\hbar^2}\sin^2\theta - m_l^2\right]\Theta = 0. \tag{7.43}$$

This equation is not easy to solve. It is called Legendre's equation, and fortunately, the solutions to the equation are known. The solutions can be expressed using the so-called Legendre polynomials (after Adrien-Marie Legendre). These are given by

$$P_l(z) = \frac{1}{2^l l!}\frac{d^l}{dz^l}(z^2 - 1)^l. \tag{7.44}$$

We see that here we have introduced a second number l (besides m_l). l can take the values $0, 1, 2, \ldots$. Using the Legendre polynomials, the (normalized) solutions to equation (7.43) are then expressed as

$$\Theta(\theta) = S_{lm_l}(\theta) = (-1)^{m_l}\sqrt{\frac{2l+1}{2}}\sqrt{\frac{(l-m_l)!}{(l+m_l)!}}\sin^{m_l}(\theta)\frac{d^{m_l}P_l(\cos(\theta))}{(d\cos(\theta))^{m_l}}. \tag{7.45}$$

One often introduces the so-called associated Legendre functions

$$P_l^{m_l}(z) = (1-z^2)^{m_l/2}\frac{d^{m_l}}{dz^{m_l}}P_l(z). \tag{7.46}$$

Then the wave functions ψ in equation (7.47) become

$$\psi(\theta,\phi) = \Theta(\theta)\cdot\Phi(\phi) \equiv Y_{lm_l}(\theta,\phi)$$

$$= \begin{cases} (-1)^{m_l}\sqrt{\frac{2l+1}{2}}\sqrt{\frac{(l-m_l)!}{(l+m_l)!}}P_l^{m_l}(\cos\theta)\frac{e^{im_l\phi}}{\sqrt{2\pi}} & m_l \geq 0 \\ (-1)^{m_l}Y_{l,-m_l}^*(\theta,\phi) & m_l < 0. \end{cases} \tag{7.47}$$

It turns out that there are solutions for only certain values of (l, m_l):

$$l = 0, 1, 2, \ldots$$
$$m_l = -l, -l+1, -l+2, \ldots, l-2, l-1, l. \tag{7.48}$$

Moreover, the Y functions are called **spherical harmonic functions** or just **spherical harmonics**.

The mathematical expressions of the spherical harmonics for $l = 0, 1, 2$, and 3 are shown in Table 7.1.

The solutions to the Schrödinger equation, equation (7.32), can therefore be expressed in terms of the spherical harmonics. This equation is a two-dimensional differential equation (we have two coordinates: θ and ϕ), and so their solution leads to the introduction of two quantum numbers, which we have labeled l and m_l, and which

Table 7.1: The spherical harmonic functions for $l = 0, 1, 2,$ and 3.

l	m_l	$Y_{lm_l}(\theta, \phi)$
0	0	$\sqrt{\frac{1}{4\pi}}$
	1	$-\sqrt{\frac{3}{8\pi}} \sin\theta e^{i\phi}$
1	0	$\sqrt{\frac{3}{4\pi}} \cos\theta$
	-1	$\sqrt{\frac{3}{8\pi}} \sin\theta e^{-i\phi}$
	2	$\sqrt{\frac{15}{32\pi}} \sin^2\theta e^{2i\phi}$
	1	$-\sqrt{\frac{15}{8\pi}} \sin\theta \cos\theta e^{i\phi}$
2	0	$\sqrt{\frac{5}{16\pi}}(3\cos^2\theta - 1)$
	-1	$\sqrt{\frac{15}{8\pi}} \sin\theta \cos\theta e^{-i\phi}$
	-2	$\sqrt{\frac{15}{32\pi}} \sin^2\theta e^{-2i\phi}$
	3	$-\frac{1}{8}\sqrt{\frac{35}{\pi}} \sin^3\theta e^{3i\phi}$
	2	$\frac{1}{4}\sqrt{\frac{105}{2\pi}} \cos\theta \sin^2\theta e^{2i\phi}$
	1	$-\frac{1}{8}\sqrt{\frac{21}{\pi}}(5\cos^2\theta - 1)\sin\theta e^{i\phi}$
3	0	$\frac{1}{4}\sqrt{\frac{7}{\pi}}(5\cos^2\theta - 3)\cos\theta$
	-1	$\frac{1}{8}\sqrt{\frac{21}{\pi}}(5\cos^2\theta - 1)\sin\theta e^{-i\phi}$
	-2	$\frac{1}{4}\sqrt{\frac{105}{2\pi}} \cos\theta \sin^2\theta e^{-2i\phi}$
	-3	$\frac{1}{8}\sqrt{\frac{35}{\pi}} \sin^3\theta e^{-3i\phi}$

can only assume certain values. This applies in general: a Schrödinger equation in P dimensions leads to P quantum numbers that can only assume certain values.

The energy we obtain is

$$E = \frac{l(l+1)\hbar^2}{2I}.$$ (7.49)

We see that E depends only on l but not on m_l. This means that all $2l + 1$ wave functions Y_{lm_l} with the same l but the different m_l values lead to the same energy: they are energetically degenerate.

At the end of this chapter, we discuss the angular momentum. In Section 7.1, we saw that the angular momentum can be written as

$$\vec{l} = (l_x, l_y, l_z) = \vec{r} \times \vec{p} = (yp_z - zp_y, zp_x - xp_z, xp_y - yp_x).$$ (7.50)

In Section 7.2, we also introduced the quantum mechanical operator for l_z,

$$\hat{l}_z = \frac{\hbar}{i}\frac{\partial}{\partial\phi}$$ (7.51)

It can also be shown that the operator for the square of the length of the angular momentum vector becomes

$$\hat{l}^2 = \hat{l}_x\hat{l}_x + \hat{l}_y\hat{l}_y + \hat{l}_z\hat{l}_z = -\hbar^2\hat{\Lambda}^2. \tag{7.52}$$

Except for a factor $2I$, $\hat{l}^2$ is therefore identical to the Hamilton operator so that

$$\hat{l}^2 Y_{lm_l} = l(l+1)\hbar^2 Y_{lm_l}. \tag{7.53}$$

Simultaneously, it is easily shown by inserting that

$$\hat{l}_z Y_{lm_l} = m_l\hbar Y_{lm_l}. \tag{7.54}$$

The spherical harmonics are thus eigenfunctions both for the operator for the square of the length of the angular momentum vector and for the operator for its z component. At the same time, the second identify in equation (7.48) must be satisfied, which we write as

$$|m_l\hbar| \le l\hbar \le \sqrt{l(l+1)}\hbar \tag{7.55}$$

(and the "=" in the second identity applies only to $l = m_l = 0$). This means that the z component of the angular momentum vector is smaller than the length of the angular momentum vector, i. e., the vector cannot lie along the z axis. This result for the angular momentum can be interpreted as shown in Figure 7.7.

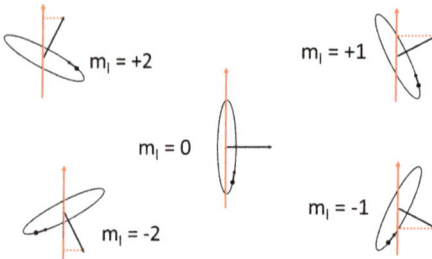

Figure 7.7: The five allowed orientations of the angular momentum vector for $l = 2$. The normal (which is parallel to the angular momentum vector) of the circular orbit of the rotating particle makes a fixed angle to the z axis (shown in red).

Thus, our results give that both the length of the angular momentum and its z component are quantized. But we know nothing about the x and y components, which in principle can take any values. For given l and m_l, the angular momentum vector can therefore lie at any point on one of the circles shown in the right part of Figure 7.8. The vector can even follow a path on this circle. One speaks then of precession.

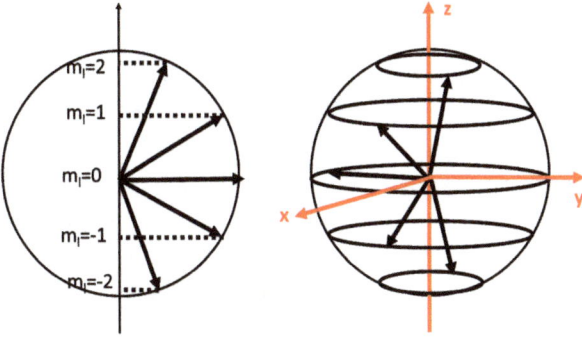

Figure 7.8: The allowed orientations of the angular momentum vector for $l = 2$. As in Figure 7.7, the normal (which is parallel to the angular momentum vector) of the circular orbit of the rotating particle makes a fixed angle to the z axis (shown in red). But because the angles to the other two directions are not fixed, all points on the circles in the right part of the figure are allowed directions of the angular momentum vector.

Precisely because of this precession, the x and y components of the angular momentum vector have no unique values. Would you try, e. g., to measure the x component after determining the z component, the precession will rotate and then take place around the x axis. This means that now, although the x component of the angular momentum vector is known, you lose any information about the z component. Ultimately, this is a consequence of the fact that the three components of the angular momentum vector do not commute,

$$[\hat{l}_x, \hat{l}_y] = i\hbar\hat{l}_z$$
$$[\hat{l}_y, \hat{l}_z] = i\hbar\hat{l}_x$$
$$[\hat{l}_z, \hat{l}_x] = i\hbar\hat{l}_y, \tag{7.56}$$

so that the exact determination of more components of the angular momentum vector is not possible [cf. equation (3.71)].

From equation (7.47), it is easy to see that $|Y_{lm_l}(\theta, \phi)|^2$ is independent of ϕ, and that

$$|Y_{lm_l}(\theta, \phi)|^2 = |Y_{l,-m_l}(\theta, \phi)|^2. \tag{7.57}$$

In order to depict the spherical harmonics $|Y_{lm_l}(\theta, \phi)|^2$, one can proceed as follows. Y_{lm_l} is a function of the two angles θ and ϕ, which describe a direction in the three-dimensional coordinate system. Therefore, one chooses a direction (i. e., values for θ and ϕ), computes $|Y_{lm_l}(\theta, \phi)|^2$, and marks a point with this value along the direction (θ, ϕ). Having done this for all values of θ and ϕ, one has generated a surface in three-dimensional space that can be depicted, as done in Figure 7.9.

The spherical harmonics we have introduced are generally complex, meaning that they take complex (rather than purely real) values. From these spherical harmonics,

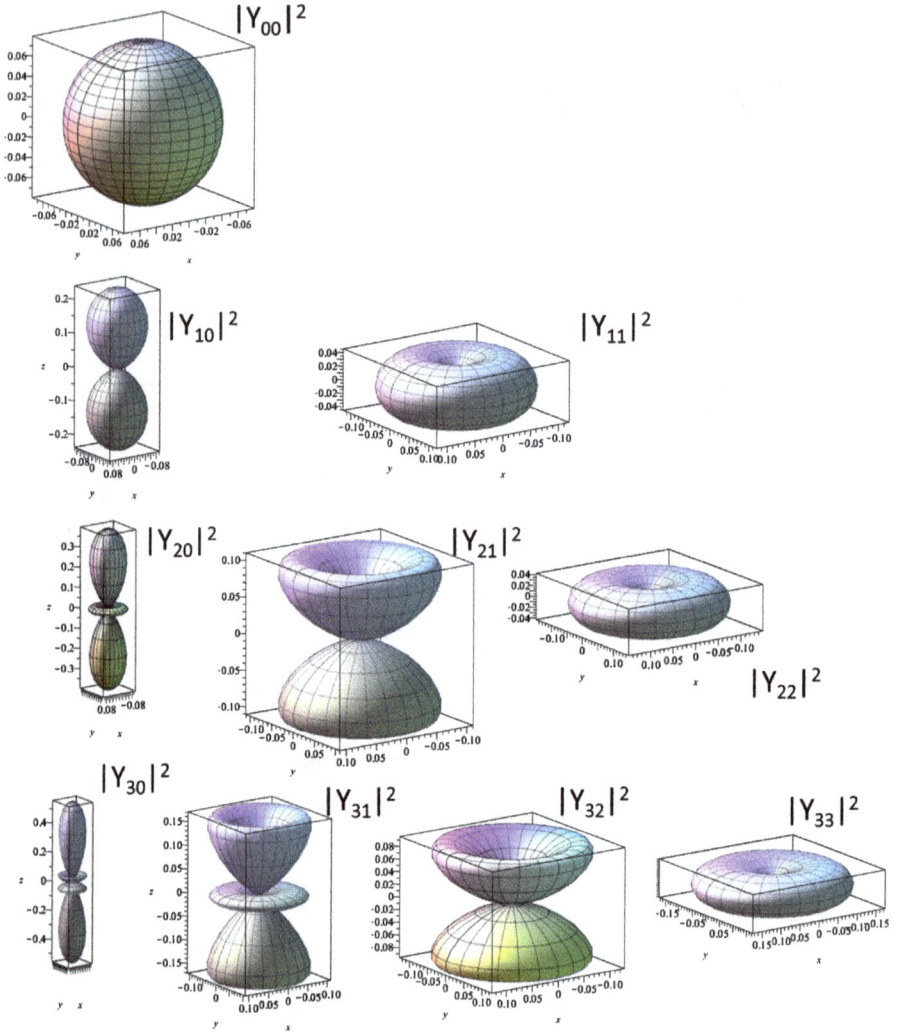

Figure 7.9: Representation of the squared absolute values of some of the complex spherical harmonics.

however, the more well-known s, p_x, ... functions can be constructed that take only real values. In general, one creates two new functions from Y_{l,m_l} and $Y_{l,-m_l}$ for $m_l \neq 0$,

$$Y_{l,m_l,\pm} = \frac{c_{l,m_l,\pm}}{\sqrt{2}} (Y_{l,m_l} \pm Y_{l,-m_l}), \tag{7.58}$$

while the functions $Y_{l,0}$ remain unchanged. By judicious choice of the constants $c_{l,m_l,\pm}$ (e. g., $c_{l,m_l,\pm} = \pm 1$ or $\pm i$), it can be achieved that the functions $Y_{l,m_l,\pm}$ are real (rather than complex). For example, p_x and p_y are formed from $Y_{1,1}$ and $Y_{1,-1}$, while $d_{x^2-y^2}$ and d_{xy} are formed from $Y_{2,2}$ and $Y_{2,-2}$. These functions are shown in Figure 7.10. Also, these

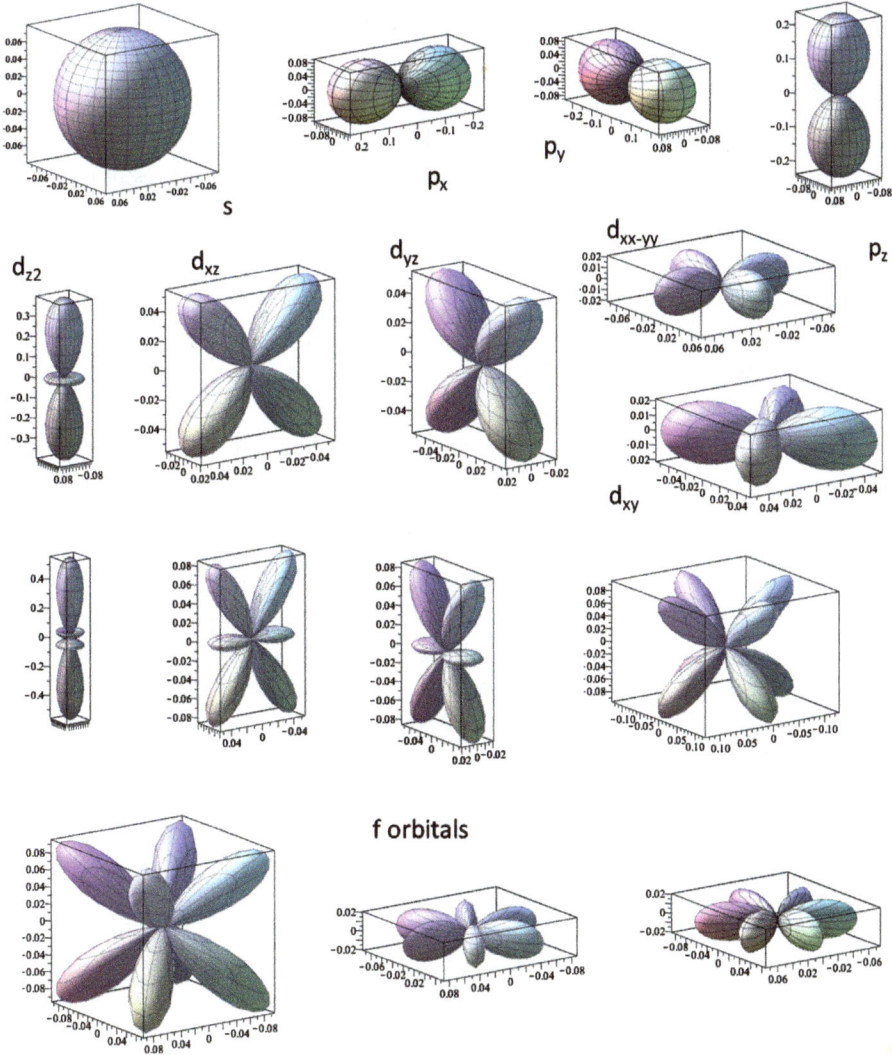

Figure 7.10: Representation of the real spherical harmonics.

functions are eigenfunctions for the Hamilton operator as well as for $\hat{l}^2$ but only for $m_l = 0$ also for $\hat{l}_z$.

That also the real spherical harmonics are eigenfunctions to $\hat{H}$ can be seen as follows. We consider two functions, ψ_1 and ψ_2, that both are eigenfunctions to some (linear) Hamilton operator $\hat{H}$ with the same eigenvalue,

$$\hat{H}\psi_1 = E\psi_1$$
$$\hat{H}\psi_2 = E\psi_2. \tag{7.59}$$

Then, for any linear combination of ψ_1 and ψ_2 we find

$$\hat{H}(c_1\psi_1 + c_2\psi_2) = c_1\hat{H}\psi_1 + c_2\psi_2 = c_1E\psi_1 + c_2E\psi_2 = E(c_1\psi_1 + c_2\psi_2), \tag{7.60}$$

i. e., this is also an eigenfunction to $\hat{H}$. In the present case where we combine Y_{l,m_l} and $Y_{l,-m_l}$ for $m_l \neq 0$, similar arguments can be used to show that the resulting linear combination is also an eigenfunction to $\hat{l}^2$.

7.4 Addition of angular momenta

For systems with several angular momenta, they are added vectorially both in the classical case and in the quantum mechanical case (see Figure 7.3). Then, in the quantum mechanical case, too, the length of the total angular momentum and its z component are quantized. In the general case, however, this does not apply to the individual angular momentum vectors, but only to their sum, if the individual motions influence each other. For example, electrons of an atom are charged and repel each other, so that the motion of each electron depends on those of the others.

For us, it is often the case that the angular momentum will be caused by electrons moving in some orbits. The motion of a single electron generates an angular momentum vector, and from the individual ones a total angular momentum vector is then formed. Then it may happen that the sum disappears.

In the general case, we have that if

$$\vec{L} = \vec{l}_1 + \vec{l}_2 + \cdots + \vec{l}_n \tag{7.61}$$

we have for the corresponding quantum mechanical operator [where $\Psi(\vec{x}_1, \vec{x}_2, \ldots, \vec{x}_n)$ is the wave function of the total system of the n particles]

$$\hat{L}^2\Psi(\vec{x}_1, \vec{x}_2, \ldots, \vec{x}_n) = L(L+1)\hbar^2\Psi(\vec{x}_1, \vec{x}_2, \ldots, \vec{x}_n) \tag{7.62}$$

with L equal to an integer. $L = 0$ is then possible. Furthermore,

$$\hat{L}_z\Psi(\vec{x}_1, \vec{x}_2, \ldots, \vec{x}_n) = M_L\hbar\Psi(\vec{x}_1, \vec{x}_2, \ldots, \vec{x}_n). \tag{7.63}$$

The integer M_L satisfies

$$-L \leq M_L \leq L. \tag{7.64}$$

Similar formulas do not apply for the operators of the individual angular momenta, $\vec{l}_1$, $\vec{l}_2, \ldots, \vec{l}_n$.

As an example, we consider two electrons in one atom. If we first assume that their motions are independent of each other, we can calculate the total angular momentum in equation (7.61) obtained by adding the angular momenta of the individual electrons,

$$\vec{L} = \vec{l}_1 + \vec{l}_2, \tag{7.65}$$

whereby we initially assume that the individual angular momentum vectors are quantized,

$$|\vec{l}_1|^2 = l_1(l_1 + 1)\hbar^2$$
$$|\vec{l}_2|^2 = l_2(l_2 + 1)\hbar^2. \tag{7.66}$$

If subsequently the motions of the individual electrons are allowed to influence each other, only the sum will be quantized,

$$|\vec{L}|^2 = L(L + 1)\hbar^2, \tag{7.67}$$

and it turns out that the possible values of L are

$$L = |l_1 - l_2|, |l_1 - l_2| + 1, \ldots, l_1 + l_2 - 1, l_1 + l_2. \tag{7.68}$$

Specifically, we first have assumed that the two electrons are in the two (different) orbitals ψ_1 and ψ_2, for which

$$\hat{l}_1^2 \psi_1 = l_1(l_1 + 1)\hbar^2 \psi_1$$
$$\hat{l}_2^2 \psi_2 = l_2(l_2 + 1)\hbar^2 \psi_2. \tag{7.69}$$

Subsequently, the total wave function of the two electrons Ψ is not an eigenfunction to the operators $\hat{l}_1^2$ and $\hat{l}_2^2$, but only to $\hat{L}^2$.

Finally, we mention that we have repeatedly discussed the z component of the angular momentum vector that is quantized while the x and y components are not. It is said that the z direction is the quantization direction. But there is no coordinate system in normal everyday life, and then the choice of the z direction is actually arbitrary. However, there may be experimental conditions that define a "special" or "preferred" direction. This is the case, e. g., when the system is in a magnetic field (a case we will discuss in the next section). Then the direction of the magnetic field is "special" or "preferred" and it is then often convenient to choose this as the z direction.

7.5 Spin of the electron

In 1922, the physicists Otto Stern and Walther Gerlach carried out an experiment in Frankfurt am Main, the results of which led to the discovery of the electron's spin. However, the exact existence of the spin was not suggested before 1925 and then by Samuel Abraham Goudsmit and George Eugene Uhlenbeck.

So far, we have considered small bodies that can move in more or less circular orbits. When these bodies carry a charge, this results in a small circuit. Such a circuit corresponds to a magnetic moment, $\vec{\mu}$, which is proportional to the angular momentum of the body,

$$\vec{\mu} \propto \vec{l}. \tag{7.70}$$

In a magnetic field with field vector $\vec{B}$, the energy of the magnetic moment equals

$$E_{magn} = \vec{\mu} \cdot \vec{B}. \tag{7.71}$$

We now place such small circuits in an inhomogeneous magnetic field as in Figure 7.11. Those for which $\vec{\mu}$ is parallel to $\vec{B}$ can reduce their energy by moving in the direction in which $\vec{B}$ becomes smaller. The reverse applies to those for which $\vec{\mu}$ is antiparallel to $\vec{B}$. Overall, the force acting on a circuit depends on the angle between $\vec{\mu}$ and $\vec{B}$, assuming that $|\vec{\mu}|$ is constant.

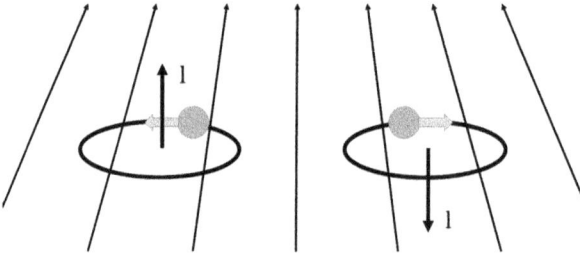

Figure 7.11: Small circuits that are in an inhomogeneous magnetic field. The figure shows two such circuits whose angular momenta point in opposite directions. The thinner lines represent the magnetic field lines. The smaller distances between the field lines in the upper part of the figure compared with the lower part illustrates the inhomogeneity of the field.

Exactly the latter, according to the discussion in this chapter, is the case because $\vec{l}$ is quantized. Furthermore, of convenience, we have introduced a quantization direction through the magnetic field, which in turn means that $\vec{\mu} \cdot \vec{B}$ can only assume discrete values.

Stern and Gerlach used Ag atoms in their experiment. For these, we have to consider the total angular momentum, but it turns out that for Ag atoms in their ground state $\vec{L} = \vec{0}$. This would mean that for all small circuits (= atoms) $\vec{\mu} \cdot \vec{B} = 0$, and that no forces act on the circuits / atoms. But the experiment revealed something else; see Figure 7.12. The Ag atoms were collected on a glass plate, so that one could see exactly where the atoms arrived after passing through the magnetic field. And it was found that there were two groups of atoms. This was explained by the fact that $\vec{\mu} \cdot \vec{B}$ can assume two different values, i. e., that the z component of the angular momentum can take two values.

This result is not easy to explain. We have seen that actually $\vec{L} = \vec{0}$, which means that M_L can only be equal to 0, so that $\vec{B} \cdot \vec{\mu} \propto BM_L$ (if making the convenient choice of placing the z axis along the magnetic field) can only assume one value, 0. It could be suggested that the outermost valence electron of the Ag atom is not in the 5s but in the 5p orbital, but this would give $L = 1$, so that M_L could assume three different values. The experimental finding that there are only two values for the magnetic moment of

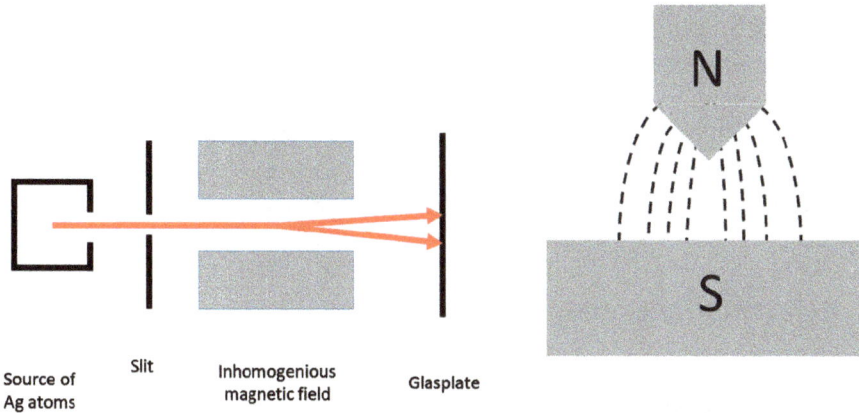

Figure 7.12: Schematic representation of the Stern-Gerlach experiment. Ag atoms enter an inhomogeneous magnetic field and are deflected in two directions and collected on a glass plate. On the right is a cross-section of the magnetic field outlined.

the individual Ag atoms cannot be explained by our previous results that L always has to be an integer, implying that $2L + 1$ is odd and can never be equal to 2.

Nevertheless, one makes the following interpretation: The splitting comes from an additional rotation, which we have not considered so far. The electron cannot only revolve around the nucleus, but also around its own axis (just as the earth revolves around the sun as well as around its own axis). This second rotation is also quantized, and for whatever reason, this motion can be described by means of an angular momentum vector whose length is always equal to $\sqrt{\frac{1}{2}(\frac{1}{2} + 1)}\hbar$. This rotational motion is called spin, and its angular momentum is denoted by $\vec{s}$. So this quantum number is always for an electron

$$s = \frac{1}{2} \tag{7.72}$$

so that

$$m_s = -\frac{1}{2}, \frac{1}{2}. \tag{7.73}$$

This interpretation is not perfect. We have found that every rotational motion is quantized and that the allowed quantum numbers can only be integers. In fact, the existence of a spin can only be explained by the combination of quantum theory and the theory of relativity, as Paul Andre Maurice Dirac showed. This combination yields a different equation than the Schrödinger equation (the so-called Dirac equation). It is easy to see that the Schrödinger equation can not be reconciled so easily with the relativity theory. The time-dependent Schrödinger equation (see Section 2.1) includes a second-order differentiation with respect to the position coordinates but only a first-order differentiation with respect to time. The theory of relativity, however, deals with time and position coordinates equivalently. To remedy this contradiction, Dirac introduced a new equation, the Dirac equation, which becomes equivalent to the Schrödinger equation in the limiting case that the speed of light approaches infinity. But as a result of the Dirac equation, one obtains that particles also have a spin which is an integral multiple of a half. As we have seen, it equals a half for electrons.

The Dirac equation will hardly play a role in this course, except for the introduction of the so-called spin-orbit coupling which will be discussed later in Section 11.4.

7.6 Relevance for experiment

By measuring the rotational energy levels of a molecule, information about the moments of inertia of the molecule is obtained. This is part of rotational spectroscopy. This information can be helpful in characterizing the molecule. If the theory in this chapter is used directly to interpret the results, it is assumed that the molecule is rigid. Then one would obtain spectra as shown in Figure 7.13. As this figure suggests, there are so-called selection rules that state that only transitions between neighboring energy levels are possible. This issue shall not be discussed further here. Since the number of degenerate states (i. e., states with the same l but different m_l) increases with l, the heights of the peaks increase initially with l, i. e., with energy. On the other hand, due to the Boltzmann distribution (for the reader who is less familiar with this, see Section 16.4), the number of molecules with a higher energy decreases and, in total, we obtain a curve like that shown in the lower part of Figure 7.13.

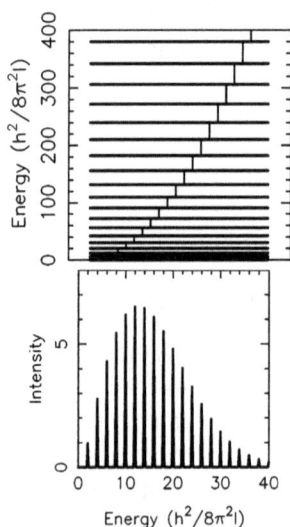

Figure 7.13: Schematic representation of the energy levels of a rigid rotor (upper part) and the accompanying absorption spectrum (lower part). The small, vertical lines in the upper panel show the allowed transitions.

The assumption that the molecule is rigid may be too simple. For example, one can imagine that the molecule expands when it rotates fast (because of the centrifugal force). Then the theory has to be modified accordingly.

Furthermore, one will often simultaneously excite both vibrational and rotational levels. This produces rotational-vibrational spectra, as shown in Figure 7.14. Figure 7.15 shows that the spectrum of the HCl molecule has a fine structure that is even finer than what is caused by the rotation. These are produced by isotope effects: if one isotope is replaced by another, the mass of the corresponding atom changes, and thus also the moment of inertia, and so does the energy level splitting. The analysis of such spectra gives further information on the structure of the molecule.

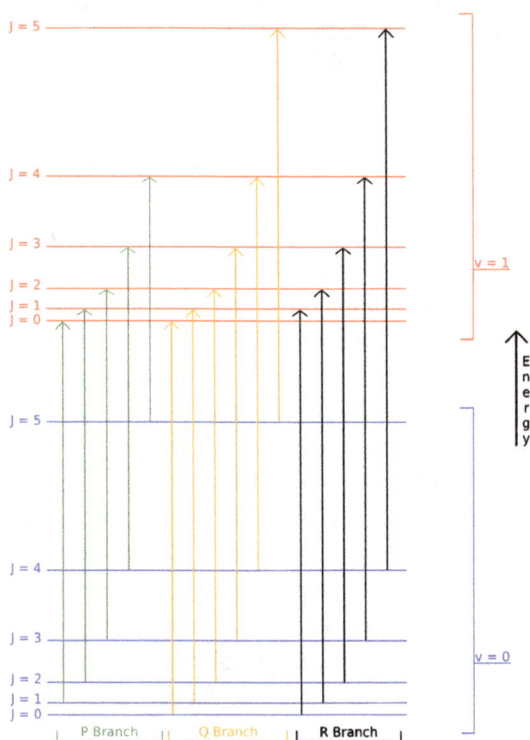

Figure 7.14: The schematic representation of the rotational-vibrational levels of a rotor. $v = 0$ and $v = 1$ represent two levels of vibration, while J represents the different levels of rotation. The vertical arrows show the possible excitations.

But one of the experimentally most important results of this chapter is the existence of spin. The spin also corresponds to an angular momentum vector and therefore couples with the other angular momentum vectors. This, as we shall see later, is important for spin-orbit coupling. Furthermore, the mere existence of a spin is important every-

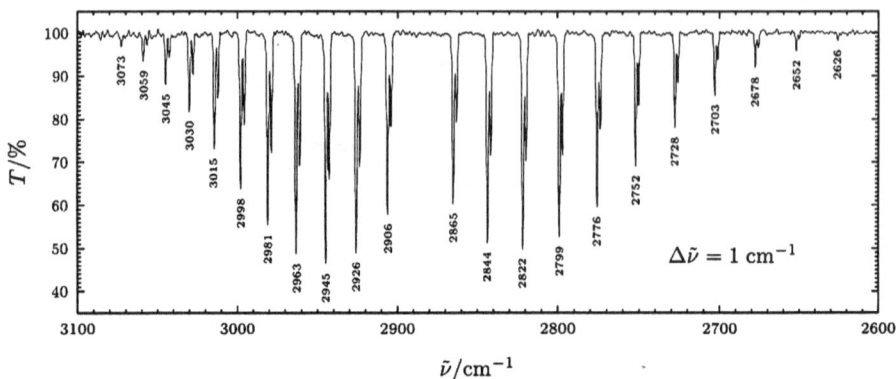

Figure 7.15: The rotation-vibration absorption spectrum of a HCl molecule. Copied on 03.02.17 from https://commons.wikimedia.org/wiki/File:Ir_hcl_rot-vib_mrtz.svg. From mrtzmrtz [CC BY-SA 2.5 (http://creativecommons.org/licenses/by-sa/2.5)], via Wikimedia Commons.

where. Thus, spin-resonance experiments are used to characterize molecules (e. g., NMR, ESR, EPR spectroscopy). The existence of magnetism can to a large part be attributed to the existence of a spin.

In such spin-resonance experiments, it is exploited that those particles that have an angular momentum vector often are charged so that their rotational motion is ultimately equivalent to a small circuit. One can influence this in an external magnetic field. In describing the resulting effect, it is convenient to choose a coordinate system so that the z axis is parallel to the magnetic field. Because the coupling between magnetic field and angular momentum is quantified by the scalar product between magnetic field vector and angular momentum vector, the fact that the z component of the angular momentum vector is quantized means that the size of the coupling can only take finite, discrete values. If one finally measures these couplings, one obtains information about the magnetic field at the place where the particle is located. This magnetic field is not only the externally applied magnetic field, but also has a component, which is due to the fact that all other particles (especially electrons) of the system respond to the externally applied magnetic field and shield it partly. Thus, the information about the magnetic field at the place where the particle is located describes the distribution of the other particles. This provides important chemical information.

Here, we have based the discussion on that the z axis lies along the externally applied magnetic field and the z axis is defined as the quantization direction. This is not necessary: in the end, you would get the same results regardless of how you choose the coordinate system. But this choice is very useful, because it makes all considerations and calculations easier.

7.7 Problems with answers

1. **Problem:** Consider the operators: $\hat{l}_x = \hat{y}\hat{p}_z - \hat{z}\hat{p}_y$, $\hat{l}_y = \hat{z}\hat{p}_x - \hat{x}\hat{p}_z$, $\hat{l}_z = \hat{x}\hat{p}_y - \hat{y}\hat{p}_x$, and $\hat{l}^2 = \hat{l}_x\hat{l}_x + \hat{l}_y\hat{l}_y + \hat{l}_z\hat{l}_z$. Determine the operators $[\hat{l}_x, \hat{l}_y]$, $[\hat{l}_y, \hat{l}_z]$, $[\hat{l}_z, \hat{l}_x]$, and $[\hat{l}_z, \hat{l}^2]$.
   Answer: We use

$$[\hat{q}_k, \hat{q}_l] = 0$$
$$[\hat{p}_k, \hat{p}_l] = 0$$
$$[\hat{q}_k, \hat{p}_l] = i\hbar\delta_{k,l}. \tag{7.74}$$

Then

$$\begin{aligned}
[\hat{l}_x, \hat{l}_y] &= \hat{l}_x\hat{l}_y - \hat{l}_y\hat{l}_x = (\hat{y}\hat{p}_z - \hat{z}\hat{p}_y)(\hat{z}\hat{p}_x - \hat{x}\hat{p}_z) - (\hat{z}\hat{p}_x - \hat{x}\hat{p}_z)(\hat{y}\hat{p}_z - \hat{z}\hat{p}_y) \\
&= \hat{y}\hat{p}_z\hat{z}\hat{p}_x - \hat{z}\hat{p}_y\hat{z}\hat{p}_x - \hat{y}\hat{p}_z\hat{x}\hat{p}_z + \hat{z}\hat{p}_y\hat{x}\hat{p}_z \\
&\quad - \hat{z}\hat{p}_x\hat{y}\hat{p}_z + \hat{z}\hat{p}_x\hat{z}\hat{p}_y + \hat{x}\hat{p}_z\hat{y}\hat{p}_z - \hat{x}\hat{p}_z\hat{z}\hat{p}_y \\
&= \hat{y}\hat{p}_z\hat{z}\hat{p}_x + \hat{z}\hat{p}_y\hat{x}\hat{p}_z - \hat{z}\hat{p}_x\hat{y}\hat{p}_z - \hat{x}\hat{p}_z\hat{z}\hat{p}_y = -\hat{y}\hat{p}_x[\hat{z}, \hat{p}_z] + \hat{x}\hat{p}_y[\hat{z}, \hat{p}_z] \\
&= \hat{l}_z[\hat{z}, \hat{p}_z] = i\hbar\hat{l}_z. \tag{7.75}
\end{aligned}$$

Subsequently, we use the cyclic permutation

$$\hat{l}_x \rightarrow \hat{l}_y$$
$$\hat{l}_y \rightarrow \hat{l}_z$$
$$\hat{l}_z \rightarrow \hat{l}_x \tag{7.76}$$

which changes equation (7.75) into

$$[\hat{l}_y, \hat{l}_z] = i\hbar\hat{l}_x \tag{7.77}$$

and subsequently

$$[\hat{l}_z, \hat{l}_x] = i\hbar\hat{l}_y. \tag{7.78}$$

From these commutator relations, we get

$$\begin{aligned}
\hat{l}_z\hat{l}_x &= \hat{l}_x\hat{l}_z + i\hbar\hat{l}_y \\
\hat{l}_x\hat{l}_z &= \hat{l}_z\hat{l}_x - i\hbar\hat{l}_y \\
\hat{l}_z\hat{l}_y &= \hat{l}_y\hat{l}_z - i\hbar\hat{l}_x \\
\hat{l}_y\hat{l}_z &= \hat{l}_z\hat{l}_y + i\hbar\hat{l}_x. \tag{7.79}
\end{aligned}$$

Then

$$\begin{aligned}
[\hat{l}_z, \hat{l}^2] &= \hat{l}_z(\hat{l}_x\hat{l}_x + \hat{l}_y\hat{l}_y + \hat{l}_z\hat{l}_z) - (\hat{l}_x\hat{l}_x + \hat{l}_y\hat{l}_y + \hat{l}_z\hat{l}_z)\hat{l}_z \\
&= \hat{l}_z\hat{l}_x\hat{l}_x - \hat{l}_x\hat{l}_x\hat{l}_z + \hat{l}_z\hat{l}_y\hat{l}_y - \hat{l}_y\hat{l}_y\hat{l}_z
\end{aligned}$$

$$= (\hat{l}_z \hat{l}_x)\hat{l}_x - \hat{l}_x(\hat{l}_x \hat{l}_z) + (\hat{l}_z \hat{l}_y)\hat{l}_y - \hat{l}_y(\hat{l}_y \hat{l}_z)$$

$$= (\hat{l}_x \hat{l}_z + i\hbar \hat{l}_y)\hat{l}_x - \hat{l}_x(\hat{l}_z \hat{l}_x - i\hbar \hat{l}_y) + (\hat{l}_y \hat{l}_z - i\hbar \hat{l}_x)\hat{l}_y - \hat{l}_y(\hat{l}_z \hat{l}_y + i\hbar \hat{l}_x)$$

$$= 0. \tag{7.80}$$

Since the commutator vanishes, it is possible to identify states that are eigenfunctions to both operators simultaneously.

2. **Problem:** Consider a system with two p electrons. What values can the total spin quantum number S and the total angular momentum quantum number L assume? Justify the answer.

Answer: It is always true that for a sum like

$$\vec{L} = \vec{l}_1 + \vec{l}_2, \tag{7.81}$$

for which the individual angular momentum vectors

$$|\vec{l}_1|^2 = l_1(l_1 + 1)\hbar^2$$
$$|\vec{l}_2|^2 = l_2(l_2 + 1)\hbar^2, \tag{7.82}$$

are quantized, the sum will also be quantized,

$$|\vec{L}|^2 = L(L + 1)\hbar^2. \tag{7.83}$$

The possible values of L are then given by

$$L = |l_1 - l_2|, |l_1 - l_2| + 1, \ldots, l_1 + l_2 - 1, l_1 + l_2. \tag{7.84}$$

For the p electrons $l_1 = l_2 = 1$, so that the values

$$L = 0, 1, 2 \tag{7.85}$$

are possible.
The same applies to the spin, where $s_1 = s_2 = \frac{1}{2}$, so that the values

$$S = 0, 1 \tag{7.86}$$

are possible.

7.8 Problems

1. Explain the Stern–Gerlach experiment.
2. Describe how the energies of the 2-dimensional rotor depend on the quantum number.
3. Sketch the five energetically lowest wave functions of the 2-dimensional rotor.

4. How do the energies of the 2-dimensional rotor change when (a) the mass is doubled or (b) the length of the rotor is doubled?

5. Compare the spherical harmonics Y_{lm_l}, $l = 1$, with the p_x, p_y, and p_z functions. Sketch the functions in (x, y), (x, z), and (y, z) planes.

6. Compare the spherical functions Y_{lm_l}, $l = 2$, with the d_{xz}, d_{yz}, d_{xy}, d_{z^2}, and $d_{x^2-y^2}$ functions. Sketch the functions in (x, y), (x, z), and (y, z) planes.

7. Why is it difficult to explain the spin as a rotation of an electron about its own axis when using the Schrödinger equation?

8. Explain the vector model of the angular momenta.

9. Explain the term "product Ansatz."

10. Consider a system with two d electrons. What values can the total spin quantum number S and the total angular momentum quantum number L assume? Justify the answer.

11. Consider a system with one p electron and one d electron. What values can the total spin quantum number S and the total angular momentum quantum number L assume? Justify the answer.

12. Show that the Y_{lm_l} functions are eigenfunctions to the operator $\hat{l}^2 + \hat{l}_z^2$. Which maximum value or minimum value can the eigenvalue have for a given l?

8 The hydrogen atom

8.1 Experimental findings

The spectrum of the hydrogen atom (see Figure 8.1) represented one of the challenges that could not be explained with the classical physics. Empirically, it had been found that the spectral lines could be divided into different groups as shown in the figure. Ultimately, it was found that the energies of the spectral lines could be described by

$$h\nu = \Delta E = R_H \left(\frac{1}{n_1^2} - \frac{1}{n_2^2} \right) \tag{8.1}$$

where n_1 and n_2 are integers, and R_H is the so-called Rydberg constant whose value could be determined experimentally. But the underlying cause of this formula was unknown. However, with the new (quantum-)theory it became possible to explain equation (8.1). Here, we shall first describe the theory of Niels Bohr (although slightly modified) that was capable of describing the origin of the spectrum in Figure 8.1, and subsequently show how also Schrödinger's equation can be used in explaining the experimental findings.

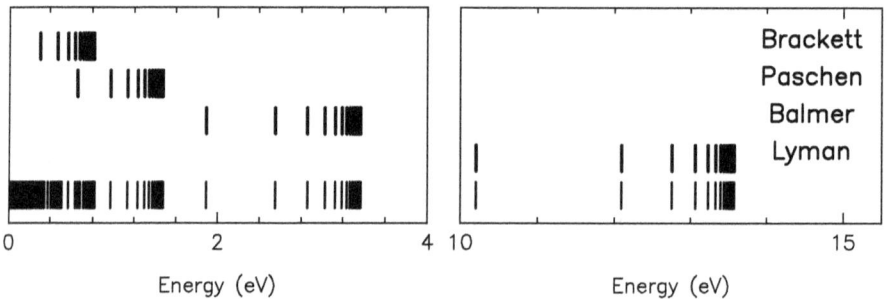

Figure 8.1: Spectral lines for the hydrogen atom. The complete spectrum is shown at the bottom, and above that, is shown its decomposition into different series whose names are given to the right.

8.2 Bohr's model for the hydrogen atom

In 1913, Niels Bohr presented a model that could qualitatively and quantitatively explain the properties of a H atom. Here, we will briefly outline the **Bohr model** in a slightly modified form, without going into too many details.

According to classical physics, an accelerated, charged body will emit electromagnetic radiation. That would also apply to an electron revolving around a nucleus, if the classical physics were also applicable to this system. By emitting electromagnetic radiation, the electron would lose energy and, therefore, approach the nucleus more

https://doi.org/10.1515/9783110742206-008

and more. Simple considerations show that the electron takes fractions of a second to reach the nucleus. However, we know that the electrons remain quite far away from the nucleus. So, other laws have to apply to such systems.

Few years before Niels Bohr presented his theory, the structure of an atom had been discovered by Ernest Rutherford. Before his work, it was known that atoms often were neutral and that the electrons were negatively charged. Therefore, an atom had to consist also of a positive charge. Previously, the so-called plum pudding model was assumed to describe the structure of atoms. According to this, suggested in 1904 by Joseph John Thomson, an atom consisted of some more or less homogeneous positively charged mass inside which the negatively charged electrons were found (like raisins in the plum pudding). Experimental studies in 1909 in the group of Ernest Rutherford were explained by him in 1911 and demonstrated that the plum pudding model was incorrect. Instead, far the largest part of the total mass of an atom was concentrated in a tiny part of the total volume of the atom. This part, the positively charged nucleus, was surrounded by the electrons. This was the background information upon which Niels Bohr developed his model for the hydrogen atom.

Niels Bohr postulated that the electron of a hydrogen atom moves in a circular orbit around the nucleus. The total energy of the electron consists of the kinetic energy as well as the potential energy resulting from the attraction between nucleus and electron:

$$E = \frac{p^2}{2m_e} + \frac{1}{4\pi\epsilon_0}\frac{-e^2}{r}. \tag{8.2}$$

Here, ϵ_0 is the dielectric constant of the vacuum, e is the elementary charge ($-e$ is then that of the electron and $+e$ that of the nucleus), and r is the radius of the circular path. Furthermore, m_e is the mass of the electron and p is its momentum.

However, the kinetic energy cannot be arbitrary: the velocity of the electron must be exactly so that the centrifugal force and the force of attraction due to the nucleus cancel each other in order to satisfy that the electron remains on the circular path. Mathematically expressed, this is

$$\frac{p^2}{m_e r} = \frac{1}{4\pi\epsilon_0}\frac{e^2}{r^2} \tag{8.3}$$

or

$$p^2 = \frac{e^2 m_e}{4\pi\epsilon_0 r}, \tag{8.4}$$

as well as

$$r = \frac{e^2 m_e}{4\pi\epsilon_0 p^2}. \tag{8.5}$$

From equations (8.2) and (8.4), we obtain then the following expression for the energy of the electron:

$$E = -\frac{1}{8\pi\epsilon_0}\frac{e^2}{r}. \tag{8.6}$$

According to the de Broglie's relation (which was actually introduced about 10 years later),

$$\lambda = \frac{h}{p}, \tag{8.7}$$

i. e., from the momentum of a particle one can determine the corresponding wavelength. Because we can determine the velocity of the electron, and thus the momentum, from the above considerations, we can also determine the wavelength that we will assign to the quantum mechanical behavior of the electron. This wavelength describes a wave that lies along the circular path followed by the electron. This wavelength must then turn into itself once we have completed a full circular orbit. Hence, the circumference of the circle must equal an integer (n) multiplied by the wavelength,

$$2\pi r = n\lambda. \tag{8.8}$$

Using this relationship, we obtain with the help of equation (8.5)

$$r = \frac{e^2 m_e}{4\pi\epsilon_0 p^2} = \frac{e^2 m_e}{4\pi\epsilon_0 h^2}\lambda^2 = \frac{e^2 m_e}{4\pi\epsilon_0 h^2}\frac{(2\pi r)^2}{n^2} \tag{8.9}$$

or that the radius of the orbit of the electron is

$$r = r_n = \frac{\epsilon_0 h^2}{\pi m_e e^2}n^2, \tag{8.10}$$

whereby the energy becomes

$$E = E_n = -\frac{m_e e^4}{8\epsilon_0^2 h^2}\frac{1}{n^2}. \tag{8.11}$$

Through this theory, we have obtained an expression for the Rydberg constant, i. e.,

$$R_H = \frac{m_e e^4}{8\epsilon_0^2 h^2}, \tag{8.12}$$

which contains no adjustable parameters. By comparison with experiments, it was discovered that the experimental results actually agreed with this theory, which earned Niels Bohr a Nobel Prize in physics.

8.3 $\hat{H}$ for H

After the Schrödinger equation was introduced (in 1926), the hydrogen atom was one of the first systems to be treated within this new theory. Here, we will discuss this quantum mechanical treatment in some detail, also because the system is one of the few real systems for which the Schrödinger equation can be solved analytically.

The hydrogen atom consists of a nucleus and an electron. We will call the position coordinates of the nucleus $\vec{R}_n$ and those of the electron $\vec{r}_e$. The wave function of the hydrogen atom, Ψ, is thus a function of $\vec{R}_n$ and $\vec{r}_e$,

$$\Psi = \Psi(\vec{R}_n, \vec{r}_e). \tag{8.13}$$

We let M_n and m_e be the masses of the nucleus and the electron, respectively. Then the Hamilton operator of the atom can be written as (see also Chapter 9)

$$\hat{H} = -\frac{\hbar^2}{2M_n}\nabla_n^2 - \frac{\hbar^2}{2m_e}\nabla_e^2 - \frac{e^2}{4\pi\epsilon_0|\vec{R}_n - \vec{r}_e|} \tag{8.14}$$

The first term represents the kinetic energy of the nucleus, the second that of the electron, and the third term represents the electrostatic attraction between the two particles. ∇_n^2 is the operator that, when applied to a function [as in equation (8.13)] this function is differentiated twice with respect to the nuclear coordinates, while the electronic coordinates are considered constant. Similarly, ∇_e^2 means: it only operates on the electronic coordinates. In equation (8.14), e is the elementary charge and $|\vec{R}_n - \vec{r}_e|$ the distance between the electron and the nucleus.

It is convenient to introduce other coordinates:

$$\vec{R} = \frac{m_e}{M_n + m_e}\vec{r}_e + \frac{M_n}{M_n + m_e}\vec{R}_n$$
$$\vec{r} = \vec{r}_e - \vec{R}_n. \tag{8.15}$$

$\vec{R}$ is the position vector for the center of gravity of the atom, and $\vec{r}$ is the vector for the position of the electron relative to that of the nucleus.

By applying the chain rule for differentiation, we obtain, e. g.,

$$\frac{\partial}{\partial x_e} = \frac{\partial X}{\partial x_e}\frac{\partial}{\partial X} + \frac{\partial x}{\partial x_e}\frac{\partial}{\partial x} = \frac{m_e}{M_n + m_e}\frac{\partial}{\partial X} + \frac{\partial x}{\partial x_e}$$
$$\frac{\partial}{\partial X_n} = \frac{\partial X}{\partial X_n}\frac{\partial}{\partial X} + \frac{\partial x}{\partial X_n}\frac{\partial}{\partial x} = \frac{M_n}{M_n + m_e}\frac{\partial}{\partial X} - \frac{\partial x}{\partial x_e} \tag{8.16}$$

and analogous expressions for the first-order differential quotients with respect to the y and z coordinates. We use the chain rule a second time (without giving the results now) and ultimately obtain

$$-\frac{\hbar^2}{2M_n}\nabla_n^2 - \frac{\hbar^2}{2m_e}\nabla_e^2 = -\frac{\hbar^2}{2(M_n + m_e)}\nabla_R^2 - \frac{\hbar^2}{2\mu}\nabla^2. \tag{8.17}$$

This operator can act on wave functions of the type

$$\Psi = \Psi(\vec{R}, \vec{r}) \tag{8.18}$$

whereby when ∇_R^2 operates on this function then we differentiate two times with respect to coordinates of the center of gravity, while the relative coordinates are considered constant. Similarly, ∇^2 means that it only operates on the relative coordinates $\vec{r}$.

In equation (8.17), μ is the reduced mass,

$$\mu = \frac{m_e M_n}{m_e + M_n} = \frac{m_e}{\frac{m_e}{M_n} + 1}. \tag{8.19}$$

Since $m_e \ll M_n$, $\mu \approx m_e$.

The Hamilton operator is now

$$\hat{H} = -\frac{\hbar^2}{2(M_n + m_e)}\nabla_R^2 - \frac{\hbar^2}{2\mu}\nabla^2 - \frac{e^2}{4\pi\epsilon_0 r}. \tag{8.20}$$

To solve the Schrödinger equation,

$$\hat{H}\Psi(\vec{R}, \vec{r}) = \tilde{E}\Psi(\vec{R}, \vec{r}) \tag{8.21}$$

we set up a product Ansatz for Ψ:

$$\Psi(\vec{R}, \vec{r}) = \psi_R(\vec{R}) \cdot \psi(\vec{r}). \tag{8.22}$$

We insert this into equation (8.21) with $\hat{H}$ from equation (8.20),

$$-\psi_R(\vec{R})\frac{\hbar^2}{2\mu}\nabla^2\psi(\vec{r}) - \psi_R(\vec{R})\frac{e^2}{4\pi\epsilon_0 r}\psi(\vec{r}) = \tilde{E}\psi_R(\vec{R}) \cdot \psi(\vec{r}) + \psi(\vec{r})\frac{\hbar^2}{2(M_n + m_e)}\nabla_R^2\psi_R(\vec{R}). \tag{8.23}$$

By dividing with the product of equation (8.22), we obtain

$$-\frac{1}{\psi(\vec{r})}\frac{\hbar^2}{2\mu}\nabla^2\psi(\vec{r}) - \frac{1}{\psi(\vec{r})}\frac{e^2}{4\pi\epsilon_0 r}\psi(\vec{r}) = \tilde{E} + \frac{1}{\psi_R(\vec{R})}\frac{\hbar^2}{2(M_n + m_e)}\nabla_R^2\psi_R(\vec{R}). \tag{8.24}$$

The left-hand side is independent of $\vec{R}$, while the right-hand side is independent of $\vec{r}$. Thus, both have to be a constant. We denote this constant E.

From this, we first get one equation for the motion of the center of mass,

$$-\frac{\hbar^2}{2(M_n + m_e)}\nabla_R^2\psi_R(\vec{R}) = (\tilde{E} - E)\psi_R(\vec{R}). \tag{8.25}$$

This is the equation for a free particle in three dimensions which we already have discussed in detail (in Chapter 4), and we shall not consider it further here.

The situation is different for the equation for the relative motion of the electron and the nucleus,

$$-\frac{\hbar^2}{2\mu}\nabla^2\psi(\vec{r}) - \frac{e^2}{4\pi\epsilon_0 r}\psi(\vec{r}) = E\psi(\vec{r}). \tag{8.26}$$

This is the equation we will solve in the next chapter.

8.4 $\hat{H}$ for e^-

We rewrite the Schrödinger equation (8.26):

$$\nabla^2 \psi(\vec{r}) + \frac{e^2 \mu}{2\pi\epsilon_0 \hbar^2 r} \psi(\vec{r}) = -\frac{2\mu E}{\hbar^2} \psi(\vec{r}). \tag{8.27}$$

As a simplification, we introduce

$$\epsilon = \frac{2\mu E}{\hbar^2}$$

$$\gamma = \frac{e^2 \mu}{2\pi\epsilon_0 \hbar^2}. \tag{8.28}$$

This allows to write equation (8.27) as

$$\nabla^2 \psi(\vec{r}) + \frac{\gamma}{r} \psi(\vec{r}) = -\epsilon\psi(\vec{r}). \tag{8.29}$$

As in Chapter 7.3, it is also here convenient to use spherical coordinates. Then (see Chapter 17)

$$\nabla^2 = \frac{\partial^2}{\partial r^2} + \frac{2}{r}\frac{\partial}{\partial r} + \frac{1}{r^2}\hat{\Lambda}^2. \tag{8.30}$$

with

$$\hat{\Lambda}^2 = \frac{1}{\sin^2\theta}\frac{\partial^2}{\partial\phi^2} + \frac{1}{\sin\theta}\frac{\partial}{\partial\theta}\sin\theta\frac{\partial}{\partial\theta}. \tag{8.31}$$

Equation (8.29) becomes then

$$\frac{\partial^2}{\partial r^2}\psi(\vec{r}) + \frac{2}{r}\frac{\partial}{\partial r}\psi(\vec{r}) + \frac{1}{r^2}\hat{\Lambda}^2\psi(\vec{r}) + \frac{\gamma}{r}\psi(\vec{r}) = -\epsilon\psi(\vec{r}). \tag{8.32}$$

As before, a product Ansatz is also applied here:

$$\psi(\vec{r}) = R(r) \cdot Y(\theta, \phi). \tag{8.33}$$

By inserting this into equation (8.32), we find

$$Y\frac{d^2 R}{dr^2} + Y\frac{2}{r}\frac{dR}{dr} + \frac{1}{r^2}R\hat{\Lambda}^2 Y + \frac{\gamma}{r}RY = -\epsilon RY. \tag{8.34}$$

As done before, we divide with the product (8.33), and the various terms are rearranged:

$$\frac{r^2}{R}\left(\frac{d^2 R}{dr^2} + \frac{2}{r}\frac{dR}{dr} + \frac{\gamma}{r}R + \epsilon R\right) = -\frac{1}{Y}\hat{\Lambda}^2 Y. \tag{8.35}$$

Because the left-hand side only depends on r, but not on θ and ϕ, whereas the opposite is the case for the right-hand side, the two sides can only be identical if they are equal to a constant C. This results in the following equation for the Y function,

$$\hat{\Lambda}^2 Y = -CY. \tag{8.36}$$

This is the equation we have already solved in Chapter 7.3. Therefore, we can immediately make use of the results from there:

$$Y(\theta, \phi) = Y_{lm_l}(\theta, \phi)$$
$$C = l(l+1)$$
$$l = 0, 1, 2, \ldots$$
$$m_l = -l, -l+1, -l+2, \ldots, l-1, l. \tag{8.37}$$

The equation for $R(r)$ becomes then

$$\frac{d^2R}{dr^2} + \frac{2}{r}\frac{dR}{dr} + \left[\frac{\gamma}{r} - \frac{l(l+1)}{r^2}\right]R = -\epsilon R. \tag{8.38}$$

It is nontrivial to solve this differential equation. But others have done so and we can use their results. Accordingly,

$$R(r) = R_{nl}(r) = \rho^l L_{nl}(\rho)e^{-\rho/2}$$
$$\rho = 2\sqrt{\epsilon}r$$
$$n = 1, 2, 3, \ldots$$
$$l = 0, 1, \ldots, n-1$$
$$E = -\frac{\mu e^4}{32\pi^2\epsilon_0^2\hbar^2}\frac{1}{n^2}. \tag{8.39}$$

The L functions are the so-called associated Laguerre polynomials (after Edmond Laguerre).

In Table 8.1, the R_{nl} functions are listed for n up to 4 and in Figure 8.2 they are shown for n up to 3.

We shall now briefly summarize the main findings that result:

– The solutions to the Schrödinger equation are labeled with three quantum numbers, n, l, and m_l. This is consistent with the fact that we are considering a three-dimensional system.
– n is a positive integer. n is called the main quantum number.
– l is a nonnegative integer, smaller than n. It is called the secondary quantum number. It is usual, instead of writing $l = 0, l = 1, l = 2, l = 3, \ldots$, to use letters: $s, p, d, f, \ldots$ for $l = 0, 1, 2, 3, \ldots$.
– m_l is an integer whose absolute value is less than or equal to l. It is called magnetic quantum number.

Table 8.1: The radial functions R_{nl} for the hydrogen atom for $n = 1$, 2, 3, and 4.

n	l	$R_{nl}(r)$
1	0	$a_0^{-3/2} 2 e^{-r/a_0}$
2	0	$a_0^{-3/2} \frac{1}{\sqrt{2}}(1 - \frac{r}{2a_0}) e^{-r/(2a_0)}$
2	1	$a_0^{-3/2} \frac{1}{2\sqrt{6}} \frac{r}{a_0} e^{-r/(2a_0)}$
3	0	$a_0^{-3/2} \frac{2}{3\sqrt{3}}(1 - \frac{2r}{3a_0} + \frac{2r^2}{27a_0^2}) e^{-r/(3a_0)}$
3	1	$a_0^{-3/2} \frac{8}{27\sqrt{6}} \frac{r}{a_0}(1 - \frac{r}{6a_0}) \frac{r}{a_0} e^{-r/(3a_0)}$
3	2	$a_0^{-3/2} \frac{4}{81\sqrt{30}} \frac{r^2}{a_0^2} e^{-r/(3a_0)}$
4	0	$a_0^{-3/2} \frac{1}{4}(1 - \frac{3r}{4a_0} + \frac{r^2}{8a_0^2} - \frac{r^3}{192a_0^3}) e^{-r/(4a_0)}$
4	1	$a_0^{-3/2} \frac{\sqrt{5}}{16\sqrt{3}}(1 - \frac{r}{4a_0} + \frac{r^2}{80a_0^2}) \frac{r}{a_0} e^{-r/(4a_0)}$
4	2	$a_0^{-3/2} \frac{1}{64\sqrt{5}}(1 - \frac{r}{12a_0}) \frac{r^2}{a_0^2} e^{-r/(4a_0)}$
4	3	$a_0^{-3/2} \frac{1}{768\sqrt{35}} \frac{r^3}{a_0^3} e^{-r/(4a_0)}$

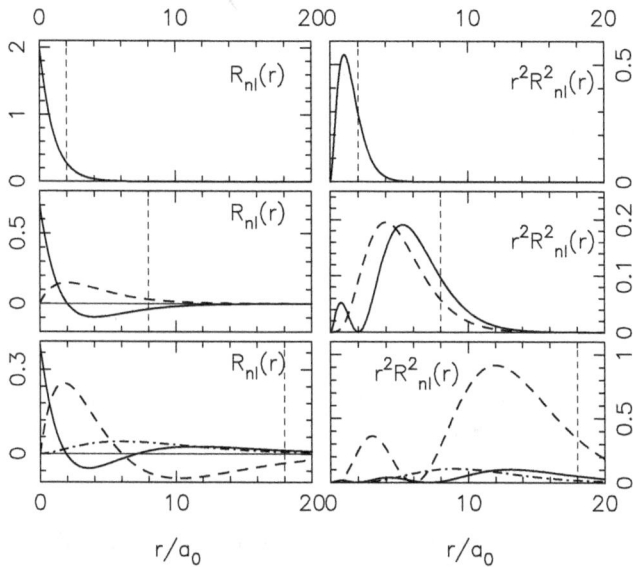

Figure 8.2: The radial functions $R_{nl}(r)$ for the hydrogen atom for $n = 1$, 2, and 3. The left part shows the functions themselves, while the right part shows the radial distributions $r^2 R_{nl}^2(r)$. The solid and dashed curves show the s and p functions, while the dot–dash curve shows the d function. The vertical, dashed lines separate the classically allowed area from the classically forbidden area.

- In addition, it is customary to assign a fourth quantum number, m_s, to the electron. m_s can only be $+\frac{1}{2}$ or $-\frac{1}{2}$ and is called spin quantum number. $m_s = \frac{1}{2}$ is often called spin-up or spin-α. $m_s = -\frac{1}{2}$ is often called spin-down or spin-β.

-   The energy depends only on n, but not on l or m_l (see Figure 8.3 and Figure 8.4). This is a special case, which applies only to hydrogen.

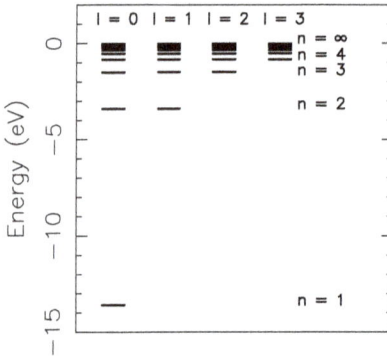

Figure 8.3: The energy levels for the hydrogen atom.

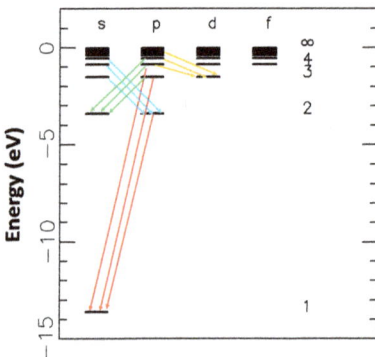

Figure 8.4: The theoretically calculated transitions for the hydrogen atom. Only a few are shown. The set of transitions of $n \to 1$ is called the Lyman series, that of $n \to 2$ Balmer series, and that of $n \to 3$ Paschen series (see Figure 8.1).

-   As Figure 8.4 indicates, the most probable transitions are those for which n changes by an arbitrary value, while l changes by ± 1. The reason for these so-called selection rules can be understood by means of the time-dependent perturbation theory, including Fermi's golden rule, which we will discuss in Chapter 9.8. However, the selection rules will not be discussed further here.
-   The experimental findings discussed in Section 8.1 are explained with great accuracy by this theory. At the same time, the theory provides a numerical value for the Rydberg constant.
-   The wave functions are complex valued for $m_l \neq 0$. To depict the orbitals, there are several possibilities. Either one chooses a particular direction (e. g., along the

z axis) and draws the wave function along that direction. Or a value is chosen, and one draws the surface in three-dimensional space, where the wave function (or its absolute magnitude) has this (constant) value. Finally, one can select any plane and display the function values either with the help of contour lines or as a three-dimensional object in this plane. In Figures 8.5, 8.6, 8.7, 8.8, and 8.9, we show some examples of such representations.

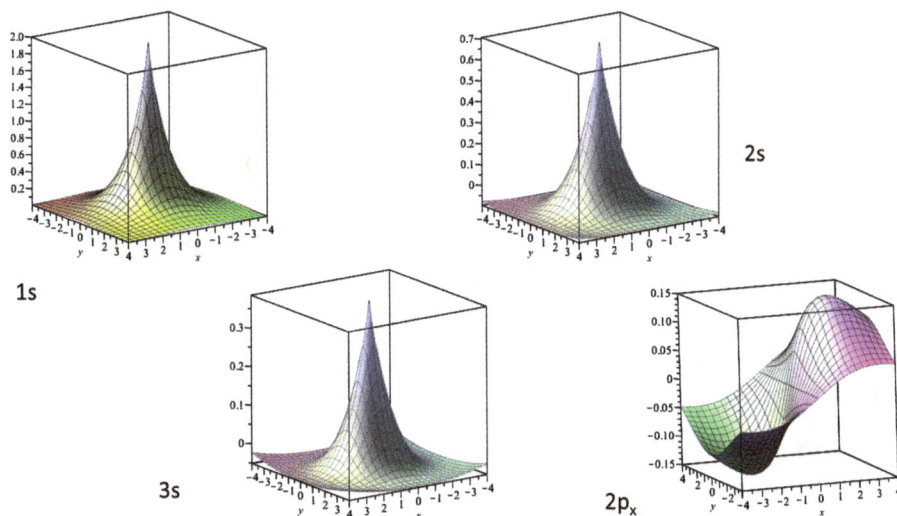

Figure 8.5: The real wave functions 1s, 2s, 3s, and 2p_x for the hydrogen atom, drawn in the plane $z = 0$.

- As we will see below, it is possible to define real-valued functions.
- In Table 8.1, we have introduced the so-called Bohr radius:

$$a_0 = \frac{4\pi\epsilon_0 \hbar^2}{m_e e^2}. \qquad (8.40)$$

Thereby, we did not consider that m_e should actually be the reduced mass μ. However, m_e and μ differ by less than 0.1 %. $a_0 \approx 0.52918$ Å.
- For the ground state (the state with the lowest energy) $n = 1$ and $l = m_l = 0$. For this the wave function is given by

$$\psi_{100}(r,\theta,\phi) = R_{10}(r) \cdot Y_{00}(\theta,\phi) = \frac{2}{\sqrt{a_0^3}} e^{-r/a_0} \cdot \frac{1}{\sqrt{4\pi}} = \frac{1}{\sqrt{\pi a_0^3}} e^{-r/a_0}, \qquad (8.41)$$

and from this it can be seen that the largest electron density is to be found at the position of the nucleus, and that the wave function possesses a kink there. The fact that the wave function is not differentiable there is consistent with the fact that at this point the potential diverges.

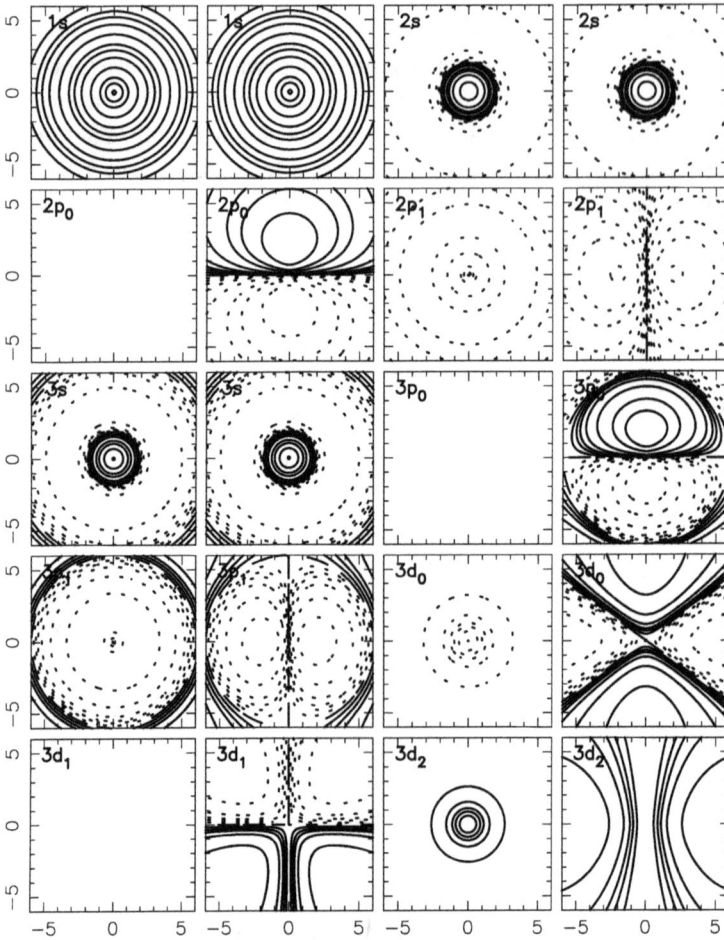

Figure 8.6: Contour curves for the complex wave functions ψ_{nlm_l} without the factor $e^{im_l\phi}$ for the hydrogen atom. Each function is shown in two different planes: (x, y) (left) and (x, z) (right). Solid, dotted, and dashed curves mark positive values, negative values, and zero, respectively. Starting at the top left, the first two panels show $(n, l, m_l) = (1, 0, 0)$, the next two $(2, 0, 0)$, then $(2, 1, 0)$, $(2, 1, 1)$, $(3, 0, 0)$, $(3, 1, 0)$, $(3, 1, 1)$, $(3, 2, 0)$, $(3, 2, 1)$, and finally $(3, 2, 2)$. As can be seen, some of the functions are zero in some of the planes. All functions shown are rotationally symmetric about the z axis. A spatial representation can be found in Figure 7.9.

– On the other hand, we can also look at the radial distribution, that is, the probability that the electron is at a certain distance from the nucleus. This probability is for $(n, l, m_l) = (1, 0, 0)$ given by

$$P(r) = r^2 [R_{10}(r)]^2 \tag{8.42}$$

and is shown in Figure 8.10.

Figure 8.7: As Figure 8.6 but for $(n, l, m_l) = (4, 0, 0), (4, 1, 0), (4, 1, 1), (4, 2, 0), (4, 2, 1), (4, 2, 2), (4, 3, 0), (4, 3, 1), (4, 3, 2),$ and $(4, 3, 3)$. A spatial representation can be found in Figure 7.9.

– Because the Hamilton operator is a linear operator, if ψ_1 and ψ_2 are orthonormal eigenfunctions of $\hat{H}$ for the same eigenvalue,

$$\hat{H}\psi_1 = E\psi_1$$
$$\hat{H}\psi_2 = E\psi_2,$$

(8.43)

then the two linear combinations

$$\psi_a = c_1\psi_1 + c_2\psi_2$$
$$\psi_b = -c_2\psi_1 + c_1\psi_2$$

(8.44)

Figure 8.8: As in Figure 8.6, but for the real wave functions $\frac{1}{\sqrt{2}}((-1)^{m_l}\psi_{n,l,m_l} + \psi_{n,l,-m_l})$ for $m_l \neq 0$, or ψ_{n,l,m_l} for $m_l = 0$ for $(n, l, ml) = (1, 0, 0), (2, 0, 0), (2, 1, 0), (2, 1, 1), (3, 0, 0), (3, 1, 0), (3, 1, 1), (3, 2, 0),$ $(3, 2, 1),$ and $(3, 2, 2)$. The ϕ-dependence, $\cos(m_l\phi)$, is also shown in this case, so that there is no rotational symmetry about the z axis. A spatial representation can be found in Figure 7.10.

are also orthonormal eigenfunctions of $\hat{H}$ for the same eigenvalue. By setting

$$|c_1|^2 + |c_2|^2 = 1 \tag{8.45}$$

also the new functions are normalized.

This can be exploited to construct pairs of new functions from the two functions ψ_{n,l,m_l} and $\psi_{n,l,-m_l}$ for $m_l \neq 0$, both of which are real. This is achieved by choosing c_1 and c_2 equal to $\frac{\pm 1}{\sqrt{2}}$ or $\frac{\pm i}{\sqrt{2}}$. The new functions are the well-known functions of type p_x, p_y, d_{xy}, etc. These are shown in Figures 8.8 and 8.9.

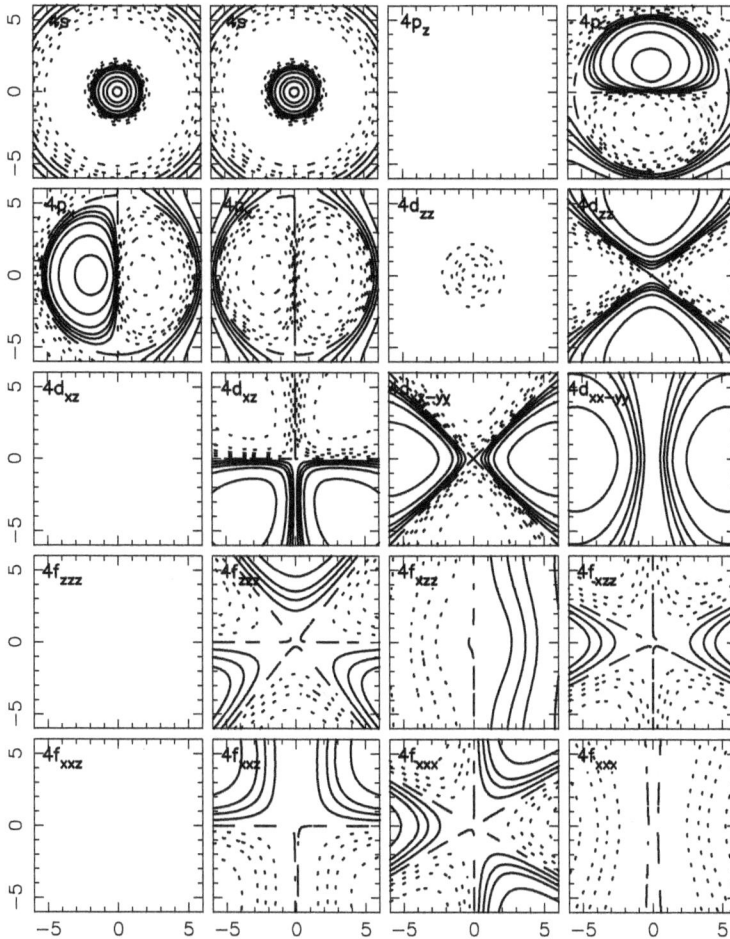

Figure 8.9: As Figure 8.8 but for $(n, l, m_l) = (4, 0, 0), (4, 1, 0), (4, 1, 1), (4, 2, 0), (4, 2, 1), (4, 2, 2), (4, 3, 0),$ $(4, 3, 1), (4, 3, 2),$ and $(4, 3, 3)$. A spatial representation can be found in Figure 7.10.

– As shown in Figure 8.2, the particle (here: an electron) in this case also has a finite probability of being in the classically forbidden area in which the potential is
   higher than the energy of the particle.
– Although this theory can reproduce the experimental findings with great accuracy, there are still small deviations between experiment and theory. Ultimately,
   the theory is not perfect, as we have seen, e. g., in the case of the existence of the
   spin. In fact, one can therefore use very accurate experimental results to analyze
   more accurate theories. If, e. g., replacing the Schrödinger equation with the Dirac
   equation, relativistic effects are taken into account and small changes in the energy levels of the hydrogen atom are obtained. With further developments that,
   e. g., are relevant if you want to treat the universe shortly after its formation, there

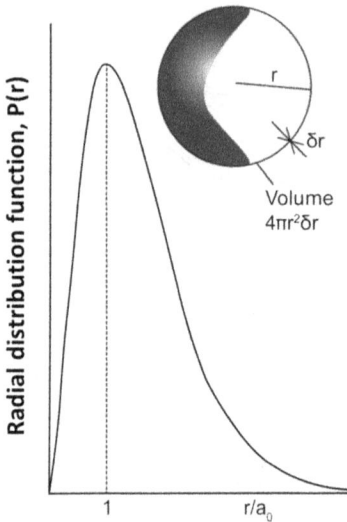

Figure 8.10: The radial distribution function for the 1s orbital of the H atom, the ground state of the hydrogen atom. Adapted from the book Peter W. Atkins, *Kurzlehrbuch Physikalische Chemie*, Wiley-VCH, 2001.

are other small changes in the energy levels of the hydrogen atom. Again, these can be compared to very accurate experimental results, with the validity of the theory ultimately being investigated. This makes the hydrogen atom being a system of relevance for the treatment of the whole universe.

– In the ground state, only the 1s function is occupied. Nevertheless, also the other states exist and are relevant when, e. g., the hydrogen atom forms chemical bonds with other atoms and when the hydrogen atom is excited.

8.5 Other centro-symmetric systems

The procedure discussed in the previous section can also be used for other centro-symmetric systems, i. e., for systems in which a particle moves in a spherically symmetric potential. In all these cases, the radial wave function will differ from that of the hydrogen atom (because the radial potential would be different), but in all cases the eigenfunctions can be expressed as the product of a radial function and a harmonic function.

As examples can be mentioned:

– Atomic ions with only one electron for which

$$V(r) = -\frac{Ze^2}{4\pi\epsilon_0 r}.$$ (8.46)

Here, Ze is the charge of the atomic nucleus.

- A free particle in a spherical wave. Then

$$V(r) = 0. \tag{8.47}$$

- Harmonium, also called Hooke's atom. This is a model system that is beloved by theorists. Here, we have

$$V(r) = \frac{1}{2}kr^2. \tag{8.48}$$

- A spherical box. Then

$$V(r) = \begin{cases} 0 & r \le R_0 \\ \infty & r > R_0. \end{cases} \tag{8.49}$$

8.6 Angular momentum

In this chapter, we will briefly discuss a few aspects concerning the angular momentum properties of the wavefunctions of the hydrogen atom—and actually also for the systems discussed in Section 8.5. In all cases, the wave functions can be written as

$$\psi(\vec{r}) = \psi(r, \theta, \phi) = \psi_{n,l,m_l}(r, \theta, \phi) = R_{nl}(r)Y_{lm_l}(\theta, \phi). \tag{8.50}$$

From Section 7.2, we know the expression of the operator for the z component of the angular momentum:

$$\hat{l}_z = \hat{x} \cdot \hat{p}_y - \hat{y} \cdot \hat{p}_x = \frac{\hbar}{i}\left(x\frac{\partial}{\partial y} - y\frac{\partial}{\partial x}\right). \tag{8.51}$$

It is easily realized that when $\hat{l}_z$ acts on the function in equation (8.50), the following is obtained:

$$\begin{aligned} \hat{l}_z\psi(\vec{r}) &= Y_{lm_l}(\theta, \phi)[\hat{l}_z R_{nl}(r)] + R_{nl}(r)[\hat{l}_z Y_{lm_l}(\theta, \phi)] \\ &= Y_{lm_l}(\theta, \phi)[\hat{l}_z R_{nl}(r)] + R_{nl}(r)[m_l\hbar Y_{lm_l}(\theta, \phi)] \\ &= Y_{lm_l}(\theta, \phi)\frac{\hbar}{i}\left(x\frac{\partial}{\partial y} - y\frac{\partial}{\partial x}\right)R_{nl}(r) + R_{nl}(r)[\hat{l}_z Y_{lm_l}(\theta, \phi)]. \end{aligned} \tag{8.52}$$

To calculate this, we need

$$\begin{aligned} \left(x\frac{\partial}{\partial y} - y\frac{\partial}{\partial x}\right)R_{nl}(r) &= x\frac{\partial R_{nl}(r)}{\partial y} - y\frac{\partial R_{nl}(r)}{\partial x} = x\frac{\partial R_{nl}(r)}{\partial r}\frac{\partial r}{\partial y} - y\frac{\partial R_{nl}(r)}{\partial r}\frac{\partial r}{\partial x} \\ &= \frac{\partial R_{nl}(r)}{\partial r}\left(x\frac{\partial r}{\partial y} - y\frac{\partial r}{\partial x}\right). \end{aligned} \tag{8.53}$$

We use that

$$r = (x^2 + y^2 + z^2)^{1/2} \tag{8.54}$$

so that

$$\frac{\partial r}{\partial x} = \frac{x}{r}$$
$$\frac{\partial r}{\partial y} = \frac{y}{r}. \tag{8.55}$$

From this, we see that the expression in equation (8.53)) vanishes, so that

$$\hat{l}_z \psi(\vec{r}) = R_{nl}(r)[\hat{l}_z Y_{lm_l}(\theta, \phi)] = m_l \hbar \psi(\vec{r}). \tag{8.56}$$

This could also have been obtained by utilizing that

$$\hat{l}_z = \frac{\hbar}{i} \frac{\partial}{\partial \phi}. \tag{8.57}$$

Similarly, one finds that even the operators $\hat{l}_x$ and $\hat{l}_y$ act only on the harmonic functions, although one then obtains that the harmonic functions are not eigenfunctions of $\hat{l}_x$ or $\hat{l}_y$.

Equation (8.56) implies that the wave functions for the hydrogen atom are eigenfunctions of $\hat{l}_z$, and that the eigenvalues are equal to $m_l \hbar$. This applies only to the complex wave functions denoted by (n, l, m_l), but not to the real functions we have constructed above.

The same can be shown for the operator

$$\hat{l}^2 = \hat{l}_x \hat{l}_x + \hat{l}_y \hat{l}_y + \hat{l}_z \hat{l}_z, \tag{8.58}$$

i. e., the wave functions for the hydrogen atom are eigenfunctions, and the eigenvalues are equal to $l(l+1)\hbar^2$. In this case, both the complex wave functions and the real wave functions are eigenfunctions.

This leads to something unexpected. For the ground state of the hydrogen atom, $l = 0$, which actually suggests that the angular momentum vector has a length equal to 0. This would mean that the electron either rests (a contradiction to Heisenberg's uncertainty principle), or that $\vec{r}$ and $\vec{p}$ are parallel. Then the circular orbits that Niels Bohr imagined would not be possible.

In fact, one can calculate the distribution of the angle between $\vec{r}$ and $\vec{p}$ (how to do that is not important here). The result is shown in Figure 8.11. As you can see, there is a maximum for the case that $\vec{r}$ and $\vec{p}$ are perpendicular to each other. This seems to contradict $l = 0$, but is consistent with Bohr's idea that the electrons move in the circular orbits.

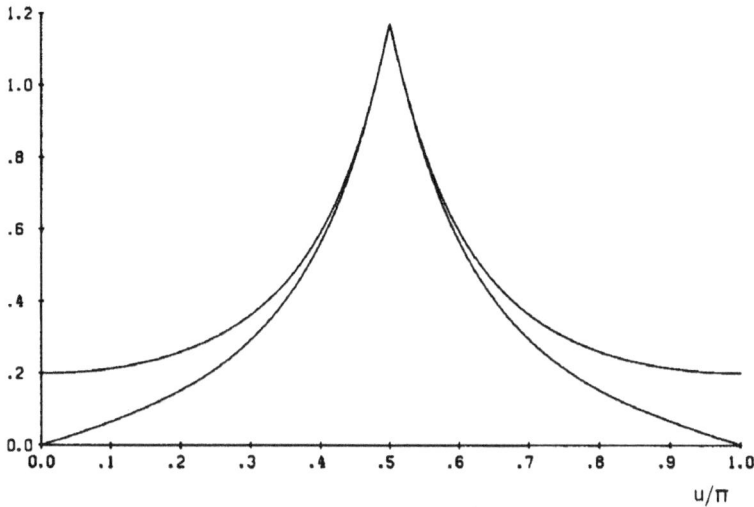

Figure 8.11: Distribution of the angle u between the position and momentum vector for the 1s state of the hydrogen atom, $P(u)$. The bottom curve shows the distribution function (upper curve) multiplied by $\sin u$. $P(u) \sin u \, du$ is the probability that the angle lies in the interval $[u, u + du]$.

The solution is that $\hat{l}^2$ is not exactly equal to the operator of the square of the length of the angular momentum. We consider, e. g.,

$$\hat{l}_z\hat{l}_z = (\hat{x}\hat{p}_y - \hat{y}\hat{p}_x)(\hat{x}\hat{p}_y - \hat{y}\hat{p}_x) = \hat{x}^2\hat{p}_y^2 + \hat{y}^2\hat{p}_x^2 - \hat{x}\hat{p}_y\hat{y}\hat{p}_x - \hat{y}\hat{p}_x\hat{x}\hat{p}_y$$
$$= \hat{x}^2\hat{p}_y^2 + \hat{y}^2\hat{p}_x^2 - \hat{x}\hat{p}_x\hat{p}_y\hat{y} - \hat{y}\hat{p}_y\hat{p}_x\hat{x}. \tag{8.59}$$

Because, e. g., $\hat{x}$ and $\hat{p}_x$ do not commute, the order is important (see Section 3.7). Therefore,

$$\widehat{l_z l_z} = \hat{x}^2\hat{p}_y^2 + \hat{y}^2\hat{p}_x^2 - \frac{1}{2}(\hat{x}\hat{p}_x + \hat{p}_x\hat{x})(\hat{y}\hat{p}_y + \hat{p}_y\hat{y}) \tag{8.60}$$

which differs from the expression of equation (8.59).

Carrying through the same analysis as in equations (8.59) and (8.60) for the other terms in equation (8.58), we find the following expression for the operator of the square of the length of the angular momentum:

$$\hat{\lambda}^2 = \hat{l}^2 + \frac{3}{2}\hbar^2. \tag{8.61}$$

This means that all eigenfunctions remain eigenfunctions, but the eigenvalues change by an additive constant of $\frac{3}{2}\hbar^2$. Thus, the length of the angular momentum vector is not equal to 0 for the 1s state either.

8.7 Problems with answers

1. **Problem:** Calculate the expectation values for x^2, y^2, and r^2 for an electron in the 1s-state of the hydrogen atom with the nucleus at the origin of the coordinate system.

    Answer: Because of the spherical symmetry of the system, we have

$$\langle x^2 \rangle = \langle y^2 \rangle = \langle z^2 \rangle = \frac{1}{3} \langle x^2 + y^2 + z^2 \rangle = \frac{1}{3} \langle r^2 \rangle. \tag{8.62}$$

We find

$$\langle r^2 \rangle = \int_0^\infty \frac{2}{\sqrt{a_0^3}} e^{-r/a_0} r^2 \frac{2}{\sqrt{a_0^3}} e^{-r/a_0} r^2 \, dr = \frac{4}{a_0^3} \int_0^\infty r^4 e^{-2r/a_0} \, dr$$

$$= \frac{4}{a_0^3} \frac{4!}{(2/a_0)^5} = \frac{4 \cdot 24 \cdot a_0^5}{a_0^3 32} = 3a_0^2 \tag{8.63}$$

using the formulas in Chapter 17.
Finally, then

$$\langle x^2 \rangle = \langle y^2 \rangle = \frac{1}{3} \langle r^2 \rangle = a_0^2. \tag{8.64}$$

8.8 Problems

1. Explain the term "atomic spectra," including the Rydberg constant.
2. Describe the mathematical dependence of the eigenfunctions of the hydrogen atom on the spherical coordinates r, θ, and ϕ.
3. How is the radial distribution function of the hydrogen atom orbitals defined? Sketch these for the 1s, 2s, 2p, 3s, and 3p orbitals.
4. Compare the radial distribution function and the probability density for the hydrogen orbitals.
5. What is the expectation value for the energy of a hydrogen atom in the $\frac{1}{\sqrt{2}}(2s + 2p_z)$ state?
6. Describe briefly how to arrive at the Schrödinger equation for the electron in the hydrogen atom from the Schrödinger equation for the entire atom by means of a product Ansatz.
7. Describe briefly the relationship between the quantum numbers of the electron in the hydrogen atom and its angular momentum vector.
8. Normalize the function $\Psi(r, \theta, \varphi) = Ne^{-r/a_0}$ for $0 \le r < \infty$, $0 \le \theta \le \pi$, and $0 \le \varphi \le 2\pi$.
9. How many quantum numbers are needed to unambiguously define an atomic orbital in the hydrogen atom? What are their names, what do they describe, and which values can they take?

10. Depict graphically the dependence of the energies on the quantum numbers for:
    (a) the particle in a box,
    (b) the harmonic oscillator, and
    (c) the hydrogen atom.
    Explain in all cases the mathematical dependence of the energies on the quantum numbers.

11. Calculate the expectation value $\langle r \rangle$ and the most probable value for r for an electron in the ground state of the H-atom.

12. Calculate the probability of finding an electron in the 1s state in the hydrogen atom outside a sphere of radius $r = a_0$ (a_0 is the Bohr radius).

13. Describe why it is not possible to create orbitals for the hydrogen atom, which are not only eigenfunctions for $\hat{l}_z$ but also for $\hat{l}_x$.

9 Foundations of the approximate methods

9.1 The problem

Shortly after the introduction of the mathematical foundations of the quantum theory by Heisenberg and Schrödinger in 1926, one began to apply this theory to different systems. Among them were some of the systems that were discussed in the previous chapters (e. g., the free particle, the particle in a box, step potentials, the rotor, the harmonic oscillator, and the hydrogen atom). However, it quickly became apparent that the Schrödinger equation could not be solved exactly for many systems of chemical or physical interest.

The very good agreement between calculated and measured quantities, especially for the spectrum of the H atom, was considered to be very encouraging, but the disillusionment soon came. If, starting with the H atom, one adds an electron or a nucleus to that atom, one obtains the He atom, or the H_2^+ molecular ion, respectively. For these two systems, it was discovered that the Schrödinger equation could hardly be solved. A few years later, precise theoretical descriptions of these systems were presented. It even took about 20 years for the hydrogen molecular ion to be treated within a numerically accurate approach, although Øyvind Burrau had laid the foundations already in 1927.

In 1929, Paul Andre Maurice Dirac had summarized the problems. [P. A. M. Dirac, Proc. Roy. Soc. (London) A **123**, 714 (1929)]:

> The general theory of quantum mechanics is now almost complete, the imperfections still remain being in connection with the exact fitting of the theory with relativity ideas. These give rise to difficulties only when high-speed particles are involved, and are therefore of no importance in the consideration of atomic and molecular structure and ordinary chemical reactions, in which it is, indeed, usually sufficiently accurate if one neglects relativity variation of mass with velocity and assumes only Coulomb forces between the various electrons and atomic nuclei. The underlying physical laws necessary for the mathematical theory of a large part of physics and the whole of chemistry are thus completely known, and the difficulty is only that the exact application of these laws leads to equations much too complex to be soluble. It therefore becomes desirable that approximate practical methods of applying quantum mechanics should be developed, which can lead to an explanation of the main features of complex atomic systems without too much computation.

Actually, first of all, this statement implies that with the help of quantum theory in chemistry, one no longer needs any experimental work, because, in principle, everything can be calculated. This statement is relativized by the fact that Dirac realized that the equations to be solved could not be solved mathematically. But also a way out of this dilemma was proposed by Dirac: he emphasized that the application of quantum theory to chemical and physical questions requires the use of approximations.

The development over the past nearly 100 years shows that many such approximate methods have been developed which further have decisively influenced our understanding of, e. g., chemical bonds. The fundamental problems that Dirac men-

https://doi.org/10.1515/9783110742206-009

tioned are circumvented by the fact that sufficiently accurate, but still approximate methods have been developed, which have become important instruments in chemistry, especially with the aid of computer programs. Because these methods represent approximations and at the same time require significant computer resources, chemical problems can not be treated 100 % exactly and the methods can only be used for idealized systems. Overall, this means that nowadays experimental studies have not been replaced by theoretical studies, nor are theoretical studies irrelevant. Instead, the theoretical methods have become an important complement to experimental work in chemistry, and are also increasingly used in industry.

In this chapter, we will present the basics for some of the approximate methods mentioned above. Later (in Chapter 10) we will use this to show that the well-known so-called orbital model often is one very good, but still "only" approximate method.

9.2 Variational principle

When using different approximate methods, there are a few key questions:
- How can one approximate the exact solution of the Schrödinger equation?
- How to compare two approximate solutions?

The variational principle can be very helpful in answering these questions. This will be discussed here in general form. For that, we use some of the basics that we presented in Chapter 3.

We consider the general eigenvalue problem

$$\hat{A}f_n = a_n f_n \tag{9.1}$$

and assume that $\hat{A}$ is a linear and hermitian operator.

Because $\hat{A}$ is a hermitian operator, all its eigenvalues a_n are real. We will now assume that there is a smallest eigenvalue (which is the case, e. g., for the Hamilton operator, but not for the position operator) so that we can sort the eigenvalues according to increasing size,

$$a_0 \leq a_1 \leq a_2 \leq a_3 \leq \cdots \leq a_{n-1} \leq a_n \leq a_{n+1} \leq \cdots. \tag{9.2}$$

We also use that the eigenfunctions form a complete set of functions, and that they can be orthonormalized,

$$\langle f_n | f_m \rangle = \delta_{n,m}. \tag{9.3}$$

We emphasize that it is not necessary for our arguments to know precisely the eigenfunctions f_n or the eigenvalues a_n, but only to know that they exist.

We will now present an approach to estimate the quality of an approximation to the eigenfunction f_0 for the lowest eigenvalue a_0:

$$f_0 \simeq \phi. \tag{9.4}$$

The idea behind this is that we can choose ϕ absolutely freely, so that it becomes also possible to compare different functions.

Because the eigenfunctions, $\{f_n\}$, form a complete set, we can expand the approximate function ϕ according to

$$\phi = \sum_n c_n f_n. \tag{9.5}$$

Then we consider

$$\frac{\langle \phi | \hat{A} | \phi \rangle}{\langle \phi | \phi \rangle} = \frac{\langle \sum_{n_1} c_{n_1} f_{n_1} | \hat{A} | \sum_{n_2} c_{n_2} f_{n_2} \rangle}{\langle \sum_{n_1} c_{n_1} f_{n_1} | \sum_{n_2} c_{n_2} f_{n_2} \rangle} = \frac{\langle \sum_{n_1} c_{n_1} f_{n_1} | \sum_{n_2} c_{n_2} \hat{A} f_{n_2} \rangle}{\langle \sum_{n_1} c_{n_1} f_{n_1} | \sum_{n_2} c_{n_2} f_{n_2} \rangle}$$

$$= \frac{\langle \sum_{n_1} c_{n_1} f_{n_1} | \sum_{n_2} c_{n_2} a_{n_2} f_{n_2} \rangle}{\langle \sum_{n_1} c_{n_1} f_{n_1} | \sum_{n_2} c_{n_2} f_{n_2} \rangle} = \frac{\sum_{n_1,n_2} \langle c_{n_1} f_{n_1} | c_{n_2} a_{n_2} f_{n_2} \rangle}{\sum_{n_1,n_2} \langle c_{n_1} f_{n_1} | c_{n_2} f_{n_2} \rangle}$$

$$= \frac{\sum_{n_1,n_2} c_{n_1}^* c_{n_2} a_{n_2} \langle f_{n_1} | f_{n_2} \rangle}{\sum_{n_1,n_2} c_{n_1}^* c_{n_2} \langle f_{n_1} | f_{n_2} \rangle} = \frac{\sum_{n_1,n_2} c_{n_1}^* c_{n_2} a_{n_2} \delta_{n_1,n_2}}{\sum_{n_1,n_2} c_{n_1}^* c_{n_2} \delta_{n_1,n_2}}$$

$$= \frac{\sum_n c_n^* c_n a_n}{\sum_n c_n^* c_n} = \frac{\sum_n |c_n|^2 a_n}{\sum_n |c_n|^2}$$

$$\geq \frac{\sum_n |c_n|^2 a_0}{\sum_n |c_n|^2} = \frac{a_0 \sum_n |c_n|^2}{\sum_n |c_n|^2} = a_0. \tag{9.6}$$

We have here used that the operator $\hat{A}$ is a linear operator as well as the properties of equations (9.1), (9.2), and (9.3).

Equation (9.6) shows that, regardless of what ϕ looks like, the expectation value $\frac{\langle \phi | \hat{A} | \phi \rangle}{\langle \phi | \phi \rangle}$ always is greater than (or equal to) the smallest eigenvalue, a_0. If we have several approximate functions, we can compare their expectation values. All are according to equation (9.6) greater than the lowest eigenvalue, so that the smallest expectation value then deviates least from a_0. Accordingly, we will take the lowest expectation value and consider it as the best approximation to a_0. This statement is well founded by the variational principle. But we will also assume that the corresponding function ϕ is then the best approximation to f_0. This is just an assumption and not founded in the variational principle. But often it is not a bad assumption.

9.3 Variation method—an example

The practical application of the variational principle is best illustrated through a simple example. We consider a particle in a box with a finite potential outside the box (see

Figure 9.1),

$$V(x) = \begin{cases} -V_0 & \text{for } |x| \leq \ell \\ 0 & \text{for } |x| > \ell. \end{cases} \tag{9.7}$$

For this potential, we seek an approximation to the ground-state energy, i. e., to the smallest energy eigenvalue of the Schrödinger equation. This equation is

$$\left[-\frac{\hbar^2}{2m} \frac{d^2}{dx^2} + V(x) \right] \Psi(x) = E\Psi(x), \tag{9.8}$$

corresponding to replacing the operator $\hat{A}$ in Section 9.2 by the Hamilton operator,

$$\hat{H} = -\frac{\hbar^2}{2m} \frac{d^2}{dx^2} + V(x), \tag{9.9}$$

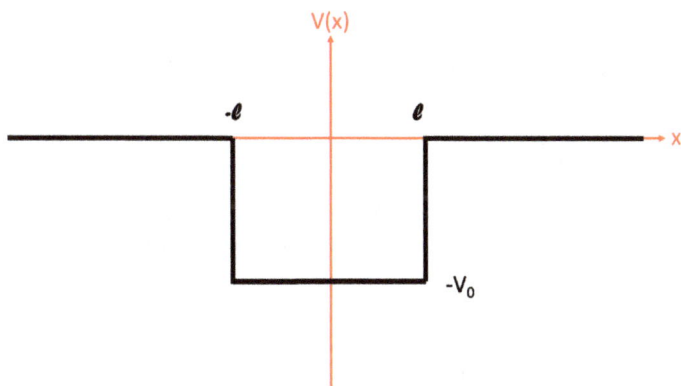

Figure 9.1: The potential of equation (9.7).

A reasonable, approximate function is a function located mainly in the range $-\ell \leq x \leq \ell$ (to make the potential energy small) and having small oscillations (to make the kinetic energy small). This could be for the case for

$$\phi_1 = \exp(-\alpha x^2). \tag{9.10}$$

Here, α is a parameter describing the width of the function: the larger α becomes, the narrower the function becomes and the larger the kinetic energy becomes, whereas the potential energy becomes smaller. This indicates that there is likely to be a "best" value for α.

To find that value, we will study

$$\frac{\langle \phi_1 | \hat{H} | \phi_1 \rangle}{\langle \phi_1 | \phi_1 \rangle} = \frac{\langle \phi_1 | -\frac{\hbar^2}{2m} \frac{d^2}{dx^2} + V(x) | \phi_1 \rangle}{\langle \phi_1 | \phi_1 \rangle}, \tag{9.11}$$

which becomes a function of α

$$\frac{\langle \phi_1 | \hat{H} | \phi_1 \rangle}{\langle \phi_1 | \phi_1 \rangle} \equiv \tilde{E}_1(\alpha). \tag{9.12}$$

Here, we will not attempt to calculate $\tilde{E}_1(\alpha)$ analytically.

For every value of α, $\tilde{E}_1(\alpha) \geq E_0$ [with E_0 being the smallest eigenvalue of equation (9.8). Therefore, we seek that value of α for which $\tilde{E}_1(\alpha)$ is the smallest. That is, α can be determined through

$$\frac{\partial}{\partial \alpha} \tilde{E}_1(\alpha) = 0. \tag{9.13}$$

Here, we have not explicitly derived the mathematical expressions, but we hope that the general approach is recognizable.

As an extension, we could have used a function with more parameters, e. g.,

$$\phi_2 = c_1 \exp(-\alpha_1 x^2) + c_2 \exp(-\alpha_2 x^2) \tag{9.14}$$

and for this minimize the expectation value

$$\frac{\langle \phi_2 | \hat{H} | \phi_2 \rangle}{\langle \phi_2 | \phi_2 \rangle} \equiv \tilde{E}_2(c_1, \alpha_1, c_2, \alpha_2) \tag{9.15}$$

by varying the parameters c_1, c_2, α_1, and α_2. Because of the variational principle, we also know that the smallest value of $\tilde{E}_2(c_1, \alpha_1, c_2, \alpha_2)$ becomes smaller than or equal to the smallest value of $\tilde{E}_1(\alpha)$.

9.4 Variational method—general

In the general case, we are searching an approximation to the smallest eigenvalue and the associated eigenfunction of the eigenvalue equation

$$\hat{A} f_0 = a_0 f_0. \tag{9.16}$$

The eigenfunction f_0 depends on the coordinates of the system $\vec{x}$ (in the example above, on the single position coordinate x). Similarly, an approximate function will also depend on $\vec{x}$. But we can also introduce additional parameters for this function, $p_1, p_2, \ldots, p_{N_p}$ [analogous to α, or $(c_1, \alpha_1, c_2, \alpha_2)$ in the example above], i. e.,

$$\phi = \phi(p_1, p_2, \ldots, p_{N_p}; \vec{x}). \tag{9.17}$$

Thus, the expectation $\frac{\langle \phi | \hat{A} | \phi \rangle}{\langle \phi | \phi}$ will become a function of these (N_p) parameters,

$$\frac{\langle \phi | \hat{A} | \phi \rangle}{\langle \phi | \phi \rangle} \equiv \tilde{a}(p_1, p_2, \ldots, p_{N_p}), \tag{9.18}$$

and we can determine the values of the parameters by requiring that $\tilde{a}(p_1, p_2, \ldots, p_{N_p})$ has a minimum,

$$\frac{\partial \tilde{a}(p_1, p_2, \ldots, p_{N_p})}{\partial p_1} = \frac{\partial \tilde{a}(p_1, p_2, \ldots, p_{N_p})}{\partial p_2} = \cdots = \frac{\partial \tilde{a}(p_1, p_2, \ldots, p_{N_p})}{\partial p_{N_p}} = 0. \qquad (9.19)$$

By introducing more and more parameters, it is often possible to obtain a very accurate approximation to a_0 (and thereby also to f_0). Unfortunately, however, the equations (9.19) are often so complex that they can hardly be solved. But there is one (very important) exception and this will be discussed in the next chapter.

9.5 Rayleigh–Ritz variational procedure—general

As mentioned above, the problem of solving the N_p equations (9.19) quickly becomes insurmountable. There is one exception, however, for which the equations, though slightly modified, can be solved. This exception shall be discussed here.

Our goal is to minimize the expectation value

$$\frac{\langle \phi | \hat{A} | \phi \rangle}{\langle \phi | \phi \rangle} \qquad (9.20)$$

with $\hat{A}$ being a linear and hermitian operator.

We will now write ϕ as a linear combination of some prechosen functions:

$$\phi(\vec{x}) = \sum_{i=1}^{N_b} c_i \chi_i(\vec{x}). \qquad (9.21)$$

The functions $\{\chi_i\}$ are so-called basis functions whose form is chosen "somehow." We denote their number by N_b and only the coefficients $\{c_i\}$ are varied, while the basis functions remain fixed. This corresponds to the case of equation (9.14) but with the restriction that α_1 and α_2 are kept fixed at prechosen values.

Eq. (9.19) then becomes

$$\frac{\partial}{\partial c_k} \frac{\langle \phi | \hat{A} | \phi \rangle}{\langle \phi | \phi \rangle} = 0 \qquad (9.22)$$

for all $k = 1, 2, \ldots, N_b$, or, by considering the complex conjugate equation,

$$\frac{\partial}{\partial c_k^*} \frac{\langle \phi | \hat{A} | \phi \rangle}{\langle \phi | \phi \rangle} = 0. \qquad (9.23)$$

We will now formulate these conditions slightly differently, arriving at an equation at the end that is relatively easy to solve (mathematically and/or numerically). To achieve this, we recognize that equation (9.23) can also be formulated in such a way that

$$\frac{\partial}{\partial c_k^*} \langle \phi | \hat{A} | \phi \rangle = 0, \quad k = 1, \ldots, N_b, \qquad (9.24)$$

if it is simultaneously demanded that the function ϕ is normalized,

$$\langle \phi | \phi \rangle = 1. \tag{9.25}$$

Compared with equation (9.23), we have replaced the denominator in the expression on the left-hand side in equation (9.23) by a constraint, equation (9.25).

Equation (9.25) represents a constraint under which the quantity $\langle \phi | \hat{A} | \phi \rangle$ should be minimized. To account for such constraints, there is a mathematical method based on so-called Lagrange multipliers (after Joseph-Louis Lagrange). These are additional parameters whose numerical values are at first unknown but can ultimately be determined through the equations.

For our purpose, in which we have only one constraint, this means that instead of the quantity in equation (9.24) we study

$$K = \langle \phi | \hat{A} | \phi \rangle - \lambda [\langle \phi | \phi \rangle - 1]. \tag{9.26}$$

The parameter λ is the Lagrange multiplier. This multiplies the quantity in the square brackets, which according to our constraint should be zero. Had we had several constraints, we would have had to introduce a Lagrange multiplier for each of them. We will treat such cases later.

Equations (9.24) and (9.25) are now written as

$$\frac{\partial K}{\partial c_k^*} = \frac{\partial K}{\partial \lambda} = 0. \tag{9.27}$$

Because λ occurs only at one place in K, it is very easy to see that the second equation in equation (9.27) is identical to the constraint, equation (9.25). But the first equation in equation (9.27) is not identical to equation (9.24), because also the Lagrangian multiplier λ appears, as we will see later.

The Lagrange multiplier initially represents only a mathematical trick to account for the constraint. First of all, it is unclear whether one can also ascribe this a chemical/physical interpretation. It is often possible to do so (including our case, as we shall see), but it is not necessarily always the case.

Thus, we insert K of equation (9.26) in the second equation in equation (9.27) and get immediately

$$\langle \phi | \phi \rangle - 1 = 0. \tag{9.28}$$

From the first equation in equation (9.27), we obtain with the help of equation (9.21),

$$\frac{\partial}{\partial c_k^*} \left[\left\langle \sum_i c_i \chi_i | \hat{A} | \sum_j c_j \chi_j \right\rangle - \lambda \left(\left\langle \sum_i c_i \chi_i | \sum_j c_j \chi_j \right\rangle - 1 \right) \right]$$

$$= \frac{\partial}{\partial c_k^*} \left(\left\langle \sum_i c_i \chi_i | \hat{A} | \sum_j c_j \chi_j \right\rangle - \lambda \left\langle \sum_i c_i \chi_i | \sum_j c_j \chi_j \right\rangle \right)$$

$$= \frac{\partial}{\partial c_k^*} \sum_{i,j} c_i^* c_j [\langle \chi_i | \hat{A} | \chi_j \rangle - \lambda \langle \chi_i | \chi_j \rangle]$$

$$= \sum_j c_j [\langle \chi_k | \hat{A} | \chi_j \rangle - \lambda \langle \chi_k | \chi_j \rangle] \equiv 0, \tag{9.29}$$

where we have as many equations as we have basis functions, i. e.,

$$k = 1, 2, \dots, N_b. \tag{9.30}$$

The unknown quantities we are looking for are, above all, the expansion coefficients $\{c_k\}$, which we derive from the linear equations [equation (9.29)].

$$\sum_j [\langle \chi_k | \hat{A} | \chi_j \rangle - \lambda \langle \chi_k | \chi_j \rangle] c_j = 0 \tag{9.31}$$

But also the Lagrange multiplier λ is unknown. If it was known, the N_b equations in equation (9.31) would be N_b linear equations with N_b unknowns. Because the right-hand side is 0 for all equations, it is very easy to solve those equations:

$$c_k = 0, \quad k = 1, 2, \dots, N_b \tag{9.32}$$

that would be the solution. However, this solution is not very useful because the associated wave function in equation (9.21) then becomes identically 0. This wave function is then also not normalizable, i. e., the second condition in equation (9.25) cannot be satisfied.

This raises the question whether there could be circumstances under which the equations in equation (9.31) have other solutions. We recognize that λ is not yet determined, so the question is whether there are values of λ in equation (9.31), for which we have solutions other than those of equation (9.32).

In the general case, we know that the solutions to P linear equations

$$\sum_j a_{kj} x_j = b_k, \quad k = 1, 2, \dots, P \tag{9.33}$$

(here the numbers a_{kj} and b_k are known, while x_j are the unknowns) can be written using determinants,

$$x_j = \frac{D_j}{D} \tag{9.34}$$

where D is the determinant of the $P \times P$ matrix with the coefficients a_{kj}, while D_j is the determinant of the matrix we get if we replace the jth column in the matrix of D by the numbers b_k.

This makes it easy to see that in our case, equation (9.31), the solutions usually disappear because the numbers on the right-hand side are all 0 so that all $D_j = 0$. An exception that would allow for other solutions than $x_j = 0$ is possible if in equation

(9.34), $D = 0$. In our case, equation (9.31), this is possible through special choices of λ. This means that we will determine λ so that

$$\det(\underline{\underline{A}} - \lambda \cdot \underline{\underline{O}}) = 0. \tag{9.35}$$

This is the determinant D for the equations (9.31).

In equation (9.35), we have introduced two matrices,

$$\underline{\underline{A}}_{kj} = \langle \chi_k | \hat{A} | \chi_j \rangle \tag{9.36}$$

and

$$\underline{\underline{O}}_{kj} = \langle \chi_k | \chi_j \rangle. \tag{9.37}$$

Equation (9.35) is the so-called secular equation. This equation is (one of) the most important ones of this manuscript.

The secular equation is the equation from which λ will be determined. It can be seen that the left-hand side of the equation can be written as a polynomial in λ. The polynomial has the order N_b, so the equation has a total of N_b solutions. So another question arises: which value is the best in our case?

But first, it should be mentioned that equation (9.31) can also be written as a so-called generalized eigenvalue equation,

$$\sum_j \langle \chi_k | \hat{A} | \chi_j \rangle c_j = \lambda \sum_j \langle \chi_k | \chi_j \rangle c_j, \tag{9.38}$$

or in matrix form

$$\underline{\underline{A}} \cdot \underline{c} = \lambda \cdot \underline{\underline{O}} \cdot \underline{c} \tag{9.39}$$

Here, $\underline{c}$ is a vector with the sought coefficients $\{c_j\}$. In case the matrix $\underline{\underline{O}}$ is equal to the unit matrix, equation (9.38) is a "normal" eigenvalue equation, but for other matrices it is called a generalized eigenvalue equation. Solving the "normal" eigenvalue equation is called diagonalizing the matrix $\underline{\underline{A}}$. This is discussed briefly in Section 16.3.

For applications, the very important point is that the matrix eigenvalue equation (9.38) easily can be solved using standard computer routines that we do not need to care about but only use. This yields all N_b eigenvalues λ and the associated coefficients $\{c_j\}$, i. e., N_b different solutions to equation (9.29). Subsequently, we shall identify which of these is the relevant one.

At first, we remind that the sought expectation value is

$$\frac{\langle \phi | \hat{A} | \phi \rangle}{\langle \phi | \phi \rangle} = \frac{\sum_{i,j} c_i^* \langle \chi_i | \hat{A} | \chi_j \rangle c_j}{\sum_{i,j} c_i^* \langle \chi_i | \chi_j \rangle c_j}. \tag{9.40}$$

We will now show that this is identical to λ. For this purpose, we multiply equation (9.38) by c_k^* and sum all the different equations for the different k. This gives

$$\sum_{j,k} c_k^* \langle \chi_k | \hat{A} | \chi_j \rangle c_j = \lambda \sum_{j,k} c_k^* \langle \chi_k | \chi_j \rangle c_j, \tag{9.41}$$

or

$$\lambda = \frac{\sum_{j,k} c_k^* \langle \chi_k | \hat{A} | \chi_j \rangle c_j}{\sum_{j,k} c_k^* \langle \chi_k | \chi_j \rangle c_j}. \tag{9.42}$$

This is exactly the expression in equation (9.40). Thus, the eigenvalue (or Lagrange multiplier) is the sought expectation value. Therefore, out of the N_b eigenvalues and eigenfunctions, we shall choose the one corresponding to the smallest eigenvalue.

9.6 Rayleigh–Ritz variation method—an example

We can illustrate the Rayleigh–Ritz method through the example of Section 9.4. We consider the test function of equation (9.14),

$$\phi = c_1 \exp(-\alpha_1 x^2) + c_2 \exp(-\alpha_2 x^2), \tag{9.43}$$

but we will keep the two nonlinear parameters α_1 and α_2 constant at prechosen values (how to choose these "reasonable" is not easy and often a matter of experience). In contrast to the example in Section 9.4, in this case, we only have to determine two parameters, c_1 and c_2.

We introduce

$$\chi_1(x) = \exp(-\alpha_1 x^2)$$
$$\chi_2(x) = \exp(-\alpha_2 x^2) \tag{9.44}$$

and then

$$
\begin{aligned}
O_{11} &= \langle \chi_1 | \chi_1 \rangle \\
O_{12} &= \langle \chi_1 | \chi_2 \rangle \\
O_{21} &= \langle \chi_2 | \chi_1 \rangle \\
O_{22} &= \langle \chi_2 | \chi_2 \rangle \\
H_{11} &= \langle \chi_1 | \hat{H} | \chi_1 \rangle \\
H_{12} &= \langle \chi_1 | \hat{H} | \chi_2 \rangle \\
H_{21} &= \langle \chi_2 | \hat{H} | \chi_1 \rangle \\
H_{22} &= \langle \chi_2 | \hat{H} | \chi_2 \rangle
\end{aligned} \tag{9.45}
$$

so that K from equation (9.26) can be written as

$$K = c_1^* c_1 H_{11} + c_1^* c_2 H_{12} + c_2^* c_1 H_{21} + c_2^* c_2 H_{22}$$
$$- \lambda[c_1^* c_1 O_{11} + c_1^* c_2 O_{12} + c_2^* c_1 O_{21} + c_2^* c_2 O_{22} - 1]. \tag{9.46}$$

In this case, equation (9.31) consists of two equations

$$[H_{11} - \lambda O_{11}]c_1 + [H_{12} - \lambda O_{12}]c_2 = 0$$
$$[H_{21} - \lambda O_{21}]c_1 + [H_{22} - \lambda O_{22}]c_2 = 0. \tag{9.47}$$

This set of equations has always the trivial solution, $c_1 = c_2 = 0$, which would, however, lead to the fact that the wave function in equation (9.43) becomes identically zero. Therefore, we seek cases for which we also have other solutions to the equations. We recognize that the value of the parameter λ is not yet fixed and that we can now determine the value by requiring that the two equations in equation (9.47) are linearly dependent. Then one equation is proportional to the other equation. That is the case when

$$[H_{11} - \lambda O_{11}] \cdot [H_{22} - \lambda O_{22}] - [H_{12} - \lambda O_{12}] \cdot [H_{21} - \lambda O_{21}] = 0 \tag{9.48}$$

i. e., equation (9.35) is satisfied.

In equation (9.48), all quantities are known except for λ, so that we can determine λ from this equation. In fact, this leads to a second-order equation in λ,

$$(O_{11}O_{22} - O_{12}O_{21})\lambda^2 - (H_{11}O_{22} - H_{12}O_{21} + O_{11}H_{22} - O_{12}H_{21})\lambda$$
$$+ (H_{11}H_{22} - H_{12}H_{21}) = 0, \tag{9.49}$$

which has two solutions. Because of equation (9.42), we choose the smaller value of these two.

Then we can substitute this value of λ into one of the two equations (9.47) (since the two equations in this case are proportional to each other, we only need to consider one of them) and obtain a relationship between the two constants c_1 and c_2,

$$c_2 = -\frac{H_{11} - \lambda O_{11}}{H_{12} - \lambda O_{12}} c_1. \tag{9.50}$$

Finally, we can then determine c_1 by normalizing the wave function,

$$1 = c_1^* c_1 O_{11} + c_1^* c_2 O_{12} + c_2^* c_1 O_{21} + c_2^* c_2 O_{22}. \tag{9.51}$$

We insert c_2 from equation (9.50) into this equation, and thereby obtain an equation according to which c_1 can be determined (almost). "Almost," because we can always change c_1 through an arbitrary phase factor,

$$c_1 \rightarrow c_1 e^{i\theta}, \tag{9.52}$$

although this leaves all the calculated values for experimental quantities unchanged. Because of equation (9.50), c_2 is also changed by this phase factor.

9.7 Time-independent perturbation theory

Above, we have presented the variational procedure in detail. With this method, an approximate wave function for the ground state of any system can be created and also systematically improved (the latter, e. g., by including more and more basis functions in the calculations). This process is the basis for much of what we will introduce later and, therefore, it has been treated in such detail. But there are other ways in which more or less accurate results can be obtained, and in this and the following section we will briefly introduce two of them. In this case, we will do so without detailed mathematical derivations.

The perturbation theory is an important theory to consider the additional effects due to small perturbations. The perturbation theory is, e. g., used to treat the effects of ligands on the orbitals of a metal atom when these orbitals are well localized or to study the effects of an electromagnetic field on a molecule. In the first case, one uses the time-independent perturbation theory, because it is assumed that the ligands occupy fixed positions, while in the second case the time-dependent perturbation theory is used because, e. g., the external electromagnetic field oscillates in time with a certain frequency. In this section, we will cover the first case and the second case in the next section.

Accordingly, we assume that the Hamilton operator can be split into two parts,

$$\hat{H} = \hat{H}_0 + \Delta\hat{H}. \tag{9.53}$$

We consider a wave function ψ_k for which the effects of $\Delta\hat{H}$ are small,

$$\langle\psi_k|\hat{H}_0|\psi_k\rangle \gg \langle\psi_k|\Delta\hat{H}|\psi_k\rangle. \tag{9.54}$$

We further assume that we know the solutions to the Schrödinger equation in the absence of the perturbation,

$$\hat{H}_0\psi_i^{(0)} = E_i^{(0)}\psi_i^{(0)} \tag{9.55}$$

[the index (0) indicates that this is the solution for the Schrödinger equation without $\Delta\hat{H}$], and that these eigenfunctions are orthonormal,

$$\langle\psi_i^{(0)}|\psi_j^{(0)}\rangle = \delta_{i,j}. \tag{9.56}$$

Ultimately, we will here treat only the so-called nondegenerate case, which means that we study a wave function $\psi_k^{(0)}$ for which there are no other wave functions with the same energy,

$$\hat{H}_0\psi_k^{(0)} = E_k^{(0)}\psi_k^{(0)} \tag{9.57}$$

with

$$E_i^{(0)} \neq E_k^{(0)} \quad \text{for } i \neq k. \tag{9.58}$$

As mentioned, we assume that the effects of the perturbation $\Delta\hat{H}$ on ψ_k are small, so that we can write

$$\psi_k = \psi_k^{(0)} + \Delta\psi_k^{(1)} + \Delta\psi_k^{(2)} + \Delta\psi_k^{(3)} + \cdots$$
$$E_k = E_k^{(0)} + \Delta E_k^{(1)} + \Delta E_k^{(2)} + \Delta E_k^{(3)} + \cdots. \tag{9.59}$$

Here, $\Delta\psi_k^{(1)}, \Delta\psi_k^{(2)}, \Delta\psi_k^{(3)}, \ldots$ are terms in $\Delta\hat{H}$ to first, second, third, ... order, which also applies to $\Delta E_k^{(1)}, \Delta E_k^{(2)}, \Delta E_k^{(3)}, \ldots$. The idea behind this is that $\Delta\hat{H}$ is small, so that terms of higher powers in $\Delta\hat{H}$ become progressively smaller.

After somewhat tedious but not very complicated mathematical derivations one arrives at the following results:

$$E_k = E_k^{(0)} + \langle\psi_k^{(0)}|\Delta\hat{H}|\psi_k^{(0)}\rangle + \sum_{i\neq k}\frac{|\langle\psi_i^{(0)}|\Delta\hat{H}|\psi_k^{(0)}\rangle|^2}{E_k^{(0)} - E_i^{(0)}} + \cdots \tag{9.60}$$

and

$$\psi_k = \psi_k^{(0)} + \sum_{i\neq k}\frac{\langle\psi_i^{(0)}|\Delta\hat{H}|\psi_k^{(0)}\rangle}{E_k^{(0)} - E_i^{(0)}}\psi_i^{(0)} + \cdots. \tag{9.61}$$

Here, we have given only the corrections of the energy to the first and second order and only those of the wave function to the first order in the perturbation.

Finally, we emphasize that this treatment is valid only for the nondegenerate case. If we consider a wave function that is one out of more, all of which have the same energy in the absence of the perturbation, the theory has to be extended slightly. However, this will not be discussed further here.

9.8 Time-dependent perturbation theory

If the perturbation is time-dependent, one has to treat the time-dependent Schrödinger equation,

$$\hat{H}\tilde{\psi}(\tilde{x}, t) = i\hbar\frac{\partial}{\partial t}\tilde{\psi}(\tilde{x}, t), \tag{9.62}$$

where $\tilde{x}$ represents all position and spin coordinates. This case is, e. g., very relevant for spectroscopy in which a system is perturbed by one or more oscillating electromagnetic fields.

We assume that $\hat{H}$ has a dominating time-independent part while the time-dependent part is small,

$$\hat{H} = \hat{H}_0 + \Delta\hat{H}(t). \tag{9.63}$$

In addition, we assume that we know the solutions to the time-independent Schrödinger equation without the perturbation,

$$\hat{H}_0 \psi_k^{(0)}(\vec{x}) = E_k^{(0)} \psi_k^{(0)}(\vec{x}). \tag{9.64}$$

From Chapter 2, we know the associated time-dependent wave functions,

$$\tilde{\psi}_k^{(0)}(\vec{x}, t) = \psi_k^{(0)}(\vec{x}) \exp\left(-\frac{iE_k^{(0)}t}{\hbar}\right). \tag{9.65}$$

In the general case (when the perturbation is turned on), we write the time-dependent wave functions as

$$\tilde{\psi}(\vec{x}, t) = \sum_i c_i(t) \tilde{\psi}_i^{(0)}(\vec{x}, t). \tag{9.66}$$

We will now treat the special case that the system is in a fixed state k until a certain time t_0,

$$c_j(t) = \delta_{j,k} \quad \text{for } t < t_0. \tag{9.67}$$

At time t_0 the perturbation is turned on and the coefficients $c_j(t)$ may then change. Specifically, this means that the occupation of the energy levels of the system can change,

$$c_i(t) = c_i^{(0)}(t) + c_i^{(1)}(t) + c_i^{(2)}(t) + \cdots, \tag{9.68}$$

where $c_i^{(1)}(t), c_i^{(2)}(t), \ldots$ depend linearly, quadratic, $\ldots$ on the perturbation $\Delta\hat{H}(t)$ (this is analogous to our approach in the last section where we treated the time-independent perturbation).

After some calculations, we arrive at

$$c_j^{(1)}(t) = \frac{1}{i\hbar} \int_{t_0}^{t} \langle \psi_j^{(0)} | \Delta\hat{H}(t') | \psi_k^{(0)} \rangle \exp(i\omega_{jk}t') \, dt' \tag{9.69}$$

with

$$\omega_{ji} = \frac{E_j^{(0)} - E_i^{(0)}}{\hbar}. \tag{9.70}$$

A very important case is that the perturbation oscillates with a fixed frequency,

$$\Delta\hat{H}(t) = \Delta H \cdot (e^{i\omega t} + e^{-i\omega t}). \tag{9.71}$$

ΔH depends only on the position coordinates but not on the time. Such cases are found, e. g., when the system is exposed to an electromagnetic field, which is relevant for spectroscopic investigations. The most common case is that in which the perturbation lasts for a very (about infinitely) long time, whereby "very long" has to be specified. The duration of the perturbation here is long for one electron, i. e., longer than picoseconds.

We may now consider the case of a resonance, i. e., $\hbar\omega$ is equal to the energy difference between two states of the undisturbed system without the time-dependent perturbation,

$$\hbar\omega = E_f - E_i. \tag{9.72}$$

In this case, there is a finite probability that the system will be excited. The excitation rate W (i. e., the number of excitations per time unit) is then given by

$$W = \frac{2\pi}{\hbar} |\langle \psi_i^{(0)} | \Delta\hat{H} | \psi_f^{(0)} \rangle|^2. \tag{9.73}$$

This is Fermi's golden rule (after Enrico Fermi).

The assumption that the experiment takes "a very long time" is very common. However, since further effects can occur if this assumption is not met, so-called femtosecond spectroscopy has been developed. One of the leading scientists in this development, Ahmed Zewail, was awarded the 1999 Nobel Prize in Chemistry for this work. More recently, attosecond spectroscopy has been developed that can provide an even more detailed understanding of the motion of electrons in molecules. Here, the time for a measurement is so short that, casually expressed, one can observe the movements of the electrons.

9.9 Problems with answers

1. **Problem:** Consider the hydrogen atom. Use the variational principle with e^{-kr} as the test function to determine an upper limit for the ground-state energy of the hydrogen atom.
    Answer: We introduce

$$\hat{H} = -\frac{\hbar^2}{2\mu}\nabla^2 - \frac{e^2}{4\pi\epsilon_0 r} \tag{9.74}$$

and

$$\phi(\vec{r}) = e^{-kr}. \tag{9.75}$$

With the help of the formulas in Chapter 17, we find

$$\left[-\frac{\hbar^2}{2\mu}\nabla^2 - \frac{e^2}{4\pi\epsilon_0 r}\right]e^{-kr} = \left[-\frac{\hbar^2}{2\mu}\frac{d^2}{dr^2} - \frac{\hbar^2}{2\mu}\frac{2}{r}\frac{d}{dr} - \frac{e^2}{4\pi\epsilon_0 r}\right]e^{-kr}$$

$$= \left[-\frac{\hbar^2}{2\mu}\left(k^2 - \frac{2k}{r}\right) - \frac{e^2}{4\pi\epsilon_0 r}\right]e^{-kr}. \tag{9.76}$$

Then

$$\langle \phi | \hat{H} | \phi \rangle = \int \left[-\frac{\hbar^2}{2\mu} \left(k^2 - \frac{2k}{r} \right) - \frac{e^2}{4\pi\epsilon_0 r} \right] e^{-2kr} \, d\vec{r}$$

$$= 4\pi \int_0^\infty \left[-\frac{\hbar^2}{2\mu} (k^2 r^2 - 2kr) - \frac{e^2}{4\pi\epsilon_0} r \right] e^{-2kr} \, dr$$

$$= 4\pi \left[-\frac{\hbar^2}{2\mu} \left(k^2 \frac{2}{(2k)^3} - 2k \frac{1}{(2k)^2} \right) - \frac{e^2}{4\pi\epsilon_0} \frac{1}{(2k)^2} \right]$$

$$= 4\pi \left[\frac{\hbar^2}{2\mu} \frac{1}{4k} - \frac{e^2}{4\pi\epsilon_0} \frac{1}{(2k)^2} \right]. \tag{9.77}$$

Similarly,

$$\langle \phi | \phi \rangle = \int e^{-2kr} \, d\vec{r} = 4\pi \int_0^\infty e^{-2kr} r^2 \, dr = 4\pi \frac{2}{(2k)^3}. \tag{9.78}$$

Thereby,

$$\frac{\langle \phi | \hat{H} | \phi \rangle}{\langle \phi | \phi \rangle} = 4k^3 \left[\frac{\hbar^2}{2\mu} \frac{1}{4k} - \frac{e^2}{4\pi\epsilon_0} \frac{1}{(2k)^2} \right] = \frac{\hbar^2}{2\mu} k^2 - \frac{e^2}{4\pi\epsilon_0} k \equiv \tilde{E}(k). \tag{9.79}$$

We seek the minimum of $\tilde{E}(k)$:

$$\frac{d}{dk} \tilde{E}(k) = \frac{\hbar^2}{\mu} k - \frac{e^2}{4\pi\epsilon_0} \equiv 0 \tag{9.80}$$

which has the solution

$$k = \frac{e^2 \mu}{4\pi\epsilon_0 \hbar^2}. \tag{9.81}$$

If we use the electron mass m_e instead of the reduced mass μ, then this value for k is actually the inverse of the Bohr radius [equation (8.40)], as it should be. For the value from equation (9.81), the estimated ground-state energy becomes

$$\tilde{E} = -\frac{\mu e^4}{32\hbar^2 \pi^2 \epsilon_0^2}. \tag{9.82}$$

This value corresponds to the expression in equation (8.39) for $n = 1$.

2. **Problem:** Consider the hydrogen atom. Use the variational principle with the test function e^{-ar^2} to determine an upper limit of the ground-state energy of the hydrogen atom.

Answer: We introduce

$$\hat{H} = -\frac{\hbar^2}{2\mu} \nabla^2 - \frac{e^2}{4\pi\epsilon_0 r} \tag{9.83}$$

and

$$\phi(\vec{r}) = e^{-\alpha r^2}. \tag{9.84}$$

Then, with the help of the formulas in Chapter 17 we find

$$\begin{aligned}
\left[-\frac{\hbar^2}{2\mu}\nabla^2 - \frac{e^2}{4\pi\epsilon_0 r}\right]e^{-\alpha r^2} &= \left[-\frac{\hbar^2}{2\mu}\frac{d^2}{dr^2} - \frac{\hbar^2}{2\mu}\frac{2}{r}\frac{d}{dr} - \frac{e^2}{4\pi\epsilon_0 r}\right]e^{-\alpha r^2} \\
&= \left[-\frac{\hbar^2}{2\mu}(-2\alpha + 4\alpha^2 r^2 - 4\alpha) - \frac{e^2}{4\pi\epsilon_0 r}\right]e^{-\alpha r^2} \\
&= \left[-\frac{\hbar^2}{2\mu}(-6\alpha + 4\alpha^2 r^2) - \frac{e^2}{4\pi\epsilon_0 r}\right]e^{-\alpha r^2}. \tag{9.85}
\end{aligned}$$

Thereby

$$\begin{aligned}
\langle\phi|\hat{H}|\phi\rangle &= \int\left[-\frac{\hbar^2}{2\mu}(-6\alpha + 4\alpha^2 r^2) - \frac{e^2}{4\pi\epsilon_0 r}\right]e^{-2\alpha r^2}\,d\vec{r} \\
&= 4\pi\int_0^\infty\left[-\frac{\hbar^2}{2\mu}(-6\alpha r^2 + 4\alpha^2 r^4) - \frac{e^2 r}{4\pi\epsilon_0}\right]e^{-2\alpha r^2}\,dr \\
&= 4\pi\frac{\hbar^2}{2\mu}\left(6\alpha\frac{1}{8\alpha}\sqrt{\frac{\pi}{2\alpha}} - 4\alpha^2\frac{3}{8(2\alpha)^2}\sqrt{\frac{\pi}{2\alpha}}\right) - 4\pi\frac{e^2}{4\pi\epsilon_0}\frac{1}{4\alpha} \\
&= 4\pi\left[\frac{\hbar^2}{2\mu}\frac{3}{8}\sqrt{\frac{\pi}{2\alpha}} - \frac{e^2}{4\pi\epsilon_0}\frac{1}{4\alpha}\right]. \tag{9.86}
\end{aligned}$$

Similarly,

$$\langle\phi|\phi\rangle = \int e^{-2\alpha r^2}\,d\vec{r} = 4\pi\int_0^\infty e^{-2\alpha r^2}r^2\,dr = 4\pi\frac{1}{8\alpha}\sqrt{\frac{\pi}{2\alpha}}. \tag{9.87}$$

In total,

$$\frac{\langle\phi|\hat{H}|\phi\rangle}{\langle\phi|\phi\rangle} = \frac{3\hbar^2\alpha}{2\mu} - \frac{2e^2}{4\pi\epsilon_0}\sqrt{\frac{2\alpha}{\pi}} \equiv \tilde{E}(\alpha). \tag{9.88}$$

We are searching for the minimum of $\tilde{E}(\alpha)$:

$$\frac{d}{d\alpha}\tilde{E}(\alpha) = \frac{3\hbar^2}{2\mu} - \frac{e^2}{4\pi\epsilon_0}\sqrt{\frac{2}{\pi\alpha}} \equiv 0 \tag{9.89}$$

which has the solution

$$\alpha = \frac{\mu^2 e^4}{18\pi^3\hbar^4\epsilon_0^2}. \tag{9.90}$$

For this value,

$$\tilde{E} = -\frac{\mu e^4}{12\hbar^2\pi^3\epsilon_0^2}. \qquad (9.91)$$

It is interesting to compare this value with the exact value for the hydrogen atom, which we also gave in equation (9.82). The ratio of the two equals

$$\frac{\mu e^4}{12\hbar^2\pi^3\epsilon_0^2} \Big/ \frac{\mu e^4}{32\hbar^2\pi^2\epsilon_0^2} = \frac{8}{3\pi} \approx 0.8488. \qquad (9.92)$$

Thus, with this approximate function we get an energy, which corresponds to almost 85 % of the exact energy.

For the function in problem 1, we have obtained that the argument of the exponential function should be $-r/a_0$ so that the energy becomes as small as possible. Similarly, one could suggest that in the present problem the argument of the exponential function should be equal to $-(r/a_0)^2$. This would mean that $a^{-1/2} \cdot k = 1$ with k from equation (9.81). In fact, one finds

$$\alpha^{-1/2} \cdot k = \sqrt{\frac{9\pi}{8}} \approx 1.88. \qquad (9.93)$$

3. **Problem:** Use the variational method to determine an upper limit of the ground-state energy for a particle (mass m) in a 1-dimensional potential, $V(x) = k|x|$. As a test function, use $\phi(x) = e^{-\alpha x^2}$.

   Answer: We have

$$\frac{d^2}{dx^2}\phi(x) = \frac{d^2}{dx^2}e^{-\alpha x^2} = (4\alpha^2 x^2 - 2\alpha)e^{-\alpha x^2}. \qquad (9.94)$$

Then

$$\left\langle \phi \left| -\frac{\hbar^2}{2m}\frac{d^2}{dx^2} \right| \phi \right\rangle = -\frac{\hbar^2}{2m}\int_{-\infty}^{\infty}(4\alpha^2 x^2 - 2\alpha)e^{-2\alpha x^2}\,dx$$

$$= -\frac{\hbar^2}{2m}2\int_{0}^{\infty}(4\alpha^2 x^2 - 2\alpha)e^{-2\alpha x^2}\,dx = -\frac{\hbar^2}{2m}\left(4\alpha^2\frac{1}{4\alpha}\sqrt{\frac{\pi}{2\alpha}} - 2\alpha\sqrt{\frac{\pi}{2\alpha}}\right)$$

$$= \frac{\hbar^2}{2m}\alpha\sqrt{\frac{\pi}{2\alpha}} \qquad (9.95)$$

by using the integrals in Chapter 17.

Equivalently, we find

$$\langle\phi|V(x)|\phi\rangle = \int_{-\infty}^{0}(-kx)e^{-2\alpha x^2}\,dx + \int_{0}^{\infty}kxe^{-2\alpha x^2}\,dx$$

$$= 2\int_{0}^{\infty}kxe^{-2\alpha x^2}\,dx = 2k\frac{1}{4\alpha} = \frac{k}{2\alpha}. \qquad (9.96)$$

Finally, we need

$$\langle \phi | \phi \rangle = \int_{-\infty}^{\infty} e^{-2\alpha x^2} \, dx = \sqrt{\frac{\pi}{2\alpha}}. \tag{9.97}$$

Then

$$\frac{\langle \phi | \hat{H} | \phi \rangle}{\langle \phi | \phi \rangle} = \frac{\frac{\hbar^2}{2m} \alpha \sqrt{\frac{\pi}{2\alpha}} + \frac{k}{2\alpha}}{\sqrt{\frac{\pi}{2\alpha}}}$$

$$= \frac{\hbar^2}{2m} \alpha + \frac{k}{2\alpha} \sqrt{\frac{2\alpha}{\pi}} = \frac{\hbar^2}{2m} \alpha + \frac{k}{\sqrt{2\pi}} \alpha^{-1/2} \equiv \tilde{E}(\alpha). \tag{9.98}$$

We are searching for the minimum of $\tilde{E}(\alpha)$:

$$0 \equiv \frac{d}{d\alpha} \tilde{E}(\alpha) = \frac{\hbar^2}{2m} - \frac{k}{2\sqrt{2\pi}} \alpha^{-3/2}, \tag{9.99}$$

which has the solution

$$\alpha = \left(\frac{km}{\hbar^2 \sqrt{2\pi}} \right)^{2/3} \equiv \alpha_0. \tag{9.100}$$

For this value is

$$\tilde{E}(\alpha_0) = \frac{\hbar^2}{2m} \left(\frac{km}{\hbar^2 \sqrt{2\pi}} \right)^{2/3} + \frac{k}{\sqrt{2\pi}} \left(\frac{\hbar^2 \sqrt{2\pi}}{km} \right)^{1/3}$$

$$= \frac{\hbar^2 k^{2/3} m^{2/3}}{2m \hbar^{4/3} (\sqrt{2\pi})^{2/3}} + \frac{k \hbar^{2/3} (\sqrt{2\pi})^{1/3}}{\sqrt{2\pi} k^{1/3} m^{1/3}} = \frac{3}{2} \left(\frac{\hbar^2 k^2}{2m\pi} \right)^{1/3}. \tag{9.101}$$

This value is the upper limit that was asked for.

4. **Problem:** Use the variational method to determine an upper limit for the ground-state energy for a particle (mass m) in a 1-dimensional potential, $V(x) = k|x|$. As a test function, use $\phi(x) = c_1 e^{-ax^2} + c_2 e^{-2ax^2}$, for which a is a given constant.
   Answer: We denote the two basic functions

$$X_1(x) = e^{-ax^2}$$
$$X_2(x) = e^{-2ax^2} \tag{9.102}$$

or in general

$$X_p(x) = e^{-apx^2}. \tag{9.103}$$

For the secular equation,

$$\begin{pmatrix} \langle X_1 | \hat{H} | X_1 \rangle & \langle X_1 | \hat{H} | X_2 \rangle \\ \langle X_2 | \hat{H} | X_1 \rangle & \langle X_2 | \hat{H} | X_2 \rangle \end{pmatrix} \begin{pmatrix} c_1 \\ c_2 \end{pmatrix} = \epsilon \begin{pmatrix} \langle X_1 | X_1 \rangle & \langle X_1 | X_2 \rangle \\ \langle X_2 | X_1 \rangle & \langle X_2 | X_2 \rangle \end{pmatrix} \begin{pmatrix} c_1 \\ c_2 \end{pmatrix} \tag{9.104}$$

we need overlap matrix elements

$$S_{pq} = \langle \chi_p | \chi_q \rangle = \int_{-\infty}^{\infty} e^{-pax^2} e^{-qax^2} \, dx = \int_{-\infty}^{\infty} e^{-(p+q)ax^2} \, dx = \sqrt{\frac{\pi}{a(p+q)}} \qquad (9.105)$$

as well as Hamilton matrix elements

$$H_{pq} = \langle \chi_p | \hat{H} | \chi_q \rangle$$

$$= \left\langle \chi_p \left| -\frac{\hbar^2}{2m}\frac{d^2}{dx^2} + k|x| \right| \chi_q \right\rangle$$

$$= \int_{-\infty}^{\infty} e^{-pax^2} \left[-\frac{\hbar^2}{2m}\frac{d^2}{dx^2} + k|x| \right] e^{-qax^2} \, dx$$

$$= \int_{-\infty}^{\infty} e^{-pax^2} \left[-\frac{\hbar^2}{2m}(4a^2q^2x^2 - 2qa) + k|x| \right] e^{-qax^2} \, dx$$

$$= 2\int_{0}^{\infty} \left[-\frac{\hbar^2}{2m}(4a^2q^2x^2 - 2qa) + kx \right] e^{-(p+q)ax^2} \, dx$$

$$= 2\left[-\frac{\hbar^2}{2m}\left(4a^2q^2 \frac{1}{4a(p+q)} \sqrt{\frac{\pi}{a(p+q)}} - 2aq\frac{1}{2}\sqrt{\frac{\pi}{a(p+q)}} \right) \right.$$

$$\left. + k\frac{1}{2a(p+q)} \right]$$

$$= \frac{\hbar^2}{m}\frac{apq}{p+q}\sqrt{\frac{\pi}{a(p+q)}} + k\frac{1}{a(p+q)}. \qquad (9.106)$$

The secular equation (9.104) has nontrivial solutions [solutions for $(c_1, c_2) \neq (0,0)$] if

$$0 = \begin{vmatrix} H_{11} - \epsilon S_{11} & H_{12} - \epsilon S_{12} \\ H_{21} - \epsilon S_{21} & H_{22} - \epsilon S_{22} \end{vmatrix}$$

$$= (S_{11}S_{22} - S_{12}S_{21})\epsilon^2 + (H_{11}S_{22} + H_{22}S_{11} - H_{12}S_{21} - H_{21}S_{12})\epsilon$$

$$+ (H_{11}H_{22} - H_{12}H_{21})$$

$$\equiv A\epsilon^2 + B\epsilon + C \qquad (9.107)$$

with

$$A = S_{11}S_{22} - S_{12}S_{21} = \sqrt{\frac{\pi^2}{2a \cdot 4a}} - \sqrt{\frac{\pi^2}{3a \cdot 3a}} = \frac{\pi}{a}\left(\frac{1}{\sqrt{8}} - \frac{1}{\sqrt{9}} \right)$$

$$B = H_{11}S_{22} + H_{22}S_{11} - H_{12}S_{21} - H_{21}S_{12}$$

$$= \left(\frac{\hbar^2}{m}\frac{a}{2}\sqrt{\frac{\pi}{2a}} + \frac{k}{2a} \right)\sqrt{\frac{\pi}{2a}} - 2\left(\frac{\hbar^2}{m}\frac{2a}{3}\sqrt{\frac{\pi}{3a}} + \frac{k}{3a} \right)\sqrt{\frac{\pi}{3a}}$$

$$+ \left(\frac{\hbar^2}{m} a \sqrt{\frac{\pi}{4a}} + \frac{k}{4a} \right) \sqrt{\frac{\pi}{4a}}$$

$$C = H_{11}H_{22} - H_{12}H_{21}$$

$$= \left(\frac{\hbar^2}{m} \frac{a}{2} \sqrt{\frac{\pi}{2a}} + \frac{k}{2a} \right) \left(\frac{\hbar^2}{m} a \sqrt{\frac{\pi}{4a}} + \frac{k}{4a} \right) - \left(\frac{\hbar^2}{m} \frac{2a}{3} \sqrt{\frac{\pi}{3a}} + \frac{k}{3a} \right)^2. \quad (9.108)$$

The solutions to equation (9.107) are

$$\epsilon = \frac{-B \pm \sqrt{B^2 - 4AC}}{A}, \quad (9.109)$$

where the "-" should be chosen to get the lower energy, and thus the upper limit of the exact energy. It is possible to calculate the values of A, B, and C from equation (9.108), which we did not do here.

5. **Problem:** Use the time-independent perturbation theory to first order to calculate the change in the ground-state energy of a particle located in a one-dimensional box, $0 \le x \le L$, and experiencing a perturbation $\Delta V(x) = V_0 - 4V_0(\frac{2x}{L} - 1)^2$ for $\frac{L}{4} \le x \le \frac{3L}{4}$, and otherwise 0. (See Figure 9.2.)

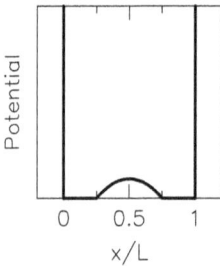

Figure 9.2: The potential including the perturbation from problem 5 in Section 9.9.

Answer: The eigenfunctions for the unperturbed system are

$$\psi_n(x) = \sqrt{\frac{2}{L}} \sin\left(\frac{n\pi x}{L} \right), \quad 0 \le x \le L \quad (9.110)$$

and the associated eigenvalues (energies)

$$E_n = \frac{\hbar^2 \pi^2}{2mL^2} n^2. \quad (9.111)$$

In both cases,

$$n = 1, 2, 3, \ldots, \quad (9.112)$$

where $n = 1$ corresponds to the ground state.

The perturbation potential is

$$\Delta V(x) = V_0\left(-\frac{16x^2}{L^2} + \frac{16x}{L} - 3\right), \quad \frac{L}{4} \le x \le \frac{3L}{4}. \tag{9.113}$$

The first-order correction in the perturbation is then

$$\langle \psi_1|\Delta V|\psi_1\rangle = \frac{2}{L}\int_{L/4}^{3L/4} \sin\left(\frac{\pi x}{L}\right)\Delta V \sin\left(\frac{\pi x}{L}\right)dx$$

$$= \frac{2}{L}\int_{L/4}^{3L/4} \frac{1}{2}\left[1 - \cos\left(\frac{2\pi x}{L}\right)\right]\Delta V\, dx$$

$$= \frac{V_0}{L}\int_{L/4}^{3L/4}\left[1 - \cos\left(\frac{2\pi x}{L}\right)\right]\left(-\frac{16x^2}{L^2} + \frac{16x}{L} - 3\right)dx$$

$$= \frac{V_0}{L}\left\{\frac{16}{L^2}\left[-\frac{2L^3}{(2\pi)^3}\sin\left(\frac{2\pi x}{L}\right) + \frac{2L^2}{(2\pi)^2}x\cos\left(\frac{2\pi x}{L}\right)\right.\right.$$

$$+ \frac{L}{2\pi}x^2\sin\left(\frac{2\pi x}{L}\right)\Big]_{L/4}^{3L/4} - \frac{16}{L}\left[\frac{L^2}{(2\pi)^2}\cos\left(\frac{2\pi x}{L}\right) + \frac{L}{2\pi}x\sin\left(\frac{2\pi x}{L}\right)\right]_{L/4}^{3L/4}$$

$$+ 3\left[\frac{L}{2\pi}\sin\left(\frac{2\pi x}{L}\right)\right]_{L/4}^{3L/4} + \left[-\frac{16x^3}{3L^2} + \frac{8x^2}{L} - 3x\right]_{L/4}^{3L/4}\right\}$$

$$= \frac{V_0}{L}\left\{\frac{16}{L^2}\left[\frac{2L^3}{(2\pi)^3}2 + \frac{L}{2\pi}\left(-\left(\frac{3L}{4}\right)^2 - \left(\frac{L}{4}\right)^2\right)\right] - \frac{16}{L}\frac{L}{2\pi}\left(-\frac{3L}{4} - \frac{L}{4}\right)\right.$$

$$- \frac{3L}{2\pi}2 - \frac{16}{3L^2}\left[\left(\frac{3L}{4}\right)^3 - \left(\frac{L}{4}\right)^3\right] + \frac{16}{2L}\left[\left(\frac{3L}{4}\right)^2 - \left(\frac{L}{4}\right)^2\right]$$

$$- 3\left[\frac{3L}{4} - \frac{L}{4}\right]\right\}$$

$$= \frac{V_0}{L}\left\{\frac{8L}{\pi^3} - \frac{5L}{\pi} + \frac{8L}{\pi} - \frac{3L}{\pi} - \frac{13L}{6} + 4L - \frac{3L}{2}\right\}$$

$$= V_0\left(\frac{8}{\pi^3} + \frac{10}{\pi} + \frac{1}{3}\right). \tag{9.114}$$

That this correction is positive for $V_0 > 0$ corresponds to the fact that the perturbation potential is positive; see Figure 9.2.
In equation (9.114), we have used:

$$\cos\left(\frac{6\pi}{4}\right) = 0$$

$$\cos\left(\frac{2\pi}{4}\right) = 0$$

$$\sin\left(\frac{6\pi}{4}\right) = -1$$

$$\sin\left(\frac{2\pi}{4}\right) = 1. \tag{9.115}$$

6. **Problem:** As in the previous problem, the time-independent perturbation theory shall be used to calculate the change in the ground-state energy of a particle located in a one-dimensional box, $0 \le x \le L$. This time, however, the correction is to be determined to second order in the perturbation. The perturbation is equal to $\Delta V(x) = V_0$ for $\frac{L}{4} \le x \le \frac{3L}{4}$, and otherwise 0.

Answer: We use expressions from the previous problem. Then the first-order correction is as in that case:

$$\langle \psi_1 | \Delta V | \psi_1 \rangle = \int_{L/4}^{3L/4} \frac{2}{L} \sin^2\left(\frac{\pi x}{L}\right) V_0 \, dx = \int_{L/4}^{3L/4} \frac{1}{L} V_0 \left[1 - \cos\left(\frac{2\pi x}{L}\right) \right] dx$$

$$= \frac{1}{L} V_0 \left[-\frac{L}{2\pi} \sin\left(\frac{2\pi x}{L}\right) + x \right]_{L/4}^{3L/4} = \frac{1}{L} V_0 \left[\frac{L}{2\pi} 2 + \frac{L}{2} \right]$$

$$= V_0 \left(\frac{1}{2} + \frac{1}{\pi} \right). \tag{9.116}$$

For the second-order correction, we need $(n \ne 1)$:

$$\langle \psi_n | \Delta V | \psi_1 \rangle = \int_{L/4}^{3L/4} \frac{2}{L} \sin\left(\frac{\pi x}{L}\right) \sin\left(\frac{n\pi x}{L}\right) V_0 \, dx$$

$$= \frac{2}{L} V_0 \int_{L/4}^{3L/4} \frac{1}{2} \left[\cos\left(\frac{(n-1)\pi x}{L}\right) - \cos\left(\frac{(n+1)\pi x}{L}\right) \right] dx$$

$$= \frac{1}{L} V_0 \left[\frac{L}{(n-1)\pi} \sin\left(\frac{(n-1)\pi x}{L}\right) - \frac{L}{(n+1)\pi} \sin\left(\frac{(n+1)\pi x}{L}\right) \right]_{L/4}^{3L/4}$$

$$= V_0 \left\{ \frac{1}{(n-1)\pi} \left[\sin\left(\frac{(n-1)\pi 3}{4}\right) - \sin\left(\frac{(n-1)\pi}{4}\right) \right] \right.$$

$$\left. - \frac{1}{(n+1)\pi} \left[\sin\left(\frac{(n+1)\pi 3}{4}\right) - \sin\left(\frac{(n+1)\pi}{4}\right) \right] \right\}$$

$$= V_0 \left\{ \frac{1}{(n-1)\pi} \left[\sin\left((n-1)\pi - (n-1)\frac{\pi}{4}\right) - \sin\left((n-1)\frac{\pi}{4}\right) \right] \right.$$

$$\left. - \frac{1}{(n+1)\pi} \left[\sin\left((n+1)\pi - (n+1)\frac{\pi}{4}\right) - \sin\left((n+1)\frac{\pi}{4}\right) \right] \right\}$$

$$= V_0 \left\{ \frac{-1}{(n-1)\pi} [(-1)^n + 1] \sin\left((n-1)\frac{\pi}{4}\right) \right.$$

$$\left. - \frac{-1}{(n+1)\pi} [(-1)^n + 1] \sin\left((n+1)\frac{\pi}{4}\right) \right\}$$

$$= -V_0 [(-1)^n + 1] \left\{ \frac{1}{(n-1)\pi} \sin\left((n-1)\frac{\pi}{4}\right) - \frac{1}{(n+1)\pi} \sin\left((n+1)\frac{\pi}{4}\right) \right\}. \tag{9.117}$$

Here, we have used

$$\cos((n-1)\pi) = -(-1)^n$$
$$\cos((n+1)\pi) = -(-1)^n$$
$$\sin((n-1)\pi) = 0$$
$$\sin((n+1)\pi) = 0.$$

(9.118)

The second-order correction is then

$$\sum_{n>1} \frac{|\langle \psi_n | \Delta V | \psi_1 \rangle|^2}{E_1 - E_n} = \sum_{n>1} \frac{|V_0[(-1)^n + 1]\{\frac{\sin((n-1)\frac{\pi}{4})}{(n-1)\pi} - \frac{\sin((n+1)\frac{\pi}{4})}{(n+1)\pi}\}|^2}{\frac{\hbar^2 \pi^2}{2mL^2}(1-n^2)}.$$

(9.119)

9.10 Problems

1. Explain the terms "variational theorem" and "variational method."
2. Explain the Rayleigh–Ritz variational procedure.
3. Explain the term "secular equation."
4. Use the variational method to determine an upper limit for the ground-state energy for a particle (mass m) in a 1-dimensional potential, $V(x) = kx^4$. As a test function, use $\phi(x) = e^{-ax^2}$.
5. Use the variational method to determine an upper limit for the ground-state energy for a particle (mass m) in a 2-dimensional potential, $V(x,y) = kx^2y^2$. As a test function, use $\phi(x,y) = e^{-a(x^2+y^2)}$.
6. Use first-order time-independent perturbation theory to calculate the change in the ground-state energy of a particle (mass m) that is moving in a harmonic potential, and that experiences a perturbation $\Delta V(x)$. $\Delta V(x) = cx$ with c being a (small) constant. Compare with the exact result for the ground-state energy of the particle, which moves in the total potential $V(x) + \Delta V(x)$.
7. Use first-order time-independent perturbation theory to calculate the change in the ground-state energy of a particle (mass m) that moves in a harmonic potential, $V(x) = \frac{1}{2}kx^2$, and experiences a perturbation $\Delta V(x)$, $\Delta V(x) = cx^4$ with c being a constant.
8. Use first-order time-independent perturbation theory to calculate the change in the ground-state energy of a particle (mass m) in a box, $0 \le x \le L$ and that experiences a perturbation $\Delta V(x)$, $\Delta V = c(\frac{L}{2} - x)^2$ with c being a constant
9. Explain briefly Fermi's golden rule.
10. Describe briefly why femto- and attosecond spectroscopy can provide interesting information.

10 The orbital model

10.1 Structure and orbitals

A model that we use with great success in chemistry is that molecules consist of nuclei and electrons, which we consider to be the elementary particles of relevance to us. The nuclei are sitting "somewhere," and the electrons are moving in orbitals, which (also) are responsible for the nuclei having their fixed positions. In fact, this idea is "only" a (usually very good) approximation.

In this book, we have seen that ultimately all the properties of a system can be described using its wave functions. The wave function is a function of all coordinates of the system, i. e., for a molecule, of the position coordinates of both the nuclei and the electrons. Furthermore, we also saw that the wave functions only give probability distributions. Specifically, this means that for the nuclei of a molecule, we can only give a probability distribution of their positions that ultimately will look more like (larger or smaller) clouds. For such distributions, it is not self-evident how to determine the distances between the nuclei, though we always use bond lengths, bond angles, dihedral angles, etc. when we discuss the structure of a molecule. A first question to be dealt with in this chapter is therefore: how can information about the structure of a molecule be brought into line with quantum theory?

The wave function is, as just mentioned, a function of all position coordinates of the electrons and nuclei. This also means that the electrons do not move independently of each other. On the other hand, the idea behind the orbital model, i. e., that electrons occupy different orbitals, implies that the electrons are more or less independent of each other. A second question to be dealt with in this chapter is therefore: where does this orbital picture come from, and how reliable is it?

10.2 The Schrödinger equation for a molecule

To answer these two questions, we start with the Schrödinger equation for a molecule. The solution Ψ to the time-independent Schrödinger equation of a molecule with M nuclei and N electrons

$$\hat{H}\Psi = E\Psi, \tag{10.1}$$

depends on the spin ($\{\sigma_i\}$) and position coordinates ($\{\vec{r}_i\}$) of all electrons, i. e., on

$$(\vec{r}_1, \sigma_1, \vec{r}_2, \sigma_2, \ldots, \vec{r}_N, \sigma_N) \equiv (\vec{x}_1, \vec{x}_2, \ldots, \vec{x}_N) \equiv \vec{x}, \tag{10.2}$$

as well as of the spin ($\{\Sigma_k\}$) and position coordinates ($\{\vec{R}_k\}$) of all nuclei, i. e.,

$$(\vec{R}_1, \Sigma_1, \vec{R}_2, \Sigma_2, \ldots, \vec{R}_M, \Sigma_M) \equiv (\vec{X}_1, \vec{X}_2, \ldots, \vec{X}_M) \equiv \vec{X}. \tag{10.3}$$

https://doi.org/10.1515/9783110742206-010

We have used a frequently used short notation here,

$$\vec{x}_i = (\vec{r}_i, \sigma_i), \tag{10.4}$$

i. e., a vector that contains both position and spin coordinates of the ith electron. Analogously, a vector with the coordinates of the kth kernel is defined as

$$\vec{X}_k = (\vec{R}_k, \Sigma_k). \tag{10.5}$$

The Hamilton operator is obtained by first considering the particles as classical particles. We then consider a situation as in Figure 10.1. For this, we write down the classical expression for the total energy and then use the translation rules between classical physics and quantum theory (Section 3.7). The latter means that every momentum of a particle is replaced by an operator of type $-i\hbar\vec{\nabla}$.

Classical description: Nuclei and electrons are moving in space and interacting electrostaticall with each other

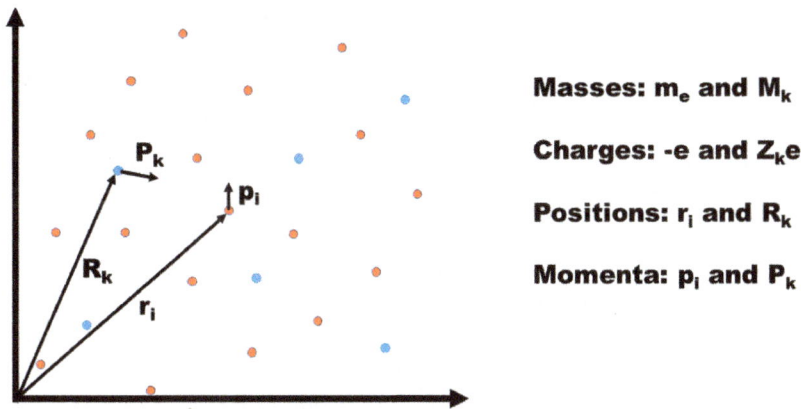

Masses: m_e and M_k

Charges: -e and $Z_k e$

Positions: r_i and R_k

Momenta: p_i and P_k

Figure 10.1: The position and momentum coordinates for a molecular system consisting of a set of nuclei (whose coordinates use capital letters) and electrons (whose coordinates use small letters).

In our case, we assume that we have M nuclei and N electrons. The position and momentum coordinates of the kth nucleus are $\vec{R}_k$ and $\vec{P}_k$ (see Figure 10.1), while its charge and mass are $Z_k e$ and M_k, respectively. Similarly, we let the position and momentum coordinates of the ith electron be $\vec{r}_i$ and $\vec{p}_i$, while its charge and mass are $-e$ and m_e, respectively.

The total kinetic energy of the system becomes then

$$E_{\text{kin}} = \sum_{k=1}^{M} \frac{P_k^2}{2M_k} + \sum_{i=1}^{N} \frac{p_i^2}{2m_e}.$$ (10.6)

The classical potential energy of this system is simply the electrostatic energy due to the interactions between the charges. It is well known that for a set of charges q_n, $n = 1, \ldots, N_q$ placed at positions $\vec{s}_n$ this energy becomes the sum over all pairs of particles

$$\sum_{n_1=1}^{N_q-1} \sum_{n_2=n_1+1}^{N_q} \frac{1}{4\pi\epsilon_0} \frac{q_{n_1} q_{n_2}}{|\vec{s}_{n_1} - \vec{s}_{n_2}|} = \frac{1}{2} \sum_{n_1 \neq n_2=1}^{N_q} \frac{1}{4\pi\epsilon_0} \frac{q_{n_1} q_{n_2}}{|\vec{s}_{n_1} - \vec{s}_{n_2}|}.$$ (10.7)

To obtain the expression on the right-hand side, we have used that the sum on the left-hand side runs over all pairs of particles. This is also the case for the sum on the right-hand side, with the difference that here each pair is counted twice. To compensate, we have multiplied by $\frac{1}{2}$.

Using this scheme for the molecule of our interest, we get

$$\begin{aligned}
E_{\text{pot}} = \ &\frac{1}{2} \sum_{k_1 \neq k_2=1}^{M} \frac{1}{4\pi\epsilon_0} \frac{Z_{k_1} Z_{k_2} e^2}{|\vec{R}_{k_1} - \vec{R}_{k_2}|} + \frac{1}{2} \sum_{i_1 \neq i_2=1}^{N} \frac{1}{4\pi\epsilon_0} \frac{e^2}{|\vec{r}_{i_1} - \vec{r}_{i_2}|} \\
&- \sum_{k=1}^{M} \sum_{i=1}^{N} \frac{1}{4\pi\epsilon_0} \frac{Z_k e^2}{|\vec{R}_k - \vec{r}_i|}.
\end{aligned}$$ (10.8)

The first term on the right-hand side is the nucleus-nucleus interactions, the second one is the electron-electron interactions, and the third one is the nucleus-electron interactions.

To obtain the Hamilton operator, we insert the gradient operators for the momenta in the total-energy expression, which gives

$$\begin{aligned}
\hat{H} = \ &- \sum_{k=1}^{M} \frac{\hbar^2}{2M_k} \nabla_{\vec{R}_k}^2 - \sum_{i=1}^{N} \frac{\hbar^2}{2m_e} \nabla_{\vec{r}_i}^2 + \frac{1}{2} \sum_{k_1 \neq k_2=1}^{M} \frac{1}{4\pi\epsilon_0} \frac{Z_{k_1} Z_{k_2} e^2}{|\vec{R}_{k_1} - \vec{R}_{k_2}|} \\
&+ \frac{1}{2} \sum_{i_1 \neq i_2=1}^{N} \frac{1}{4\pi\epsilon_0} \frac{e^2}{|\vec{r}_{i_1} - \vec{r}_{i_2}|} - \sum_{k=1}^{M} \sum_{i=1}^{N} \frac{1}{4\pi\epsilon_0} \frac{Z_k e^2}{|\vec{R}_k - \vec{r}_i|}.
\end{aligned}$$ (10.9)

The kinetic-energy operator involves terms like $-\frac{\hbar^2}{2M_k} \nabla_{\vec{R}_k}^2$. When it operates on the wave function,

$$\Psi = \Psi(\vec{X}, \vec{x})$$ (10.10)

we differentiate twice with respect to the position coordinates of the kth nucleus, while all other position coordinates are considered constant.

The total Hamilton operator consists of five terms: the kinetic-energy operator for the nucleus, the kinetic-energy operator for the electrons, the potential-energy operator for nucleus-nucleus interactions, the potential-energy operator for electron-electron interactions, and the potential-energy operator for nucleus-electron interactions. We write accordingly

$$\hat{H} = \hat{H}_{k,n} + \hat{H}_{k,e} + \hat{H}_{p,n-n} + \hat{H}_{p,e-e} + \hat{H}_{p,n-e}. \tag{10.11}$$

Thus, the Schrödinger equation becomes

$$\hat{H}\Psi = (\hat{H}_{k,n} + \hat{H}_{k,e} + \hat{H}_{p,n-n} + \hat{H}_{p,e-e} + \hat{H}_{p,n-e})\Psi(\vec{X}, \vec{x}) = E \cdot \Psi(\vec{X}, \vec{x}). \tag{10.12}$$

10.3 Born–Oppenheimer approximation

The Born–Oppenheimer approximation (after Max Born and Robert Oppenheimer) provides the basis for the notion of the structure of a molecule. Simultaneously, it makes it possible to reduce the complexity of the Schrödinger equation for the molecule. We will explain this here in more detail.

The physical idea behind this approximation is that the electrons move much faster than the nuclei, so that for a given set of positions of the nuclei the electrons adjust their positions "immediately" to these independently of the movements of the nuclei. Moreover, it is assumed that the kinetic energy of the nuclei can be ignored. As a consequence of the latter, the operator $\hat{H}_{k,n}$ in equation (10.12) can be ignored. The consequence of the former is that if there are electrons in a molecule whose nuclei are moving, they will be distributed exactly as what would have been found when keeping the nuclei fixed at the positions they have in the middle of their movements. Therefore, one can determine the electronic properties just as well by determining them for the situation that the nuclei are fixed at their positions. Of course, the distribution of electrons will depend on where exactly the nuclei are located and therefore also change when the nuclei are moving.

Mathematically, this is formulated by letting the wave function of the nuclei and electrons be factorized,

$$\Psi(\vec{X}, \vec{x}) = \Psi_n(\vec{X}) \cdot \Psi_e(\vec{X}; \vec{x}). \tag{10.13}$$

The first part depends solely on the nuclear coordinates whereas the second part also depends on the electronic ones. But because Ψ_e depends indirectly on the nuclei coordinates (this means that as the positions of the nuclei change, so does the electronic distribution; see Figure 10.2), there is also a so-called parametric dependence on the nuclear coordinates in Ψ_e.

We insert equation (10.13) into equation (10.12) and set

$$\hat{H}_{k,n} = 0. \tag{10.14}$$

Figure 10.2: The parametric dependence of the electronic wave function on the nuclear coordinates: if the positions of the nuclei change (black circles), the distribution of the electrons (red curve) changes, too; compare the left and right part of the figure.

This gives

$$[(\hat{H}_{k,n} + \hat{H}_{p,n-n}) + (\hat{H}_{k,e} + \hat{H}_{p,e-e} + \hat{H}_{p,n-e})]\Psi_n(\vec{X}) \cdot \Psi_e(\vec{X};\vec{x})$$

$$= (\hat{H}_{k,n} + \hat{H}_{p,n-n})\Psi_n(\vec{X}) \cdot \Psi_e(\vec{X};\vec{x})$$

$$+ (\hat{H}_{k,e} + \hat{H}_{p,e-e} + \hat{H}_{p,n-e})\Psi_n(\vec{X}) \cdot \Psi_e(\vec{X};\vec{x})$$

$$\approx \Psi_e(\vec{X};\vec{x})(\hat{H}_{k,n} + \hat{H}_{p,n-n})\Psi_n(\vec{X})$$

$$+ \Psi_n(\vec{X})(\hat{H}_{k,e} + \hat{H}_{p,e-e} + \hat{H}_{p,n-e})\Psi_e(\vec{X};\vec{x})$$

$$\approx \Psi_e(\vec{X};\vec{x})\hat{H}_{p,n-n}\Psi_n(\vec{X}) + \Psi_n(\vec{X})(\hat{H}_{k,e} + \hat{H}_{p,e-e} + \hat{H}_{p,n-e})\Psi_e(\vec{X};\vec{x})$$

$$\equiv E \cdot \Psi_n(\vec{X}) \cdot \Psi_e(\vec{X};\vec{x}). \tag{10.15}$$

In this equation, the first approximation involves ignoring the effects of $\hat{H}_{k,n}$ on $\Psi_e(\vec{X};\vec{x})$ and in the second approximation setting $\hat{H}_{k,n}$ equal to 0.

We proceed by dividing both sides of this equation by the function of equation (10.13), which gives

$$\frac{\hat{H}_{p,n-n}\Psi_n(\vec{X})}{\Psi_n(\vec{X})} + \frac{(\hat{H}_{k,e} + \hat{H}_{p,e-e} + \hat{H}_{p,n-e})\Psi_e(\vec{X};\vec{x})}{\Psi_e(\vec{X};\vec{x})} = E \tag{10.16}$$

or

$$\frac{(\hat{H}_{k,e} + \hat{H}_{p,e-e} + \hat{H}_{p,n-e})\Psi_e(\vec{X};\vec{x})}{\Psi_e(\vec{X};\vec{x})} = E - \frac{\hat{H}_{p,n-n}\Psi_n(\vec{X})}{\Psi_n(\vec{X})}. \tag{10.17}$$

Analogous to the procedure we have already used several times in this manuscript, we realize that the right-hand side is independent of the electronic coordinates $\vec{x}$, although it may depend on the nuclear coordinates $\vec{X}$. Both sides must then be independent of $\vec{x}$, which gives

$$\frac{(\hat{H}_{k,e} + \hat{H}_{p,e-e} + \hat{H}_{p,n-e})\Psi_e(\vec{X};\vec{x})}{\Psi_e(\vec{X};\vec{x})} = E - \frac{\hat{H}_{p,n-n}\Psi_n(\vec{X})}{\Psi_n(\vec{X})} \equiv E_e(\vec{X}). \tag{10.18}$$

Notice that $E_e(\vec{X})$ depends on $\vec{X}$, including the structure of the molecule.

This gives us the Schrödinger equation for the electrons,

$$(\hat{H}_{k,e} + \hat{H}_{p,e-e} + \hat{H}_{p,n-e})\Psi_e(\vec{X};\vec{x}) = E_e(\vec{X})\Psi_e(\vec{X};\vec{x}). \tag{10.19}$$

This equation is when written out

$$\hat{H}_e\Psi_e(\vec{X};\vec{x})$$

$$\equiv \left[-\sum_{i=1}^{N} \frac{\hbar^2}{2m_e}\nabla_{\vec{r}_i}^2 + \frac{1}{2}\sum_{i_1\neq i_2=1}^{N} \frac{1}{4\pi\epsilon_0} \frac{e^2}{|\vec{r}_{i_1} - \vec{r}_{i_2}|} - \sum_{k=1}^{M}\sum_{i=1}^{N} \frac{1}{4\pi\epsilon_0} \frac{Z_k e^2}{|\vec{R}_k - \vec{r}_i|} \right]\Psi_e(\vec{X};\vec{x})$$

$$= E_e(\vec{X}) \cdot \Psi_e(\vec{X};\vec{x}). \tag{10.20}$$

This shows that the dependence of Ψ_e on the nuclear coordinates is "only" due to the electrostatic potential of the nuclei. This potential is

$$V(\vec{r}) = -\sum_{k=1}^{M} \frac{1}{4\pi\epsilon_0} \frac{Z_k e^2}{|\vec{R}_k - \vec{r}|}. \tag{10.21}$$

Finally, we can also calculate the total energy E from equation (10.18),

$$E = E_e(\vec{X}) + \frac{\hat{H}_{p,n-n}\Psi_n(\vec{X})}{\Psi_n(\vec{X})} = E_e(\vec{X}) + \frac{1}{2}\sum_{k_1\neq k_2=1}^{M} \frac{1}{4\pi\epsilon_0} \frac{Z_{k_1} Z_{k_2} e^2}{|\vec{R}_{k_1} - \vec{R}_{k_2}|}. \tag{10.22}$$

The total energy E is a function of the structure of the molecule,

$$E = E(\vec{R}) = E(\vec{R}_1, \vec{R}_2 \cdots, \vec{R}_M), \tag{10.23}$$

i. e., it is a function of the $3M$ nuclear coordinates. E as a function of $\vec{R}$ defines the so-called potential energy surface, PES. For different electronic states [i. e., different solutions to the electronic Schrödinger equation, equations (10.19) and (10.20)], there are different potential energy surfaces.

10.4 An example

For a diatomic molecule, $\vec{X}$ represents the two vectors that describe the positions of the two nuclei if we ignore the spins of the nuclei. For the total energy of the molecule, only the distance d between the two nuclei is important, while the other five coordinates describe the rotation and translation of the complete molecule. Equation (10.22) becomes then

$$E(d) = E_e(d) + \frac{Z_1 Z_2 e^2}{4\pi\epsilon_0 d}. \tag{10.24}$$

The second term on the right-hand side describes the repulsive, electrostatic interaction between the two nuclei, which is proportional to $1/d$. This means that this term is smallest when $d \rightarrow \infty$. Consequently, $E(d)$ will have a minimum as a function of d

only if $E_e(d)$ is a growing function of d, at least for d in the range in which the minimum of the total energy lies. If this is the case, the value of d for which $E(d)$ has a minimum is the theoretically determined bond length. This is outlined in Figure 10.3.

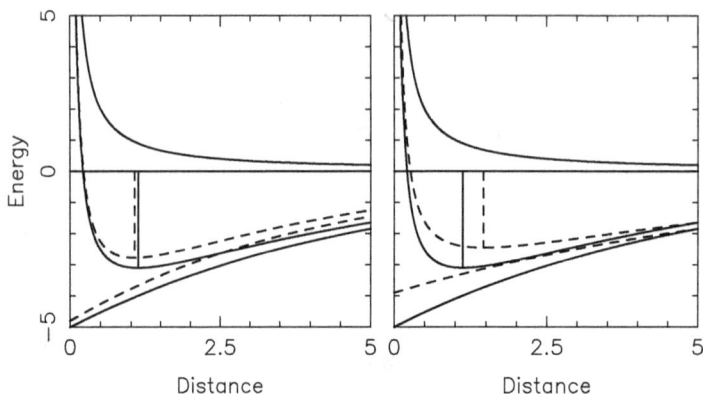

Figure 10.3: Schematic representation of various energies as a function of the bond length of a diatomic molecule. The solid curves show exact results, while the dashed curves show approximate results. The solid curve with positive energies shows the energy of the nucleus-nucleus repulsion, while the bottom, solid curve shows the electronic energy. The middle solid curve is the total energy, i. e., the sum of the other two curves, while the vertical short line shows the position of the minimum of the total energy. The dashed curves show an example of what the approximated electronic energy and the resulting approximated total energy could look like. Again, the vertical dashed line shows the position of the minimum of the approximate total energy. The two panels differ in accuracy of the approximate total energy. In the picture, the more accurate, approximated results in the left panel lead to a smaller underestimation of the bond length, while the less accurate, approximate results to the right lead to an overestimation of the bond length. We emphasize that in other examples it may be different.

However, it is almost never possible to determine E_e exactly. Instead, we will apply the variational principle, as further explained below, to determine an approximation to the lowest electronic energy for each bond length,

$$\frac{\langle\Phi|\hat{H}_e|\Phi\rangle}{\langle\Phi|\Phi\rangle} \geq E_{e0}(d) \tag{10.25}$$

with $E_{e0}(d)$ being the lowest, exact electronic energy for the molecule when the bond length is equal to d. If the difference between the expression on the left-hand side in equation (10.25) and the exact value is largely constant (i. e., d-independent), one can still determine a good approximation to the bond length, while in the case of a stronger d-dependence of the difference inaccurate bond lengths may be obtained; see Figure 10.3.

Exactly how to construct an approximate Φ in equation (10.25) is the subject of the following chapters and will, in addition, be discussed further later (in Chapter 14).

10.5 Spin-orbitals

We are used to explain chemical bonding with the help of the concept that each electron occupies an orbital. Such an electron has both a spatial distribution and a spin. In the general case, the spin depends on the position, and the orbital of the electron can then be written as

$$\psi(\vec{x}) = \psi(\vec{r}, \sigma) = \psi_\alpha(\vec{r})\alpha + \psi_\beta(\vec{r})\beta. \tag{10.26}$$

Here, α and β denote the two possible spin functions, which we also may characterize with $m_s = +\frac{1}{2}$ and $m_s = -\frac{1}{2}$, respectively.

The function in equation (10.26) represents a spin orbital. For an electron in a system (molecule, for example), this depends on the distribution of all other electrons which, however, is not relevant here.

A case that very often occurs is that $\psi(\vec{x})$ can be factorized,

$$\psi(\vec{x}) = \tilde{\psi}(\vec{r})\sigma(m_s), \tag{10.27}$$

so that $\psi_\alpha(\vec{r})$ and $\psi_\beta(\vec{r})$ have an $\vec{r}$-independent relationship. This case includes that either $\psi_\alpha(\vec{r})$ or $\psi_\beta(\vec{r})$ is identically zero, which will be the case we meet very frequently in this manuscript.

But we return to the general case of equation (10.26). We will not be able to write down how the two functions α and β actually look, but utilize that

$$\langle \alpha | \alpha \rangle = \langle \beta | \beta \rangle = 1$$
$$\langle \alpha | \beta \rangle = \langle \beta | \alpha \rangle = 0. \tag{10.28}$$

For the calculation of any expectation value such as $\langle \psi | \hat{a} | \psi \rangle$ ($\hat{a}$ is a one-particle operator whose form is irrelevant here, but that does not operate on the spin coordinate), we will "integrate" over all variables:

$$\langle \psi | \hat{a} | \psi \rangle = \int \psi^*(\vec{x})\hat{a}\psi(\vec{x}) \, d\vec{x} = \sum_{m_s} \int \psi^*(\vec{x})\hat{a}\psi(\vec{x}) \, d\vec{r}$$
$$= \int [\psi_\alpha^*(\vec{r})\hat{a}\psi_\alpha(\vec{r})\langle \alpha | \alpha \rangle + \psi_\alpha^*(\vec{r})\hat{a}\psi_\beta(\vec{r})\langle \alpha | \beta \rangle$$
$$+ \psi_\beta^*(\vec{r})\hat{a}\psi_\alpha(\vec{r})\langle \beta | \alpha \rangle + \psi_\beta^*(\vec{r})\hat{a}\psi_\beta(\vec{r})\langle \beta | \beta \rangle] \, d\vec{r}$$
$$= \int [\psi_\alpha^*(\vec{r})\hat{a}\psi_\alpha(\vec{r}) + \psi_\beta^*(\vec{r})\hat{a}\psi_\beta(\vec{r})] \, d\vec{r}$$
$$= \langle \psi_\alpha | \hat{a} | \psi_\alpha \rangle + \langle \psi_\beta | \hat{a} | \psi_\beta \rangle. \tag{10.29}$$

Another case, which will be important below, is that we consider two different spin orbitals, $\psi_p(\vec{x})$ and $\psi_q(\vec{x})$, both of which can be factorized

$$\psi_p(\vec{x}) = \tilde{\psi}_p(\vec{r})\gamma_p$$

$$\psi_q(\vec{x}) = \tilde{\psi}_q(\vec{r})\gamma_q \tag{10.30}$$

with γ_p being either equal to α or β, and the same holds true for γ_q. In fact, this is the only case that will be relevant in this manuscript and corresponds to the fact that the electrons can be split into spin-up and spin-down electrons. In this case, we consider a matrix element for an operator $\hat{a}$, which does not operate on spin coordinates,

$$\langle\psi_p|\hat{a}|\psi_q\rangle = \langle\tilde{\psi}_p|\hat{a}|\tilde{\psi}_q\rangle\langle\gamma_p|\gamma_q\rangle = \langle\tilde{\psi}_p|\hat{a}|\tilde{\psi}_q\rangle\delta_{\gamma_p,\gamma_q} \tag{10.31}$$

because of equation (10.28). Thus, the matrix element disappears if the two spin orbitals have different spin functions (i. e., for which one function is $\gamma = \alpha$, and the other is $\gamma = \beta$). This result will be important below in Section 10.8.

10.6 Hartree approximation

The orbital picture is in effect an approximation to the solution to the electronic Schrödinger equation,

$$\hat{H}_e\Psi_e = E_e\Psi_e, \tag{10.32}$$

where $\hat{H}_e$ is the sum of one- and two-electron operators:

$$\begin{aligned}
\hat{H}_e &= -\sum_{i=1}^{N}\frac{\hbar^2}{2m_e}\nabla_{\vec{r}_i}^2 + \frac{1}{2}\sum_{i_1\neq i_2=1}^{N}\frac{1}{4\pi\epsilon_0}\frac{e^2}{|\vec{r}_{i_1} - \vec{r}_{i_2}|} - \sum_{k=1}^{M}\sum_{i=1}^{N}\frac{1}{4\pi\epsilon_0}\frac{Z_k e^2}{|\vec{R}_k - \vec{r}_i|} \\
&= \sum_{i=1}^{N}\left[-\frac{\hbar^2}{2m_e}\nabla_{\vec{r}_i}^2 - \sum_{k=1}^{M}\frac{1}{4\pi\epsilon_0}\frac{Z_k e^2}{|\vec{R}_k - \vec{r}_i|}\right] + \frac{1}{2}\sum_{i_1\neq i_2=1}^{N}\frac{1}{4\pi\epsilon_0}\frac{e^2}{|\vec{r}_{i_1} - \vec{r}_{i_2}|} \\
&\equiv \sum_{i=1}^{N}\hat{h}_1(\vec{r}_i) + \frac{1}{2}\sum_{i\neq j=1}^{N}\hat{h}_2(\vec{r}_i, \vec{r}_j). \tag{10.33}
\end{aligned}$$

The one-electron operators include the kinetic energy of the electrons as well as that part of the potential energy that is caused by the interactions with the nuclei,

$$\hat{h}_1(\vec{r}) = -\frac{\hbar^2}{2m_e}\nabla^2 - \sum_{k=1}^{M}\frac{1}{4\pi\epsilon_0}\frac{Z_k e^2}{|\vec{R}_k - \vec{r}|}, \tag{10.34}$$

while $\hat{h}_2$ describes the electron-electron interactions,

$$\hat{h}_2(\vec{r}_i, \vec{r}_j) = \frac{e^2}{4\pi\epsilon_0}\frac{1}{|\vec{r}_i - \vec{r}_j|}. \tag{10.35}$$

The orbital picture assumes that each electron occupies its own orbital. This picture is used repeatedly with success in chemistry, e. g., to explain, rationalize, and predict chemical reactions, structures, and stability properties.

Mathematically, the approximation implies that each electron has its own wave function, independent of where exactly the other electrons are at a particular point in time, although the average distribution of all electrons will influence the individual orbitals. Mathematically expressed, this approximation can be written as

$$\Psi_e(\vec{x}_1, \vec{x}_2, \ldots, \vec{x}_N) \simeq \Phi(\vec{x}_1, \vec{x}_2, \ldots, \vec{x}_N) = \psi_1(\vec{x}_1) \cdot \psi_2(\vec{x}_2) \cdots \psi_N(\vec{x}_N). \tag{10.36}$$

Here, $\psi_i(\vec{x}_i)$ is the wave function for the ith electron. Equation (10.36) is the mathematical formulation of the Hartree approximation (after Douglas Rayner Hartree).

There can be only one electron in each orbital, so the wave functions must be orthonormal,

$$\langle \psi_i | \psi_j \rangle = \delta_{i,j}. \tag{10.37}$$

$\delta_{i,j}$ is the Kronecker δ,

$$\delta_{i,j} = \begin{cases} 1 & i = j \\ 0 & i \neq j. \end{cases} \tag{10.38}$$

To determine the wave functions $\{\psi_i\}$, we use the variational method including the N^2 constraints of equation (10.37). We have already seen earlier (Section 9.5) that such constraints can at best be taken into account with the help of Lagrange multipliers. While we so far have had only one constraint, in this case we have N^2 constraints. Therefore, we study

$$F = \langle \Phi | \hat{H}_e | \Phi \rangle - \sum_{i,j} \lambda_{ij} [\langle \psi_i | \psi_j \rangle - \delta_{i,j}], \tag{10.39}$$

where $\{\lambda_{ij}\}$ are the Lagrange multipliers.

The use of the variational method means that we require that the wave functions $\{\psi_i\}$ are constructed so that no matter which function we consider (e. g., the function ψ_k), and which small change we introduce for this,

$$\psi_k(\vec{x}) \rightarrow \psi_k(\vec{x}) + \delta\psi_k(\vec{x}), \tag{10.40}$$

the change in F is 0,

$$\delta F = 0. \tag{10.41}$$

This is equivalent to what we do when we identify the minimum of a function $f(x)$ by requiring $f'(x) = 0$. If that is the case, $f(x)$ does not change if we let x change slightly.

After some derivations (which will not be presented here), one finally obtains the following equation:

$$\hat{h}\psi_k = \sum_{i=1}^{N} \lambda_{ki}\psi_i \tag{10.42}$$

with

$$\hat{h}\psi_k(\vec{r}) = \hat{h}_1\psi_k(\vec{r}) + \sum_{i=1\,(i\neq k)}^{N} \int \frac{e^2}{4\pi\epsilon_0} \frac{|\psi_i(\vec{r}_1)|^2}{|\vec{r}_1 - \vec{r}|} d\vec{r}_1 \psi_k(\vec{r}), \qquad (10.43)$$

which the wave functions have to fulfill. These equations are the Hartree equations.

The most important aspects of these equations and their solutions will now be briefly summarized:

- Solving the Hartree equations is not easy. First, one realizes that they represent not a "normal" eigenvalue equations, but that on the right-hand side in equation (10.42) we have not only the same function as on the left-hand side, but a linear combination of all wave functions.
- The i summation in equation (10.43) runs over all occupied orbitals. This means that equation (10.42) has, in principle, infinitely many solutions, from which we choose N as those occupied by the electrons. How this is done will be discussed in more detail in the next section when we consider the Hartree–Fock method.
- The operator $\hat{h}$ depends on the solutions, as we see in equation (10.43). Here, the wave functions of the occupied orbitals enter. This means that one must solve the equations iteratively: one assumes something for the wave functions, inserts this into equation (10.42) to create the operator $\hat{h}$, then solve with this equation (10.42), and thereby obtains new wave functions. This procedure is repeated until the wave functions no longer change. Such a method is referred to as an SCF (Self-Consistent Field) method.
- In equation (10.43), the term $i = k$ is excluded in the sum on the right-hand side, since an electron is supposed to not interact with itself. This means that different wave functions satisfy different equations. This considerably complicates solving the equations, and the orthonormality conditions equation (10.37) are not automatically fulfilled.
- The wavefunction in equation (10.36) is an approximation. Therefore, it should not be expected that it meets all physical and chemical conditions exactly, but may be so good that the deviations are acceptable. However, it has turned out that this is not the case. Because of this, the Hartree approximation is not used in chemistry. But it is a good starting point for an improved approximation, that will be discussed in the next section.
- The approximated wave function in equation (10.36) does not fulfill a fundamental principle of quantum theory: the electrons must be indistinguishable: they are fermions (see Section 16.4). Specifically, this means that when we interchange two electrons i and j, the following must hold:

$$\hat{P}_{ij}[\Phi(\vec{x}_1, \vec{x}_2, \ldots, \vec{x}_{i-1}, \vec{x}_i, \vec{x}_{i+1}, \ldots, \vec{x}_{j-1}, \vec{x}_j, \vec{x}_{j+1}, \ldots, \vec{x}_N)]$$
$$= \Phi(\vec{x}_1, \vec{x}_2, \ldots, \vec{x}_{i-1}, \vec{x}_j, \vec{x}_{i+1}, \ldots, \vec{x}_{j-1}, \vec{x}_i, \vec{x}_{j+1}, \ldots, \vec{x}_N)$$
$$\equiv -\Phi(\vec{x}_1, \vec{x}_2, \ldots, \vec{x}_{i-1}, \vec{x}_i, \vec{x}_{i+1}, \ldots, \vec{x}_{j-1}, \vec{x}_j, \vec{x}_{j+1}, \ldots, \vec{x}_N). \qquad (10.44)$$

Here, $\hat{P}_{ij}$ is the permutation operator that interchanges the two electrons i and j.

10.7 Hartree–Fock approximation

The fact that the wave function of the Hartree approximation, equation (10.36), does not satisfy the indistinguishability of the electrons, equation (10.44), can be seen as the reason why this approximation is not considered sufficiently accurate. However, to retain the orbital picture behind the Hartree approximation, the wave function of the Hartree approximation can be improved to preserve the orbital picture on one side and the indistinguishability of the electrons on the other. How to do this is best illustrated through simple examples.

For a two-electron system, the wave function of the Hartree approximation is

$$\psi_1(\vec{x}_1)\psi_2(\vec{x}_2). \tag{10.45}$$

If we interchange the two electrons, we get the wave function

$$\psi_1(\vec{x}_2)\psi_2(\vec{x}_1), \tag{10.46}$$

which is not the negative wave function of the one of equation (10.45). But we may combine these two terms to satisfy the antisymmetry condition of equation (10.44):

$$\Phi(\vec{x}_1,\vec{x}_2) = \psi_1(\vec{x}_1)\psi_2(\vec{x}_2) - \psi_1(\vec{x}_2)\psi_2(\vec{x}_1). \tag{10.47}$$

This wave function consists of terms as in equation (10.45), where each electron occupies an orbital, but here we have considered all the combinations that result from arbitrarily distributing the two electrons in the two orbitals. We therefore cannot say which electron is in which orbital. Furthermore, we have introduced the minus such that the antisymmetry condition of equation (10.44) is satisfied.

For three electrons, starting with the Hartree approximation,

$$\psi_1(\vec{x}_1)\psi_2(\vec{x}_2)\psi_3(\vec{x}_3), \tag{10.48}$$

we can generate a new wave function according to the orbital picture and that simultaneously satisfies the antisymmetry condition of equation (10.44). This new function becomes

$$\begin{aligned}\Phi(\vec{x}_1,\vec{x}_2,\vec{x}_3) = {}&\psi_1(\vec{x}_1)\psi_2(\vec{x}_2)\psi_3(\vec{x}_3) + \psi_1(\vec{x}_2)\psi_2(\vec{x}_3)\psi_3(\vec{x}_1)\\ &+ \psi_1(\vec{x}_3)\psi_2(\vec{x}_1)\psi_3(\vec{x}_2) - \psi_1(\vec{x}_1)\psi_2(\vec{x}_3)\psi_3(\vec{x}_2)\\ &- \psi_1(\vec{x}_3)\psi_2(\vec{x}_2)\psi_3(\vec{x}_1) - \psi_1(\vec{x}_2)\psi_2(\vec{x}_1)\psi_3(\vec{x}_3).\end{aligned} \tag{10.49}$$

We can now see that the two wave functions of equations (10.47) and (10.49) can be written as determinants. We can then generalize this when setting up the wave function for N electrons,

$$\begin{vmatrix} \psi_1(\vec{x}_1) & \psi_2(\vec{x}_1) & \cdots & \psi_N(\vec{x}_1) \\ \psi_1(\vec{x}_2) & \psi_2(\vec{x}_2) & \cdots & \psi_N(\vec{x}_2) \\ \vdots & \vdots & \ddots & \vdots \\ \psi_1(\vec{x}_N) & \psi_2(\vec{x}_N) & \cdots & \psi_N(\vec{x}_N) \end{vmatrix}. \tag{10.50}$$

It turns out that it is useful to multiply this function by the constant factor $\frac{1}{\sqrt{N!}}$ so that it is normalized. This gives us the wave function,

$$\Phi(\vec{x}_1, \vec{x}_2, \ldots, \vec{x}_N) = \frac{1}{\sqrt{N!}} \begin{vmatrix} \psi_1(\vec{x}_1) & \psi_2(\vec{x}_1) & \cdots & \psi_N(\vec{x}_1) \\ \psi_1(\vec{x}_2) & \psi_2(\vec{x}_2) & \cdots & \psi_N(\vec{x}_2) \\ \vdots & \vdots & \ddots & \vdots \\ \psi_1(\vec{x}_N) & \psi_2(\vec{x}_N) & \cdots & \psi_N(\vec{x}_N) \end{vmatrix}, \tag{10.51}$$

i.e., the so-called Slater determinant (after John C. Slater), which represents the wavefunction of the Hartree–Fock approximation (after Douglas Rayner Hartree and Vladimir Alexandrovich Fock).

From the rules for determinants, it is easy to see that this wave function satisfies the antisymmetry condition: the interchange of two electrons corresponds to the interchange of two rows of the determinant, giving the determinant the same value, except that the sign changes. Furthermore, we learn from these rules that the Slater determinant contains in total $N!$ terms. Each term has the form

$$(-1)^{P(i_1, i_2, \ldots, i_N)} \psi_{i_1}(\vec{x}_1) \psi_{i_2}(\vec{x}_2) \cdots \psi_{i_N}(\vec{x}_N), \tag{10.52}$$

so that every orbital and every electron occurs exactly once. The prefactor $(-1)^{P(i_1, i_2, \ldots, i_N)}$ equals $+1$ or -1.

Also for this wave function, we require that

$$\langle \psi_i | \psi_j \rangle = \delta_{i,j}, \tag{10.53}$$

i.e., the wave functions of the individual electrons are orthonormal. Subsequently, the orbitals are determined by searching the minimum for the quantity

$$F = \langle \Phi | \hat{H}_e | \Phi \rangle - \sum_{i,j} \lambda_{ij} [\langle \psi_i | \psi_j \rangle - \delta_{i,j}], \tag{10.54}$$

where $\{\lambda_{ij}\}$ are Lagrange multipliers for the constraints of equation (10.53). This is completely analogous to the Hartree approach, except that the N-electron wave function now is the Slater determinant.

After a tedious calculation, one finally obtains the Hartree–Fock equations, which determine the individual orbitals,

$$\hat{F} \psi_k = \sum_{i=1}^{N} \lambda_{ki} \psi_i. \tag{10.55}$$

$\hat{F}$ is the so-called Fock operator,

$$\hat{F} = \hat{h}_1 + \sum_{i=1}^{N} (\hat{J}_i - \hat{K}_i). \tag{10.56}$$

Here, $\hat{h}_1$ is the operator of the kinetic energy and the potential energy caused by the nuclei. The $\hat{J}_i$ and $\hat{K}_i$ operators are operators that derive from the interactions among the electrons,

$$\hat{J}_i\psi_k(\vec{x}_1) = \int \psi_i^*(\vec{x}_2)\hat{h}_2\psi_i(\vec{x}_2)\psi_k(\vec{x}_1)\,d\vec{x}_2 = \int \frac{e^2}{4\pi\epsilon_0} \frac{|\psi_i(\vec{x}_2)|^2\psi_k(\vec{x}_1)}{|\vec{r}_2 - \vec{r}_1|}\,d\vec{x}_2$$

$$= \int \frac{e^2}{4\pi\epsilon_0} \frac{|\psi_i(\vec{x}_2)|^2}{|\vec{r}_2 - \vec{r}_1|}\,d\vec{x}_2\psi_k(\vec{x}_1), \tag{10.57}$$

and

$$\hat{K}_i\psi_k(\vec{x}_1) = \int \psi_i^*(\vec{x}_2)\hat{h}_2\psi_i(\vec{x}_1)\psi_k(\vec{x}_2)\,d\vec{x}_2 = \int \frac{e^2}{4\pi\epsilon_0} \frac{\psi_i^*(\vec{x}_2)\psi_k(\vec{x}_2)\psi_i(\vec{x}_1)}{|\vec{r}_2 - \vec{r}_1|}\,d\vec{x}_2$$

$$= \int \frac{e^2}{4\pi\epsilon_0} \frac{\psi_i^*(\vec{x}_2)\psi_k(\vec{x}_2)}{|\vec{r}_2 - \vec{r}_1|}\,d\vec{x}_2\psi_i(\vec{x}_1). \tag{10.58}$$

The $\hat{J}_i$ operators are called Coulomb operators, while the $\hat{K}_i$ operators are called exchange operators. The first correspond to the concepts of classical physics that every electron forms a charged cloud that interacts electrostatically with every other electron. However, a similar classical interpretation is not possible for the exchange operators, which can be completely attributed to the indistinguishability of the electrons (i. e., quantum effects). The Coulomb operators are the ones that also occur in the Hartree approximation while the exchange operators have not yet occurred.

Instead of discussing the equations in detail, we briefly list the key features and results:

– Like the Hartree equations, the Hartree–Fock equations are not that easy to solve. First, one realizes that they represent not only "normal" eigenvalue equations, but that on the right-hand side of equation (10.55) not only the same function as on the left-hand side is found, but a linear combination of all wave functions.

– As with the Hartree approximation, the i summation runs in equation (10.56) over all occupied orbitals. This means that equation (10.55) has in principle infinitely many solutions, of which we choose N as those occupied by the electrons. How this is done will be discussed below.

– The Fock operator $\hat{F}$ depends on the solutions, as can be seen from equations (10.57) and (10.58). This is where the wave functions of the occupied orbitals enter. This also means that one must solve the equations iteratively: one makes an Ansatz for the occupied wave functions, uses this in equation (10.56) to generate the operator $\hat{F}$, solves with this equation (10.55), and thereby obtains new wave functions. This procedure is repeated until the wave functions no longer change. Such a method is referred to as an SCF (Self-Consistent Field) method.

– In contrast to the Hartree method, in the summation on the right-hand side of equation (10.56), the term $i = k$ is not excluded. This is possible because, as one easily recognizes from equations (10.57) and (10.58), the term $i = k$ in equation

(10.55) for the Coulomb and the exchange interactions of an electron with itself are identical,

$$\hat{J}_k \psi_k = \hat{K}_k \psi_k, \tag{10.59}$$

and, therefore, cancel.

– It is possible to choose

$$\lambda_{ki} = \delta_{k,i} \epsilon_k. \tag{10.60}$$

Then the Hartree–Fock equations become standard eigenfunction equations.

– With the choice of equation (10.60), one obtains one set of orbitals, the so-called canonical orbitals. Other possibilities can be created by forming new ones from the orbitals so obtained,

$$\tilde{\psi}_l = \sum_{k=1}^{N} U_{lk} \psi_k, \tag{10.61}$$

where

$$\sum_{k} U_{lk}^* U_{mk} = \delta_{l,m} \tag{10.62}$$

has to be satisfied. The new functions are also orthogonal. The orbitals $\{\tilde{\psi}_k\}$ will most likely look very different from the orbitals $\{\psi_k\}$. A matrix $\underline{U}$ with elements that satisfy equation (10.62) is called a unitary matrix, and, accordingly, the transformation (10.61) is called a unitary transformation.

– This ultimately means that the individual orbitals are not unique. In fact, it will never be possible to design an experiment with which the individual orbitals become visible. On the other hand, when applying the unitary transformation of equation (10.61) for the occupied canonical orbitals, equation (10.55) with equation (10.60), the expectation value of any experimentally accessible quantity remains unchanged.

– The orbitals that satisfy the Hartree–Fock equations together with equation (10.60) are very easy to interpret and these equations are easier to solve. That is why usually only these are considered. See, however, also Section 15.9.

– For any experimental quantity that can be written as a sum of identical operators of the individual electrons,

$$\hat{A} = \sum_{i=1}^{N} \hat{a}(i), \tag{10.63}$$

the expectation value is

$$\langle \Phi | \hat{A} | \Phi \rangle = \sum_{i=1}^{N} \langle \psi_i | \hat{a} | \psi_i \rangle. \tag{10.64}$$

Thus the contributions of the individual orbitals are simply added together.

- A special case is that of the electron density. This becomes

$$\rho(\vec{r}) = \sum_{i=1}^{N} |\psi_i(\vec{r})|^2. \tag{10.65}$$

- With the Hartree–Fock approach, the electronic energy is approximated through

$$E_e \simeq \sum_{k=1}^{N} \langle \psi_k | \hat{h}_1 | \psi_k \rangle + \frac{1}{2} \sum_{k,l=1}^{N} [\langle \psi_k \psi_l | \hat{h}_2 | \psi_k \psi_l \rangle - \langle \psi_l \psi_k | \hat{h}_2 | \psi_k \psi_l \rangle]$$

$$= \sum_{k=1}^{N} \epsilon_k - \frac{1}{2} \sum_{k,l=1}^{N} [\langle \psi_k \psi_l | \hat{h}_2 | \psi_k \psi_l \rangle - \langle \psi_l \psi_k | \hat{h}_2 | \psi_k \psi_l \rangle] \tag{10.66}$$

The first term in the last expression shows the importance of the so-called orbital energies ϵ_i and the second term is a correction, which results from the fact that in the first term the interactions between the electrons are counted twice.
- Equation (10.66) shows that a substantial part of the total energy comes from the sum of the orbital energies. Accordingly, in order to minimize the total energy, it makes sense that the orbitals with the lowest orbital energies are occupied by electrons. Thereby we obtain a procedure with which one can construct the Slater determinant that gives the lowest total energy: This is that Slater determinant containing the orbitals of the lowest ϵ_i.
- Assuming that the orbitals do not change when electrons are added or removed, one obtains

$$E_e(N-1) - E_e(N) \simeq -\epsilon_n \tag{10.67}$$

and

$$E_e(N+1) - E_e(N) \simeq \epsilon_m. \tag{10.68}$$

In the first case, an electron has been removed from the occupied orbital n, and in the second case an electron has been added to the unoccupied orbital m. The first case corresponds to the ionization potential and the second case to the electron affinity. The fact that the other orbitals must not change if these formulas are to be valid means that so-called (electronic) relaxation effects are ignored. The result of equations (10.67) and (10.68) is the theorem of Tjalling Charles Koopmans.
- This theorem provides another argument for why the lowest energy is obtained by occupying the energetically lowest orbitals. If (electronic) relaxation effects are ignored, it will cost energy to excite electrons into other orbitals.

10.8 RHF and UHF

We will now discuss those contributions to the total energy, equation (10.66), which arise from the electron-electron interactions ($\hat{h}_2$) and explicitly consider effects due to the spin of the electrons. We assume that the electronic spin orbitals can be written as in equation (10.30),

$$\psi_k(\vec{x}) = \tilde{\psi}_k(\vec{r})\gamma_k$$
$$\psi_l(\vec{x}) = \tilde{\psi}_l(\vec{r})\gamma_l \qquad (10.69)$$

with γ_k and γ_l each equal to an α or β spin function. We take advantage of the fact that $\hat{h}_2$ is independent of the spin, so that we have for the Coulomb interactions

$$\langle\psi_k\psi_l|\hat{h}_2|\psi_k\psi_l\rangle = \frac{e^2}{4\pi\epsilon_0}\int\int \frac{\psi_k^*(\vec{x}_1)\psi_l^*(\vec{x}_2)\psi_k(\vec{x}_1)\psi_l(\vec{x}_2)}{|\vec{r}_1 - \vec{r}_2|}\, d\vec{x}_1\, d\vec{x}_2$$

$$= \frac{e^2}{4\pi\epsilon_0}\int\int \frac{\tilde{\psi}_k^*(\vec{r}_1)\tilde{\psi}_l^*(\vec{r}_2)\tilde{\psi}_k(\vec{r}_1)\tilde{\psi}_l(\vec{r}_2)}{|\vec{r}_1 - \vec{r}_2|}\, d\vec{r}_1\, d\vec{r}_2 \langle\gamma_k|\gamma_k\rangle\langle\gamma_l|\gamma_l\rangle$$

$$= \frac{e^2}{4\pi\epsilon_0}\int\int \frac{\tilde{\psi}_k^*(\vec{r}_1)\tilde{\psi}_l^*(\vec{r}_2)\tilde{\psi}_k(\vec{r}_1)\tilde{\psi}_l(\vec{r}_2)}{|\vec{r}_1 - \vec{r}_2|}\, d\vec{r}_1\, d\vec{r}_2, \qquad (10.70)$$

which shows that the Coulomb interactions are independent of spin.

But it is different for the exchange interactions. In this case, we have

$$\langle\psi_l\psi_k|\hat{h}_2|\psi_k\psi_l\rangle = \frac{e^2}{4\pi\epsilon_0}\int\int \frac{\psi_l^*(\vec{x}_1)\psi_k^*(\vec{x}_2)\psi_k(\vec{x}_1)\psi_l(\vec{x}_2)}{|\vec{r}_1 - \vec{r}_2|}\, d\vec{x}_1\, d\vec{x}_2$$

$$= \frac{e^2}{4\pi\epsilon_0}\int\int \frac{\tilde{\psi}_l^*(\vec{r}_1)\tilde{\psi}_k^*(\vec{r}_2)\tilde{\psi}_k(\vec{r}_1)\tilde{\psi}_l(\vec{r}_2)}{|\vec{r}_1 - \vec{r}_2|}\, d\vec{r}_1\, d\vec{r}_2 \langle\gamma_l|\gamma_k\rangle\langle\gamma_k|\gamma_l\rangle$$

$$= \frac{e^2}{4\pi\epsilon_0}\int\int \frac{\tilde{\psi}_k^*(\vec{r}_1)\tilde{\psi}_l^*(\vec{r}_2)\tilde{\psi}_k(\vec{r}_1)\tilde{\psi}_l(\vec{r}_2)}{|\vec{r}_1 - \vec{r}_2|}\, d\vec{r}_1\, d\vec{r}_2\delta_{\gamma_k,\gamma_l}. \qquad (10.71)$$

Thus, two spin orbitals ψ_k and ψ_l of type (10.69) have nonzero exchange interactions only if they have the same spin dependence.

This can have consequences for the calculations. In Figure 10.4, we show schematically the energies and occupations of spin orbitals, as they can be obtained with a Hartree–Fock approximation. If the total number of electrons is even (left part), it can often be assumed that the spin-up and spin-down orbitals are identical. An electron in a spin-up orbital experiences the Coulomb interactions from all electrons, while it experiences exchange interactions from the electrons with the same spin. An electron in the equivalent spin-down orbital is exposed to equivalent interactions, although it in this case then experiences exchange interactions only from the spin-down electrons. Therefore, it can often be assumed that the spin-up and spin-down orbitals are spatially identical and have the same energies.

This assumption corresponds to the so-called Restricted Hartree–Fock approximation (RHF). It should be noted that there are cases where the system, despite having an even number of electrons, can find a state of lower energy in which the spin-up and spin-down orbitals are different.

Orbital energy

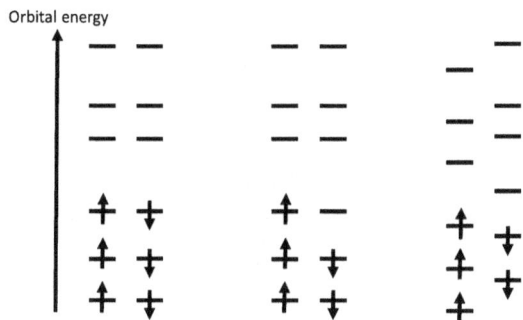

Figure 10.4: The occupation and energies of orbitals in the Hartree–Fock approximation. In the left case, the total number of electrons is even, and there is no difference between spin-up and spin-down orbitals. In the middle case, the total number of electrons is odd, but it is nonetheless assumed that there is no difference between spin-up and spin-down orbitals. In the right case, this assumption is removed.

If the number of electrons is odd, these arguments fail. If, e. g., the number of spin-down electrons is less than that of the spin-up electrons, there is less exchange interactions between the spin-down electrons than between the spin-up electrons. So, there is no reason why the spin-up and spin-down orbitals should be spatially identical and have the same energies. However, if one makes this assumption, one applies still the RHF approximation, as shown in the middle case in Figure 10.4.

Ultimately, one can give up this assumption, so that the spin-up and spin-down orbitals can be spatially different and have different energies. This is the so-called Unrestricted Hartree–Fock approximation (UHF). This is shown schematically in the right part of Figure 10.4.

As the name implies, the UHF approximation is an approximation, so it is unlikely that everything will be described perfectly. In fact, it can be shown (will not be presented here) that the UHF approximation can lead to inaccuracies in the description of spin effects, so that this approximation is not always optimal. Finally, it should be emphasized that systems with an odd number of electrons are better treated more accurately by other methods, as discussed in Chapter 14.

10.9 Bye bye, orbital picture?

In many areas of chemistry, orbitals are used to explain, rationalize, or predict chemical phenomena. The orbital picture is such a fundamental part of chemistry that the existence of orbitals is often taken for granted.

And in fact, the orbital picture is a very useful tool for understanding chemistry. Nevertheless, orbitals are not real objects, and there will never be experiments with which individual orbitals can be made visible. This statement does not imply that the orbital picture should no longer be used, only that there are cases where it may be relevant to be aware of the limits of the orbital picture. In this chapter, we will explain this statement in more detail.

The idea that each electron occupies a particular orbital has been formulated mathematically using the Hartree (Section 10.6) or the Hartree–Fock approximation (Section 10.7). The two approximations differ in whether the electrons are considered distinguishable or indistinguishable, but are otherwise equivalent. Here, we will therefore discuss only the Hartree–Fock approximation, also because the Hartree approximation has little practical importance in chemistry.

The Hartree–Fock approach describes the N-electron wave function for the N electrons of our system by means of a single Slater determinant, which contains N 1-electron wave functions for the individual orbitals. If we had approximated the N-electron wave function by means of several Slater determinants, it would no longer be possible to identify N orbitals each containing exactly one electron. So, if we give up the Hartree–Fock approximation, the orbital picture is no longer valid.

A first question is whether it will ever be relevant to write the N-electron wave function using several Slater determinants. We will see later that this can indeed be the case. For example, In Section 11.3 we will see that the spin properties of a system can lead to this. And in Section 12.6 it is shown that the dissociation of H_2 is not properly described using the Hartree–Fock approximation. So, there are actually (quite relevant) cases where the Hartree–Fock approximation fails.

But even though the Hartree–Fock approximation represents a good (= accurate) approximation, there are aspects that make it relevant to question the existence of orbitals. The Lagrange multipliers $\{\lambda_{ki}\}$ in equation (10.55) are not unique. We made the choice of equation (10.60), not because it is more correct, but "only" because the equations are easier to solve. With a different choice, we would get other orbitals that could look very different. The spatial distribution of the orbitals is anything but unique, which makes it impossible to uniquely define orbitals.

With the choice of equation (10.60), we obtain values for the orbital energies. With any other possibility, there would be no such simple interpretation of the Lagrange multipliers. This makes the choice of equation (10.60) more helpful for the general understanding of chemical issues, but not more correct.

Furthermore, it can be shown that the choice of equation (10.60) often causes the orbitals to be delocalized over the complete molecule, making it difficult to assign orbitals to specific bonds. On the other hand, we use such an assignment repeatedly when we describe chemical bonds, and, as we will see in Section 15.9, it is actually possible to create such localized orbitals. This will allow us to use a transformation as in equation (10.61) (though this is rarely explicitly stated or even mentioned). However, such a transformation has as a consequence that no orbital energies can be assigned to the new orbitals.

Overall, the orbital picture is a great help to rationalize much of the chemical knowledge. That is why the orbital picture is important and relevant and should not be dismissed. Nevertheless, it should not be forgotten that there are cases where the orbital picture becomes (too) inaccurate.

That we use such simple model concepts, we have seen earlier in this script. In Chapter 4, we discussed the simple model of a particle in a box. Nobody will doubt that this model is a simplification and can never be realized exactly. Nevertheless, we have seen that with this model, we can rationalize experimental results on the electrons in a chain of Pd atoms (Section 4.12). Furthermore, we have seen that with this model we can also obtain a fairly accurate representation of the optical properties of conjugated molecules (Section 4.11).

And in Section 7.5 we introduced the spin of the electron. We imagine that the electron rotates about its own axis, and that this movement, like all other movements, is quantized. The fact that we have to accept a half-integral spin quantum number, which actually is in contradiction to the treatment in Section 7.3, is considered as the price for obtaining an easy to understand interpretation of the spin.

Everywhere we use simple model ideas because they are helpful in understanding complex issues. The orbital picture is one, and a very good and very helpful one. But not exact and, therefore, not experimentally provable.

There are some experimental approaches to determine something like orbitals. In one case, the orbitals that then are determined are called Dyson orbitals, and will not be treated here in mathematical detail but will be briefly explained. If we assume that the orbital picture is accurate, the ionization of an N-electron molecule will cause an electron to leave some orbital. The remaining $N - 1$ electrons will most certainly respond to the new situation so that their orbitals change (i. e., there are electronic relaxation effects). In some sense, the "difference" between the system before and after the ionization is therefore the orbital from which the electron has been removed, plus the relaxation effects. We will not discuss in detail what exactly is meant by "difference," but just emphasize that this difference is related to the Dyson orbital. So, in such an experiment the best possible approximations to experimentally accessible orbitals are investigated.

The precise definition of the Dyson orbitals is

$$\Psi_I(\vec{x}_1) = \sqrt{N} \int \int \cdots \int [\Psi_I^{(N-1)}(\vec{x}_2, \vec{x}_3, \ldots, \vec{x}_N)]^*$$
$$\cdot \Psi_0^{(N)}(\vec{x}_1, \vec{x}_2, \vec{x}_3, \ldots, \vec{x}_N) \, d\vec{x}_2 \, d\vec{x}_3 \cdots d\vec{x}_N. \tag{10.72}$$

Here, $\Psi_0^{(N)}$ is the N-electron wave function of the system in the ground state before ionization. $\Psi_I^{(N-1)}$ is the wave function of the system in state I after ionization. I is not necessarily the ground state of the $(N - 1)$-electron system. Here, however, this expression shall not be discussed in further detail.

10.10 Hartree–Fock–Roothaan method

If you actually tried to solve the Hartree–Fock equations, you would have to determine the function values of each orbital at each point in position space. This is much more information than what can ever be determined. Therefore, a further approximation is introduced: each orbital is expanded in a finite number of basis functions,

$$\psi_l(\tilde{x}) = \sum_{p=1}^{N_b} \chi_p(\tilde{x})c_{pl}. \tag{10.73}$$

Here, N_b is the number of basis functions. The basis functions $\{\chi_p\}$ are chosen and fixed, and only the coefficients $\{c_{pl}\}$ are determined. This is the basis of the Hartree–Fock–Roothaan method (after Douglas Rayner Hartree, Vladimir Alexandrovich Fock, and Clemens C. J. Roothaan).

In order to set up the equations that the coefficients $\{c_{pl}\}$ shall fulfill, we proceed as described earlier in the present chapter, so we consider the quantity F from equation (10.54) and demand that it has a minimum. This means that

$$\frac{\partial F}{\partial c_{qm}} = \frac{\partial F}{\partial c_{qm}^*} = 0 \tag{10.74}$$

where c_{qm} is an arbitrary coefficient of equation (10.73). Without derivation, we give the final result:

$$\sum_{m=1}^{N_b} \left\{ \langle \chi_p|\hat{h}_1|\chi_m\rangle + \sum_{i=1}^{N}\sum_{n,q=1}^{N_b} c_{ni}c_{qi}^* [\langle \chi_p\chi_q|\hat{h}_2|\chi_m\chi_n\rangle - \langle \chi_q\chi_p|\hat{h}_2|\chi_m\chi_n\rangle] \right\} c_{ml}$$

$$= \epsilon_l \sum_{m=1}^{N_b} \langle \chi_p|\chi_m\rangle c_{ml}. \tag{10.75}$$

Here are

$$\langle \chi_p|\hat{h}_1|\chi_m\rangle = \int \chi_p^*(\tilde{x})\hat{h}_1\chi_m(\tilde{x})d\tilde{x}$$

$$= \int \chi_p^*(\tilde{x})\left[-\frac{\hbar^2}{2m}\nabla^2 - \sum_k \frac{Z_k e^2}{4\pi\epsilon_0|\vec{R}_k - \vec{r}|}\right]\chi_m(\tilde{x})\,d\tilde{x}$$

$$\langle \chi_p\chi_q|\hat{h}_2|\chi_m\chi_n\rangle = \int\int \chi_p^*(\tilde{x}_1)\chi_q^*(\tilde{x}_2)\hat{h}_2\chi_m(\tilde{x}_1)\chi_n(\tilde{x}_2)\,d\tilde{x}_1\,d\tilde{x}_2$$

$$= \int\int \chi_p^*(\tilde{x}_1)\chi_q^*(\tilde{x}_2)\frac{e^2}{4\pi\epsilon_0|\vec{r}_1 - \vec{r}_2|}\chi_m(\tilde{x}_1)\chi_n(\tilde{x}_2)\,d\tilde{x}_1\,d\tilde{x}_2. \tag{10.76}$$

Equation (10.75) can also be written as a matrix eigenvalue equation,

$$\underline{\underline{F}} \cdot \underline{c}_l = \epsilon_l \cdot \underline{\underline{O}} \cdot \underline{c}_l, \tag{10.77}$$

where $\underline{\underline{F}}$ is the Fock matrix with the elements

$$F_{pm} = \langle \chi_p | \hat{h}_1 | \chi_m \rangle + \sum_{i=1}^{N} \sum_{n,q=1}^{N_b} c_{ni} c_{qi}^* [\langle \chi_p \chi_q | \hat{h}_2 | \chi_m \chi_n \rangle - \langle \chi_q \chi_p | \hat{h}_2 | \chi_m \chi_n \rangle] \tag{10.78}$$

and $\underline{\underline{O}}$ is the overlap matrix with the elements

$$O_{pm} = \langle \chi_p | \chi_m \rangle. \tag{10.79}$$

Finally, $\underline{c_l}$ contains the coefficients of the lth orbital.

Except that the equations look slightly different from the Hartree–Fock equations, the same comments that were listed there apply here.

10.11 The orbital picture and atomic interactions

By means of the orbital picture, different types of interactions between two systems can be classified, even if the orbital picture is not always able to describe them quantitatively. The two systems can, e. g., be two atoms, and the interactions may then be those that hold the two atoms together.

Figure 10.5 shows schematically how the orbitals of the two systems interact with each other. For a covalent bond, the occupied orbitals of the two systems mix—possibly through the formation of hybrid orbitals as will be described later in Section 13.4—and the interaction between the two systems can be considered to be an interaction between the occupied orbitals. This is indicated by the arrow a in the figure.

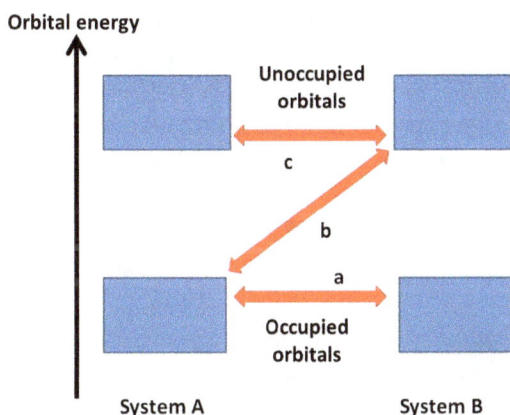

Figure 10.5: Schematic representation of the orbital interactions between two systems A and B responsible for (a) the covalent bond, (b) the ionic bond, and (c) the van der Waals interaction.

For an ionic bond, electrons are transferred from the occupied orbitals of one system to the unoccupied orbitals of the other system. This is shown by the arrow b in Figure 10.5.

Ultimately, the electrons are not distributed statically, but their charge distribution shows (time-dependent) fluctuations. As a result of these fluctuations of one system, the charge distribution of the other system is polarized, which in turn cause the charge distribution of the first system to be polarized, too. This interaction may result in induced dipoles that produces a weak binding interaction, the so-called van der Waals bond. The induced dipole of the single system is described by means of its unoccupied orbitals, so that the van der Waals interaction can be regarded as an interaction between the unoccupied orbitals, which is represented by the arrow c in Figure 10.5.

By means of such a so-called Morokuma analysis (according to Keiji Morokuma), this separation of interactions between two molecules (or atoms) can also be examined quantitatively.

Within the Hartree–Fock approximation, the occupied orbitals are accurately described, but the unoccupied ones are not. Therefore, the van der Waals interaction is described in a very inaccurate way with this approximation.

10.12 How many orbitals are needed?

The canonical orbitals obey the Hartree–Fock equations

$$\hat{F}\psi_k = \epsilon_k\psi_k, \tag{10.80}$$

i. e., the Lagrange multipliers obey equation (10.60). In setting up $\hat{F}$, we have assumed that a certain set of the orbitals is occupied, whereas the other orbitals are empty. This set of orbitals contains as many orbitals as we have electrons, N. It might therefore be assumed that in solving equation (10.80) we would need to identify only those N orbitals that are occupied whereas the other ones are irrelevant, although we, in principle, could identify infinitely many orbitals from equation (10.80). When we instead had used the Hartree–Fock–Roothaan approach, we could not identify infinitely many orbitals but only as many as we have basis functions, N_b, that is (or: should be) larger than N. Usually, the N orbitals that we consider relevant are those for the N lowest orbital energies, ϵ_k.

However, already in our treatment of the H atom (Chapter 8), we have used an approach that actually is identical to the Hartree–Fock approximation, although in this case it is exact. There, we have calculated not only the energetically lowest orbital but infinitely many orbitals, and we have also seen that the orbitals that are not occupied in the ground state are relevant when explaining, e. g., experimental findings.

Also for any other atomic or molecular system, the orbitals that are not occupied in the ground state are of relevance for a number of situations. At first, as for the hydrogen atom, experimental results require most often that also energetically higher-lying orbitals are taken into account. Furthermore, as we shall see, when we attempt to go beyond the Hartree–Fock approximation, those orbitals that in the ground state within the Hartree–Fock approximation are empty become important. Thus, it is rarely the case that for some system only the N energetically lowest orbitals are needed.

10.13 Problems

1. Explain the Hartree–Fock self-consistent field method. What does this method have to do with the variational method? Is this method an exact method?
2. Explain the theorem of Koopmans. What approximation is made, and how accurate is it?
3. Explain the term "orbital approximation."
4. Describe the physical/chemical ideas behind the Slater determinant.
5. Explain the Born–Oppenheimer approximation.
6. Explain the term "overlap integral."
7. Compare the Hartree approximation and the Hartree–Fock approximation. Which one is used in practice and why?
8. Explain briefly the Coulomb and the exchange interactions of the electrons within the Hartree–Fock approximation. Which physical effects are behind this?
9. Explain briefly why the Lagrange multipliers are introduced in the Hartree–Fock theory. Which set of multipliers is used in practice and how is it interpreted?
10. Explain what is meant by an SCF procedure and what it has to do with electronic structure calculations.
11. Are the Hartree–Fock–Roothaan equations equivalent to the Hartree–Fock equations? When are they equivalent and when are they not? Which method is more accurate? Justify the answers.
12. Explain briefly how van der Waals interactions arise.
13. Explain briefly why it will never be possible to see individual orbitals experimentally.
14. Explain why it can be useful to determine (many) more orbitals than those that are occupied in the ground state.

11 Atoms

11.1 He

The helium atom consists of a nucleus and two electrons. In the quantum-theoretical treatment of this system, we proceed as in the treatment of the hydrogen atom, and in accordance with the Born–Oppenheimer approximation: the effect of the nucleus is reduced to that of generating an electrostatic potential in which the electrons move. Because there is only one nucleus, there will be no contribution to the total energy from the nucleus-nucleus interaction.

Because of convenience, we place a coordinate system with the origin at the position of the nucleus. Then the Hamilton operator for the electrons becomes (see Figure 11.1)

$$\hat{H}_e = -\frac{\hbar^2}{2m_e}\nabla_1^2 - \frac{\hbar^2}{2m_e}\nabla_2^2 - \frac{2e^2}{4\pi\epsilon_0 r_1} - \frac{2e^2}{4\pi\epsilon_0 r_2} + \frac{e^2}{4\pi\epsilon_0|\vec{r}_1 - \vec{r}_2|}$$

$$= \left[-\frac{\hbar^2}{2m_e}\nabla_1^2 - \frac{2e^2}{4\pi\epsilon_0 r_1}\right] + \left[-\frac{\hbar^2}{2m_e}\nabla_2^2 - \frac{2e^2}{4\pi\epsilon_0 r_2}\right] + \frac{e^2}{4\pi\epsilon_0|\vec{r}_1 - \vec{r}_2|}$$

$$\equiv \hat{h}_1(\vec{r}_1) + \hat{h}_1(\vec{r}_2) + \hat{h}_2(\vec{r}_1, \vec{r}_2). \tag{11.1}$$

Here, we have separated the operator into three terms to emphasize the connection to the content of Chapter 10. $\hat{h}_1(\vec{r}_1)$ and $\hat{h}_1(\vec{r}_2)$ contain all one-electron contributions of the 1st and 2nd electron, respectively, while $\hat{h}_2(\vec{r}_1, \vec{r}_2)$ represents the two-electron contribution.

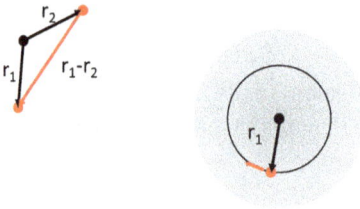

Figure 11.1: The He Atom. The left part shows the coordinates of the two electrons relative to the nucleus and relative to each other. In the right part, the motion of one electron is indicated as it moves in the field of the nucleus (black circle in the center) and in the field of the other electron (gray cloud representing the probability distribution of that electron).

If we ignore the last term, i. e., $\hat{h}_2$, the wave function of the Hartree approximation,

$$\Psi_e(\vec{x}_1, \vec{x}_2) = \psi_1(\vec{x}_1)\psi_2(\vec{x}_2) \tag{11.2}$$

is an exact solution to the electronic Schrödinger equation

$$\hat{H}_e\Psi_e = E_e\Psi_e. \tag{11.3}$$

https://doi.org/10.1515/9783110742206-011

However, as discussed in the last chapter, this function would violate the indistinguishability of the electrons. But also the wave function of the Hartree–Fock approximation

$$\Psi_e(\vec{x}_1, \vec{x}_2) = \frac{1}{\sqrt{2}}[\psi_1(\vec{x}_1)\psi_2(\vec{x}_2) - \psi_1(\vec{x}_2)\psi_2(\vec{x}_1)] \tag{11.4}$$

would be an exact solution to equation (11.3) if $\hat{h}_2$ could be ignored.

In both cases, the two wave functions ψ_1 and ψ_2 must satisfy the one-particle equations

$$\hat{h}_1(\vec{r})\psi_i(\vec{r}) = \epsilon_i\psi_i(\vec{r}) \tag{11.5}$$

To solve these equations, we proceed as in Section 8.4 for the H atom. The only changes [cf. equation (8.39)] are

$$\rho \rightarrow Z \cdot \rho$$
$$E \rightarrow Z^2 \cdot E. \tag{11.6}$$

Here, Z is the atomic number of the atom (i. e., $Z = 2$ for helium).

When we include the interactions between the electrons, we can still use the Hartree–Fock approximation, which, however, then no longer is exact. We will then have to solve the Hartree–Fock equations (10.56). For the ground state of the He atom, we will assume that the two orbitals ψ_1 and ψ_2 differ only in their spin dependence. Then it can easily be shown that in the Hartree–Fock equations (10.56) only one additional term is included compared to the simpler equations (11.5): each electron experiences an additional potential due to the distribution of the other electron. Because the electrons are negatively charged, this represents a repulsion. In other words, the potential felt by each electron is less attractive than that of the bare nucleus. Thus, the potential of the nucleus is shielded. This is shown schematically in Figure 11.1.

An approximate approach to consider these effects is to replace the nuclear charge Z in equation (11.6) with an effective nuclear charge Z_{eff}. We must have

$$Z_{\text{eff}} \leq Z. \tag{11.7}$$

Then one may still use the hydrogen-like wave functions. In principle, Z_{eff} is a variational parameter, so that its value can be optimized by variation.

In fact, such a method can also be used for heavier atoms. Often one then finds that Z_{eff} should have different values for different orbitals.

Here, however, we will treat the properties of the atoms by means of the Aufbau principle. This is the content of the next section.

11.2 The periodic table

An individual, isolated atom is spherically symmetric (see also Section 11.7). To understand how the electronic properties change throughout the periodic table, one uses the Aufbau principle. This is based on the Hartree–Fock approximation.

Due to the spherical symmetry, all orbitals can be written as

$$\psi(\vec{x}) = R_{nl}(r)Y_{lm}(\theta, \phi)\sigma(m_s), \tag{11.8}$$

where σ describes the spin dependence. The orbitals thus have the same structure as for the hydrogen atom, but with some differences. As discussed above, the radial functions $R_{nl}(r)$ can often be compared with those of the hydrogen atom, although we then introduce effective nuclear charges that depend not only on the atom but also on (n, l). To obtain the orbitals for the lowest total energy, one assumes that the orbitals are energetically arranged as indicated in Figure 11.2. This means above all that the orbital energies depend not only on n but also on l, which makes a difference to the hydrogen atom. Subsequently, the orbitals are occupied in increasing energetic order whereby it is used that the $s, p, d, \ldots$ orbitals can accommodate 2, 6, 10, $\ldots$ electrons if the spin is taken into account. In occupying the orbitals, the following procedure is used:

-   The orbitals are arranged energetically as shown in Figure 11.2.
-   Because electrons are fermions (see Chapter 16 if this concept is less well known), each orbital can only be occupied by one electron. This is also called the Pauli principle (after Wolfgang Pauli).
-   The electrons occupy the orbitals energetically from below.
-   If several spatially distinct, energetically degenerate orbitals can be occupied (e. g., the p_x, p_y, and p_z orbitals), spatially distinct orbitals are first occupied. For example, for the C atom, a $2p_x$ and a $2p_y$ orbital can each be occupied by one electron, and only for the O atom will a second electron be added to the $2p_x$ orbital. This is the first half-part of Hund's rule (after Friedrich Hund).
-   In the case discussed above, the spatially distinct orbitals with the same spin dependence are first occupied. That is the second half-part of Hund's rule.

With this approach, the periodic table, Figure 11.3, and especially the properties of the atoms can be explained. For example, the periodicity of the properties (see Figure 11.4) can be explained. If we consider that part of the electronic energy that results from the sum of the energies of the occupied orbitals, we see in Figure 11.2 that this proportion increases abruptly each time a set of orbitals with the same (n, l) is fully occupied. This is especially true when all orbitals for the same n are occupied. The latter happens at $Z = 2, 10, 18, \ldots$, i. e., for He, Ne, Ar, $\cdots$ while the first is seen also for $Z = 4, 12, 20, \ldots$, i. e. for Be, Mg, Ca, $\cdots$. This effect can be seen in Figure 11.4 as jumps in the electron affinity and the first ionization potential. Further, it can be seen that smaller jumps

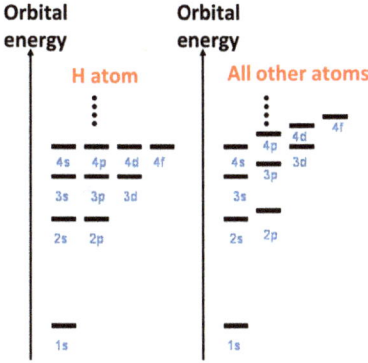

Figure 11.2: Schematic representation of the energies of the orbitals for (left) the hydrogen atom and (right) all other atoms.

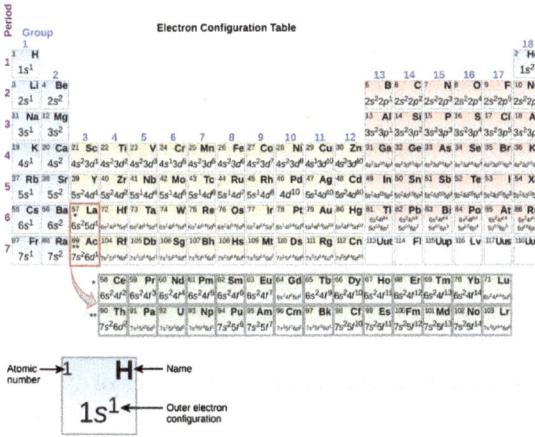

Figure 11.3: The periodic table.

occur when a set of orbitals with the same (n, l) is half-full, that is, all these orbitals with the same spin dependence are occupied. This is for instance the case for N and P. Finally, Figure 11.2 shows that the energies of the $4s$ and $3d$ orbitals are comparable, so that these orbitals are filled essentially in parallel.

11.3 Angular momentum

In Section 7.4, we discussed the addition of angular momentum vectors. Starting from the angular momenta of individual electrons we can determine a total angular momentum,

$$\vec{L} = \vec{l}_1 + \vec{l}_2 + \cdots + \vec{l}_N. \tag{11.9}$$

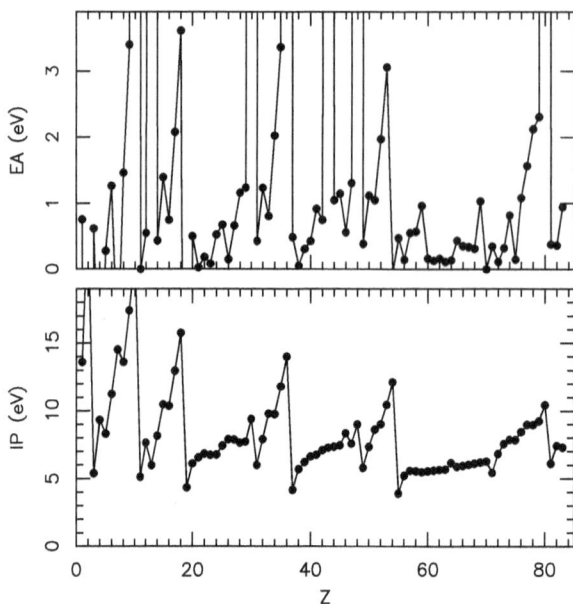

Figure 11.4: Electron affinity (EA) and first ionization potential (IP) of isolated atoms as a function of atomic number.

Of course, this also applies to the spins of the electrons,

$$\vec{S} = \vec{s}_1 + \vec{s}_2 + \cdots + \vec{s}_N. \tag{11.10}$$

And furthermore, it is also possible to obtain the total angular momentum of an electron by adding its angular momentum and spin,

$$\vec{j}_i = \vec{l}_i + \vec{s}_i. \tag{11.11}$$

We also have

$$\vec{J} = \vec{j}_1 + \vec{j}_2 + \cdots + \vec{j}_N = \vec{L} + \vec{S}. \tag{11.12}$$

The fact that the $\vec{l}_i$ couple with each other can be understood by the fact that the movements of the individual electrons are not independent of each other. Because of their electrical charge, they try to avoid each other. Similar arguments apply to the $\vec{s}_i$, where rather the Hund's rule indicate that these are not independent of each other, so that also they couple with each other.

If there was no coupling between spins and angular momenta (i. e., $\vec{L}$ and $\vec{S}$ are independent), $\vec{L}$ and $\vec{S}$ would both be quantized. This would mean that

$$\hat{L}^2 \Psi_e = L(L + 1)\hbar^2 \Psi_e$$
$$\hat{S}^2 \Psi_e = S(S + 1)\hbar^2 \Psi_e$$

$$\hat{L}_z \Psi_e = M_L \hbar \Psi_e$$
$$\hat{S}_z \Psi_e = M_S \hbar \Psi_e. \tag{11.13}$$

Here L is an integer while S is half-integral or integral. Furthermore, M_L and M_S can take the values

$$M_L = -L, -L+1, \ldots, L-1, L$$
$$M_S = -S, -S+1, \ldots, S-1, S. \tag{11.14}$$

On the other hand, when $\vec{L}$ and $\vec{S}$ couple, we only have

$$\hat{J}^2 \Psi_e = J(J+1)\hbar^2 \Psi_e$$
$$\hat{J}_z \Psi_e = M_J \hbar \Psi_e \tag{11.15}$$

(which also is true when $\vec{L}$ and $\vec{S}$ do not couple) and J can be half-integral or integral and M_J is one of the values

$$M_J = -J, -J+1, \ldots, J-1, J. \tag{11.16}$$

The whole field of quantization and coupling of angular momenta is not that simple, and we will explain it only briefly through an example.

We consider the He Atom. In the ground state, there are two electrons in the 1s orbital, which differ only in spin dependence. We can therefore write the two-electron wave function as

$$\Psi_e(\vec{x}_1, \vec{x}_2) = \frac{1}{\sqrt{2}}[\psi_{1s}(\vec{r}_1)\alpha(1)\psi_{1s}(\vec{r}_2)\beta(2) - \psi_{1s}(\vec{r}_2)\alpha(2)\psi_{1s}(\vec{r}_1)\beta(1)]$$
$$= \frac{1}{\sqrt{2}}\psi_{1s}(\vec{r}_1)\psi_{1s}(\vec{r}_2)[\alpha(1)\beta(2) - \alpha(2)\beta(1)] \tag{11.17}$$

Here, for instance, $\psi_{1s}(\vec{r}_i)\alpha(i)$ is the 1s wave function in position space for electron i multiplied by the α spin function for the same electron. A similar notation is used when α is replaced by β.

For the 1s function, the l quantum number is 0. Therefore (see Section 7.4) L can only be 0. On the other hand, the spin quantum number for each electron is $\frac{1}{2}$, so S can be both 0 and 1. However, it can be shown (which is not the content of this book) that for the wave function in equation (11.17) $S = 0$. Thus, M_S can only be 0, which can also be seen by inserting:

$$\hat{S}_z \Psi_e(\vec{x}_1, \vec{x}_2) = \frac{1}{\sqrt{2}}\psi_{1s}(\vec{r}_1)\psi_{1s}(\vec{r}_2)[\hat{s}_{1z} + \hat{s}_{2z}][\alpha(1)\beta(2) - \alpha(2)\beta(1)]$$
$$= \frac{1}{\sqrt{2}}\psi_{1s}(\vec{r}_1)\psi_{1s}(\vec{r}_2)\frac{\hbar}{2} \cdot [\alpha(1)\beta(2) - \alpha(1)\beta(2) - \alpha(2)\beta(1) + \alpha(2)\beta(1)]$$
$$= 0. \tag{11.18}$$

Here, we used that

$$\hat{s}_{iz}\alpha(j) = \delta_{i,j}\frac{\hbar}{2}\alpha(i)$$

$$\hat{s}_{iz}\beta(j) = -\delta_{i,j}\frac{\hbar}{2}\beta(i). \tag{11.19}$$

Because $S = 0$, there is only one possible value for M_S. Therefore, the state with the wave function of equation (11.17) is called a singlet. If $S = \frac{1}{2}, S = 1, \ldots$, you would have two, three, $\ldots$, possible values for M_S and the state would then be called a doublet, triplet, $\ldots$.

If we in equation (11.17) replace one of the two 1s functions with a 2s function, we describe an (electronically) excited state. Depending on which spin functions the two electrons have, we can set up four different Slater determinants,

$$\Psi_{e1}(\vec{x}_1,\vec{x}_2) = \frac{1}{\sqrt{2}}\begin{vmatrix} \psi_{1s}(\vec{r}_1)\alpha(1) & \psi_{2s}(\vec{r}_1)\alpha(1) \\ \psi_{1s}(\vec{r}_2)\alpha(2) & \psi_{2s}(\vec{r}_2)\alpha(2) \end{vmatrix}$$

$$\Psi_{e2}(\vec{x}_1,\vec{x}_2) = \frac{1}{\sqrt{2}}\begin{vmatrix} \psi_{1s}(\vec{r}_1)\beta(1) & \psi_{2s}(\vec{r}_1)\beta(1) \\ \psi_{1s}(\vec{r}_2)\beta(2) & \psi_{2s}(\vec{r}_2)\beta(2) \end{vmatrix}$$

$$\Psi_{e3}(\vec{x}_1,\vec{x}_2) = \frac{1}{\sqrt{2}}\begin{vmatrix} \psi_{1s}(\vec{r}_1)\alpha(1) & \psi_{2s}(\vec{r}_1)\beta(1) \\ \psi_{1s}(\vec{r}_2)\alpha(2) & \psi_{2s}(\vec{r}_2)\beta(2) \end{vmatrix}$$

$$\Psi_{e4}(\vec{x}_1,\vec{x}_2) = \frac{1}{\sqrt{2}}\begin{vmatrix} \psi_{1s}(\vec{r}_1)\beta(1) & \psi_{2s}(\vec{r}_1)\alpha(1) \\ \psi_{1s}(\vec{r}_2)\beta(2) & \psi_{2s}(\vec{r}_2)\alpha(2) \end{vmatrix}. \tag{11.20}$$

It can now be shown (but we will not do so here) that

$$\hat{S}^2\Psi_{e1} = 2\hbar^2\Psi_{e1}$$
$$\hat{S}_z\Psi_{e1} = \hbar\Psi_{e1}$$
$$\hat{S}^2\Psi_{e2} = 2\hbar^2\Psi_{e2}$$
$$\hat{S}_z\Psi_{e2} = -\hbar\Psi_{e2}$$
$$\hat{S}_z\Psi_{e3} = 0$$
$$\hat{S}_z\Psi_{e4} = 0. \tag{11.21}$$

On the other hand, neither Ψ_{e3} nor Ψ_{e4} is an eigenfunction to $\hat{S}^2$. But we may construct

$$\Psi_{e5}(\vec{x}_1,\vec{x}_2) = \frac{1}{\sqrt{2}}[\Psi_{e3}(\vec{x}_1,\vec{x}_2) + \Psi_{e4}(\vec{x}_1,\vec{x}_2)]$$

$$= \frac{1}{\sqrt{2}}[\psi_{1s}(\vec{r}_1)\psi_{2s}(\vec{r}_2) - \psi_{1s}(\vec{r}_2)\psi_{2s}(\vec{r}_1)] \cdot \frac{1}{\sqrt{2}}[\alpha(1)\beta(2) + \alpha(2)\beta(1)]$$

$$\Psi_{e6}(\vec{x}_1,\vec{x}_2) = \frac{1}{\sqrt{2}}[\Psi_{e3}(\vec{x}_1,\vec{x}_2) - \Psi_{e4}(\vec{x}_1,\vec{x}_2)]$$

$$= \frac{1}{\sqrt{2}}[\psi_{1s}(\vec{r}_1)\psi_{2s}(\vec{r}_2) + \psi_{1s}(\vec{r}_2)\psi_{2s}(\vec{r}_1)] \cdot \frac{1}{\sqrt{2}}[\alpha(1)\beta(2) - \alpha(2)\beta(1)], \tag{11.22}$$

for which

$$\hat{S}^2 \Psi_{e5} = 2\hbar^2 \Psi_{e5}$$
$$\hat{S}_z \Psi_{e5} = 0$$
$$\hat{S}^2 \Psi_{e6} = 0$$
$$\hat{S}_z \Psi_{e6} = 0. \tag{11.23}$$

Thus, from this equation together with equation (11.21), we see that Ψ_{e1}, Ψ_{e2}, and Ψ_{e5} form the three energetically degenerate wave functions that make up the triplet state, while Ψ_{e6} forms a singlet state.

We emphasize that with the wave functions in equation (11.22) we have left the orbital picture: none of these wave functions can be written as a single Slater determinant.

The example we have discussed here is very simple. We have only considered two electrons, so that only two angular momenta would have to be added. Furthermore, the individual orbitals have $l = 0$, so that only $L = 0$ was possible. For other values of l_1 and l_2, one would have to treat the angular momenta exactly as we did with the spins. This usually gives rise to several different possible values of the length of the total angular momentum, L. Moreover, $L = 0$ means that $J = L + S$ can take only one value, $J = S$. This, too, makes the example particularly simple.

11.4 Spin-orbit coupling

When we introduced the electron spin (Section 7.5), we mentioned that the frequently used interpretation that the spin describes the rotation of the small electron about its own axis is only a convenient concept, but not completely true. Instead, the existence of spin can be explained only by using the theory of relativity.

That relativistic effects are important for electrons can be illustrated in the following way. In our everyday lives, we use the SI units to describe the phenomena that we experience. This means that we, e. g., specify lengths, masses, and time in units of m, kg, and s. Then all other quantities are expressed in appropriate units and, what is important here, the numerical values that we then give are usually neither very large nor particularly small. Rather, they are all in the range of 0.01–100. The speed of light c is $3 \cdot 10^8$ (m/s), which is so large that one can expect that this has no significance for our everyday life. That this is correct is well known.

For an electron, the SI units are not practical. Instead, one uses the so-called atomic units (a. u. = atomic units). Then the Planck constant $\hbar$, the elementary charge $|e|$, the dielectric constant of the vacuum $4\pi\epsilon_0$, and the mass of the electron m_e are all set equal to 1,

$$\hbar = 1$$
$$m_e = 1$$

$$|e| = 1$$

$$4\pi\epsilon_0 = 1. \tag{11.24}$$

Energies are then given in Hartrees (1 Hartree = 27.21 eV) and lengths in bohr (1 bohr = 0.52918 Å). The energy of the 1s electron in a hydrogen atom is then -0.5 a. u., and the average distance between the electron and the nucleus is 1 a. u. In these units, you get

$$c = 137.036. \tag{11.25}$$

This numerical value is actually the inverse of the so-called fine structure constant. Its numerical value is significantly smaller than the value in SI units, which also means that an 1s electron in the hydrogen atom typically has a speed of about 1 % of the speed of light. For heavier elements, in particular the core electrons are attracted to the nucleus more strongly, and thus become faster, so that their speed can come close to the speed of light. Because we know that the relativistic effects can occur for bodies with velocities not much smaller than c, we must conclude that this is also the case for the electrons. In order to recognize that relativistic effects may be important for heavier atoms, we can use the simple model of Niels Bohr to describe the H atom. If we were to consider a heavier element, we would in, e. g., equations (8.3) and (8.4) replace e^2 by $Z_{\text{eff}}e^2$, where Z_{eff} is supposed to be an effective nuclear charge, which may not be quite as large as the atomic number of the atom, but it is considerably larger than 1. Equation (8.4) then shows that the momentum of the electron increases.

In order to treat the relativistic effects for quantum objects, the formally correct way is to replace the Schrödinger equation with the Dirac equation. This is a very different equation, but if you let artificially $c \rightarrow \infty$, the equation becomes equivalent to the Schrödinger equation. Therefore, it is also possible to consider the additional effects arising from the theory of relativity as smaller terms that can be included directly in the Schrödinger equation. One proceeds by writing the relativistic effects in a series in $\frac{1}{c}$, because, as equation (11.25) shows, this quantity is not that large. For a particle moving in a potential $V(\vec{r})$, one can derive the following additional terms for the Hamilton operator of a particle (or for the Fock operator)

$$\hat{H}_{\text{rel}} = -\frac{\hbar^4}{8m^3c^2}\nabla^4 + \frac{1}{8m^2c^2}\nabla^2 V(\vec{r}) + \frac{\hbar}{4m^2c^2}(-i\vec{\nabla}V \times \vec{\nabla}) \cdot \vec{s}$$

$$\equiv \hat{H}_{\text{mass-velocity}} + \hat{H}_{\text{Darwin}} + \hat{H}_{\text{soc}}. \tag{11.26}$$

The first term is the so-called mass-velocity term and the second term is the so-called Darwin term. The last term is the so-called spin-orbit coupling (spin-orbit coupling = soc).

For a particle that moves in a spherically symmetric potential,

$$V(\vec{r}) = V(r), \tag{11.27}$$

like an electron in an atom, the spin-orbit coupling can be written as

$$\hat{H}_{\text{soc}} = \frac{\hbar}{4m^2c^2}(-i\vec{\nabla}V \times \vec{\nabla}) \cdot \vec{s} = \frac{1}{4m^2c^2}\frac{1}{r}\frac{dV}{dr}(\vec{s} \cdot \vec{l}). \tag{11.28}$$

If one ignored the spin-orbit coupling, the Schrödinger equation with the additional operator of equation (11.26) would have the same symmetry properties as in the absence of this additional operator. Only the spin-orbit coupling leads to a change (reduction) of the symmetry properties. With this, the angular momentum $\vec{l}$ and the spin $\vec{s}$ are no longer independent of each other because, as equation (11.28) shows, they couple with each other.

It can be seen that the larger the $\frac{dV}{dr}$ becomes, the greater the effect becomes. For an electron in an atom, V consists of the potential of the nucleus and that of the other electrons. The first is equal to $\frac{-Ze^2}{r}$, and it can then easily be seen that the heavier the atom, the larger becomes this term (and its derivative with respect to r). For the valence electrons (i. e., those of the energetically upper-most, occupied orbitals), the other electrons ensure that the potential of the nucleus is partially shielded, but nevertheless, the effects of the spin-orbit coupling are greatest for heavy elements.

In fact, it is often claimed that gold is "the most relativistic stable element." If the speed of light is artificially set to be not equal to 137 but much larger (which can be done, e. g., in theoretical calculations), the properties of gold would change markedly. For example, then the color that also is responsible for the popularity of gold would be more like that of copper.

11.5 Couplings, term symbols, and good quantum numbers

In Section 11.3, we discussed how the individual $\vec{l}$ of the electrons lead to a total $\vec{L}$ of the atom, which also applies to the individual $\vec{s}$ of the electrons, which can be combined into a total $\vec{S}$. That this is necessary can be seen from the fact that the movements of the individual electrons are not independent of each other: the electrostatic interactions between the electrons cause the electrons to move in concert; see Figure 11.5, so that the individual $\vec{l}_i$ of the individual electrons depend on each other, and one should rather consider the total orbital momentum $\vec{L}$. Similarly, Hund's rules indicate that the spins of individual electrons also depend on each other. Finally, we have just learnt that, at least for heavier atoms, the coupling between spin and orbital angular momentum may be important.

For lighter elements, for which the spin-orbit coupling is weak, the entire electronic wave function $\Psi_e(\vec{x}_1, \vec{x}_2, \ldots, \vec{x}_n)$ will not be an eigenfunction for each of the individual $\hat{l}_i^2$, $\hat{l}_{iz}$, $\hat{s}_i^2$, or $\hat{s}_{iz}$ operators, but approximately (if relativistic effects are ignored) for the operators of the total orbital angular momentum and the total spin. This means that equation (11.13) is satisfied, and that L, M_L, S and S_z have well-defined values. The

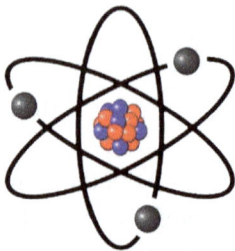

Figure 11.5: Schematic representation of the movements of the electrons in an atom.

latter is often formulated so that $\hat{H}, \hat{L}^2, \hat{L}_z, \hat{S}^2$, and $\hat{S}_z$ are said to be good quantum numbers (they have well-defined values). The existence of the good quantum numbers can also be explained with the help of commuting operators. Without spin-orbit coupling, $\hat{H}, \hat{L}^2, \hat{L}_z, \hat{S}^2$, and $\hat{S}_z$ commute. Therefore, it is possible to identify wave functions that are eigenfunctions to all these operators simultaneously.

If the wave function of a state has L and S as good quantum numbers, this wave function can also be described (in part) by these numerical values. For such a description, one uses the so-called term symbols. A term is a set of states that all have the same values of L and S, but may have different values of M_L and M_S. The term symbols have the structure ^{2S+1}X. The upper index, $2S + 1$, is the numerical value of $2S + 1$ and describes the number of different spin states with the same L and S. When $2S + 1$ equals $1, 2, 3, \ldots$, one speaks of singlet, doublet, triplet, $\ldots$. $2S + 1$ is called the multiplicity. The symbol X is equal to S, P, D, F, G, $\ldots$ when $L = 0, 1, 2, 3, 4, \ldots$. Occasionally, the term symbol is extended by a lower right index: $^{2S+1}X_J$. This index gives the quantum number J (see below).

Throughout this chapter, we have mentioned several times that one generates a total angular momentum from individual angular momenta and/or spins, but hardly described how. Here, we just mention that it is not easy, but with the help of the so-called Clebsch–Gordan coefficients it becomes possible (after Alfred Clebsch and Paul Gordan).

But there are two special cases for the addition of several angular momenta, which should be mentioned here. As we saw in the last chapter, the relativistic effects make $\vec{L}$ and $\vec{S}$ couple with each other. This means that L, M_L, S, and S_z are not good quantum numbers, but only J and M_J, which arise from the sum,

$$\vec{J} = \vec{L} + \vec{S}. \tag{11.29}$$

Then

$$\hat{J}^2\Psi_e = J(J + 1)\hbar^2\Psi_e$$
$$\hat{J}_z\Psi_e = M_J\hbar\Psi_e. \tag{11.30}$$

Due to the spin-orbit coupling, $\hat{H}$ no longer commutes with $\hat{L}^2$, $\hat{L}_z$, $\hat{S}^2$, and $\hat{S}_z$, but with $\hat{J}^2$ and $\hat{J}_z$. Because of this, the term symbols partly lose their meaning.

How strong the coupling between $\vec{L}$ and $\vec{S}$ is depends on the strength of the spin-orbit coupling. As we have seen, this is weaker for lighter elements. Then it is a reasonable approximation to first identify the term symbols, and then assume that the spin-orbit coupling leads to only minor corrections. Thus, one first determines $\vec{L}$ and $\vec{S}$ for the total set of the electrons and subsequently generates $\vec{J}$. This corresponds to the so-called Russell–Saunders coupling (after Henry Norris Russell and Frederick Albert Saunders).

For heavier elements, however, the spin-orbit coupling is strong, and its strength may even exceed the strength of the electrostatic interactions between the electrons. This means that the vectors $\vec{l}_i$ and $\vec{s}_i$ strongly couple for the individual electrons, and only as a secondary effect one may consider the couplings between these $\vec{j}_i = \vec{l}_i + \vec{s}_i$. Then it makes more sense first to form the sum of $\vec{l}_i$ and $\vec{s}_i$ of the individual electrons,

$$\vec{j}_i = \vec{l}_i + \vec{s}_i, \tag{11.31}$$

and then to form $\vec{J}$ from all $\vec{j}_i$.

$$\vec{J} = \vec{j}_1 + \vec{j}_2 + \cdots + \vec{j}_N. \tag{11.32}$$

This procedure is called jj coupling.

As said, the optimal coupling scheme depends on the relative strength between spin-orbit coupling and the electrostatic interaction between the electrons. This ratio can be quantified with a parameter χ whose exact definition is not relevant here. $\chi = 0$ means that the spin-orbit coupling can be neglected, while $1/\chi = 0$ means that the electrostatic interactions can be ignored. In Figure 11.6, we show how (on the left side) the term symbols, for which the spin-orbit coupling is not taken into account, change into the states of the jj coupling as χ increases. This example focuses on a system with two valence electrons occupying p orbitals. The vertical lines also indicate where different atoms with two p valence electrons lie on the χ scale.

11.6 On the spatial distribution and the energies of the atomic orbitals

With the Aufbau principle, we have seen how we occupy the orbitals of atoms with electrons. From the basics of quantum theory, we can get further information on the orbitals, which will be briefly discussed here.

For the H and He atoms, the first two electrons enter the $1s$ orbital. For the Li atom, the $2s$ orbital is being occupied for the first time. From quantum theory (Section 3.5), we know that the $2s$ orbital must be orthogonal to the $1s$ orbital. The two orbitals

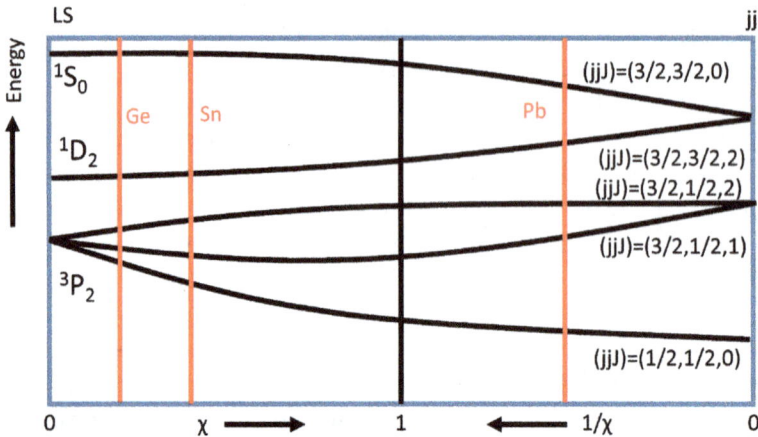

Figure 11.6: Schematic representation of the relationship between coupling schemes and relative strength of the spin-orbit coupling relative to the interelectronic interaction for atoms with two p valence electrons. $\chi = E_{soc}/E_{ee}$ is the ratio of the energy of the spin-orbit coupling and the energy of the electron-electron interaction. Typical values for Ge, Sn, and Pb are also given.

have the same angular dependence and can therefore differ only in their radial dependences. Because the $1s$ orbital is positive everywhere and also has a maximum at the nucleus, the $2s$ orbital must have both positive-valued and negative-valued spatial regions, and is accordingly made spatially more extended; hence it has larger values in areas further away from the nucleus (this can be seen in Figure 8.2). Because of this, an electron in such an orbital does not feel the electrostatic attraction through the nucleus as much as an $1s$ electron does. A further effect is therefore that the orbital has a higher energy.

Starting at B, we begin to fill electrons into p orbitals. The angular dependence of the p orbitals is different from that of the s orbitals and, therefore, the $2p$ orbitals are automatically orthogonal to the $1s$ and $2s$ orbitals. This explains why the $2p$ orbitals can approach the atomic nuclei relatively close (see Figure 8.2). Therefore, their energy is relatively low—so the energy difference between $2s$ and $2p$ orbital is not that large.

Then the $3s$ orbitals are being filled. These must be orthogonal to the $1s$ and $2s$ orbitals, while they are orthogonal to the $2p$ orbitals (due to different angular dependences). This leads to an even more complex radial behavior than for the $1s$ and $2s$ orbitals.

The $3p$ orbitals must be orthogonal to the $2p$ orbitals. As with the $1s$ and $2s$ orbitals, this causes the $3p$ orbitals to be pushed away from the nucleus and, therefore, the orbital energies of the $3p$ orbitals are not as close to those of the $3s$ orbital as we previously had for the $2s$ and $2p$ orbitals.

Ultimately, this results in hybrid orbitals (which we will discuss later in Section 13.4) not being as easily formed for elements of the $3s$ and $3p$ series, as for the elements of the 2s and 2p series. This has consequences for the chemical behavior of the atoms.

If we go further, we will eventually come to the 3d orbitals. Because of their angular dependence, these are orthogonal to the previously occupied 1s, 2s, 2p, 3s, 3p, and 4s orbitals and can thus get closer to the nucleus and are therefore often strongly localized to the region closest to the atomic nuclei. The same applies to the elements of the 4f series.

11.7 Spherical atoms

In this subsection, we shall discuss the carbon atom in some details but emphasize that the discussion easily can be adapted to any other atom with partially filled electronic shells, i. e., when applying the Aufbau principle, not all orbitals with the same (n, l) are occupied.

In order to simplify the discussion, we shall ignore spin dependences. Then we may construct the three electronic configurations that are shown in Figure 11.7. In all cases, two electrons occupy the 1s orbitals, another two occupy the 2s orbitals, and the three configurations differ in which two out of the three 2p orbitals that are occupied.

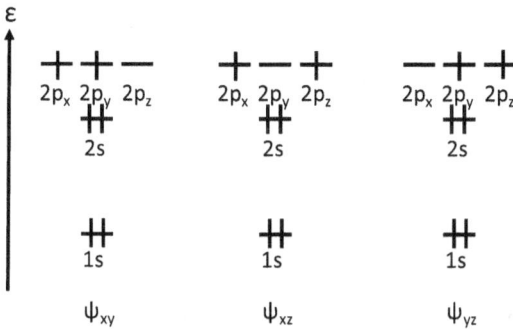

Figure 11.7: Schematic representation of the three configurations that we can imagine for the carbon atom. The corresponding wave functions are called Ψ_{xy}, Ψ_{xz}, and Ψ_{yz}, respectively. Spin has not been shown, and the orbital energies are not given to scale.

We notice that the three wave functions describing the three configurations of Figure 11.7 are energetically degenerate and in fact any (normalized) linear combination of those has the same energy and corresponds to occupying two 2p functions that are oriented along two orthogonal directions in space. Thus, we may construct a time-dependent wave function for the carbon atom according to

$$\Psi(t) = c_{xy}(t)\Psi_{xy} + c_{xz}(t)\Psi_{xz} + c_{yz}(t)\Psi_{yz},\tag{11.33}$$

where we have specified only the time dependence (t) but not the position and spin coordinates of the electrons. Except for requiring

$$|c_{xy}(t)|^2 + |c_{xz}(t)|^2 + |c_{yz}(t)|^2 = 1, \tag{11.34}$$

the three coefficients $c_{xy}(t)$, $c_{xz}(t)$, and $c_{yz}(t)$ are in principle arbitrary. Since all wave functions of the form of equation (11.33) are energetically degenerate, these three coefficients can vary freely without the need for providing energy to the system (the carbon atom).

We can now discuss different cases. At first, if the three coefficients are constants we obtain a charge density as that shown schematically in the left part of Figure 11.8. In this case, we have set

$$c_{xy} = 1, \quad c_{xz} = 0, \quad c_{yz} = 0 \tag{11.35}$$

and the electron density becomes

$$\rho(\vec{r}) = 2|\psi_{1s}(\vec{r})|^2 + 2|\psi_{2s}(\vec{r})|^2 + |\psi_{2p_x}(\vec{r})|^2 + |\psi_{2p_y}(\vec{r})|^2 \tag{11.36}$$

and the density is shown in Figure 11.8 in the (x, z) plane. In producing this figure, we have used the electronic wave functions for the hydrogen atom, which certainly is a very crude approximation. However, the main result becomes clear from an inspection of the left-most part of Figure 11.8: the electronic density is not spherical. For other constant values of the three coefficients, similar results will be found except they will be rotated compared to the case of Figure 11.8.

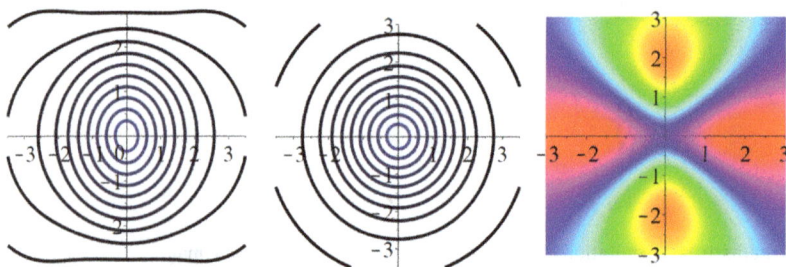

Figure 11.8: Approximate electron densities for the carbon atom obtained by using the electronic wave functions for the hydrogen atom. The left-most part shows the density of equation (11.36) and the middle part shows that of equation (11.37). The right-most part shows the difference between the two.

This will also be the case if the three coefficients possess the same time dependence. Only if they have different, uncorrelated time dependences the electron density will become

$$\rho(\vec{r}) = 2|\psi_{1s}(\vec{r})|^2 + 2|\psi_{2s}(\vec{r})|^2 + \frac{2}{3}[|\psi_{2p_x}(\vec{r})|^2 + |\psi_{2p_y}(\vec{r})|^2 + |\psi_{2p_z}(\vec{r})|^2], \tag{11.37}$$

which is shown in the middle part of Figure 11.8 and which is spherically symmetric. Comparing with the left part of that figure, small differences are visible. These differences are shown in the right part of Figure 11.8.

Thus, in order to obtain a spherical carbon atom, we need all the three configurations of Figure 11.7 and it is no longer possible to specify six different orbitals that are occupied by the six electrons: we need more orbitals.

11.8 Problems with answers

1. **Problem:** Consider a system consisting of two d electrons. Use both the Russell–Saunders and the jj coupling to determine the possible values of J for this system.
   Answer: We generally consider two angular momenta $\vec{v}_1$ and $\vec{v}_2$, which later will turn into angular and/or spin momenta. It is then always true that

$$\vec{V} = \vec{v}_1 + \vec{v}_2. \tag{11.38}$$

If the individual angular momentum vectors are quantized,

$$|\vec{v}_1|^2 = v_1(v_1 + 1)\hbar^2$$
$$|\vec{v}_2|^2 = v_2(v_2 + 1)\hbar^2, \tag{11.39}$$

then the sum will also be quantized,

$$|\vec{V}|^2 = V(V + 1)\hbar^2. \tag{11.40}$$

The possible values of V are then given by

$$V = |v_1 - v_2|, |v_1 - v_2| + 1, \ldots, v_1 + v_2 - 1, v_1 + v_2. \tag{11.41}$$

We now use this first for the Russell–Saunders coupling for the two d electrons. First, let $\vec{v}_i = \vec{l}_i$ and then obtain ($l_1 = l_2 = 2$) that

$$L = 0, 1, 2, 3, 4 \tag{11.42}$$

while for the spin ($\vec{v}_i = \vec{s}_i$ and $s_1 = s_2 = \frac{1}{2}$) the possible values are

$$S = 0, 1 \tag{11.43}$$

Finally, we set $\vec{v}_1 = \vec{L}$ and $\vec{v}_2 = \vec{S}$ and then get

$$J = 0, 1, 2, 3, 4, 5 \tag{11.44}$$

as possible values.

For the jj coupling, we first set $\vec{v}_1 = \vec{l}$ and $\vec{v}_2 = \vec{s}$ (with $v_1 = 2$, $v_2 = \frac{1}{2}$) and then obtain in a similar way that

$$j_1, j_2 = \frac{3}{2}, \frac{5}{2} \tag{11.45}$$

are possible. Then we get with $\vec{v}_i = \vec{j}_i$ as possible values for J:

$$J = 0, 1, 2, 3, 4, 5 \tag{11.46}$$

in agreement with the result of equation (11.44) as should be.

11.9 Problems

1. Write down the electronic wave function for the Be atom according to the Hartree–Fock approximation.
2. Is the two-electron wave function

$$\psi(1, 2) = 1s(1)\alpha(1)2s(2)\alpha(2) - 1s(2)\alpha(2)2s(1)\alpha(1) \tag{11.47}$$

   an eigenfunction to the operator $\hat{S}_z$? Prove the answer.
3. Explain the rules of Hund.
4. Explain the term "spin-orbit coupling."
5. Explain the term "term."
6. What does the Hamilton operator look like for the Li atom? What does the electronic Hamilton operator look like for this system?
7. Consider an atom with N electrons. What happens to the electronic wave function of this system when applying the various spin and angular momentum operators to it: (a) if there is no spin-orbit interaction, and (b) if there is a non-vanishing spin-orbit interaction?
8. To which additional terms in the Hamilton operator does the theory of relativity lead? How do the individual terms change the symmetry properties of the system?
9. Compare Russell–Saunders and jj coupling.
10. Consider a system consisting of one d and one p electron. Use both the Russell–Saunders and the jj coupling to determine the possible values of J for this system.
11. Explain how the commutator relations can be used to explain why the electronic eigenfunctions to the Hamilton operator for an N-electron system are not simultaneously eigenfunctions to the individual $\hat{l}_i^2$, $\hat{s}_i^2$, and $\hat{j}_i^2$ operators and explain when they are not eigenfunctions to the $\hat{L}^2$ and $\hat{S}^2$ operators.

12 The smallest molecules

12.1 The problem

As mentioned in Section 9.1, it took just little more than 3 years after the introduction of the mathematical theory of quantum theory by Heisenberg and Schrödinger in 1926, when Paul Andre Maurice Dirac summarized the problems in the application of this theory:

> The fundamental laws necessary for the mathematical treatment of large parts of physics and the whole of chemistry are thus fully known, and the difficulty lies only in the fact that application of these laws leads to equations that are too complex to be soluble.

He continued:

> It therefore becomes desirable that approximate practical methods of applying quantum mechanics should be developed, which can lead to an explanation of the main features of complex atomic systems without too much computation.

We presented the basics of such approximate methods in Chapter 10, and in Chapter 11 we saw how these approximations can provide accurate results for the isolated (charged or neutral) atoms. But we have also seen that, if you look very carefully, also the inaccuracies are recognizable. This was the case, e. g., when we demanded that the wave functions should be eigenfunctions to the spin operators at the same time. But in many cases, the obtained, approximate wave functions provide results that are meaningful and helpful in developing an understanding of chemical and physical phenomena.

The approximate methods mentioned above are based on mathematical arguments that we have discussed in part in the earlier chapters and that will be discussed in more detail later. Here, we need only to emphasize that reasonably developed approximate approaches can provide accurate information. In fact, we used these arguments in the last chapter to explain the structure of the periodic table. In this chapter, we will use similar techniques to analyze the electronic orbitals of very small diatomic molecules. In the following chapters, we will deal with more complex molecular systems.

12.2 The H_2^+ molecular ion

We start our discussion with the H_2^+ molecular ion, i. e., the simplest system with more than one nucleus. For this system, we have $M = 2$ nuclei and $N = 1$ electrons, and the single electron moves in the field of the two nuclei. We will use the Born–Oppenheimer approximation and search accordingly the corresponding wave function for the electron moving in the field of the two nuclei that are kept at fixed positions.

https://doi.org/10.1515/9783110742206-012

When the two nuclei are far apart, the electron experiences the potential of only one nucleus. In that case, the wave function of the electron must be equal to that of the isolated hydrogen atom. This must be true regardless of whether the electron is near one or the other nucleus. Therefore, a meaningful, approximate wave function is (see Figures 12.1 and 12.2)

$$\psi(\vec{r}) = N \cdot [\chi_a(\vec{r}) + \chi_b(\vec{r})] \tag{12.1}$$

with the 1s orbitals χ_a and χ_b, centered on the left (χ_a) and right (χ_b) hydrogen atom, respectively. N is a constant that shall ensure that the function is normalized and that will be discussed further below.

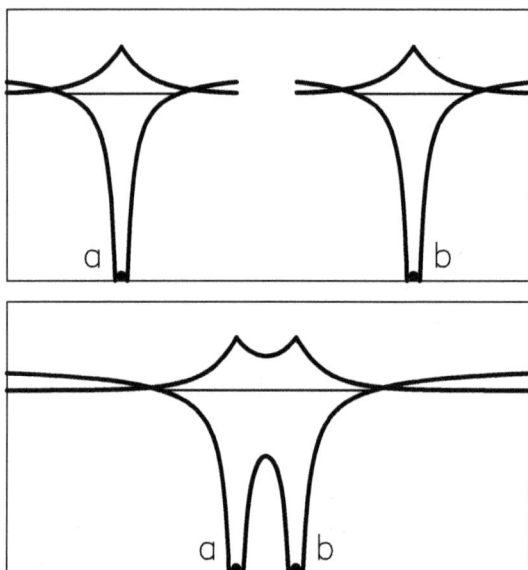

Figure 12.1: The potential and the orbital for the H_2^+ molecular ion for (above) a large and (below) a small distance between the two nuclei (shown as dark circles). The horizontal thin lines show the energies of the orbitals relative to the potential. These lines also represent the axes for the orbitals.

We will now assume that we can use this wave function even for small distances between the two nuclei. If the electron is in the immediate vicinity of one nucleus, the potential of this nucleus is so dominating that the potential of the other nucleus at first can be ignored, so that in this region the 1s orbitals of the corresponding hydrogen atom provides a good approximation to the exact orbital. Combining these two 1s functions smoothly gives again a function as in equation (12.1). This function corresponds to a chemically motivated, meaningful approximation to the exact wave function.

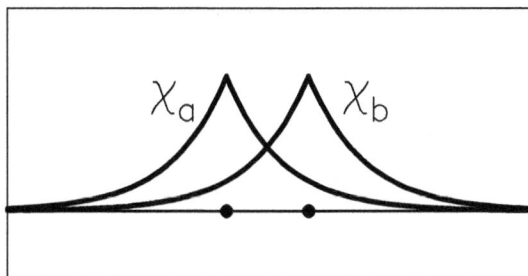

Figure 12.2: The two functions χ_a and χ_b, which are centered on the individual hydrogen atoms, and which are used for the hydrogen molecule ion to generate molecular orbitals.

If we then calculate the total energy as a function of the distance between the nucleus, we obtain a curve that has a minimum for some value of the interatomic distance, d, (see Figure 12.3). The position of this minimum corresponds to the theoretically predicted equilibrium distance between the two cores.

We will write the wave function we have generated above as

$$\psi_b(\vec{r}) = N_b \cdot [\chi_a(\vec{r}) + \chi_b(\vec{r})]. \tag{12.2}$$

Occupying this leads to an increase in the electron density between the nucleus. This creates a binding interaction between the nucleus and, accordingly, we call the wave function binding. Although this description is not entirely accurate, it is very helpful. According to this description, if we increase the electron density between the two nuclei compared to the case of noninteracting atoms, we have a binding interaction between two atoms (nuclei). In some sense, there is a stabilizing interaction because the negatively charged electron is attracted not by one but by two positively charged nuclei, thereby experiencing a decrease in energy.

We can also interpret the stabilization of the bonding orbital with the help of the kinetic energy, and in fact the kinetic energy is just as important for the chemical bond as the potential energy. In Section 4.5, we have seen that the kinetic energy can be identified in the form of strong, rapid spatial changes of the wave function. Because the binding orbital of the molecule (ion) is spatially more extended than the 1s orbital of the isolated H atom, and at the same time has no nodes, the spatial changes of the wave function in position space become smaller, and the kinetic energy decreases. Again, this is a stabilization of the chemical bond.

The constant N_b ensures that the wave function is normalized,

$$1 = \langle \psi_b | \psi_b \rangle = N_b^2 (\langle \chi_a | \chi_a \rangle + \langle \chi_b | \chi_b \rangle + \langle \chi_a | \chi_b \rangle + \langle \chi_b | \chi_a \rangle) = N_b^2 (2 + 2S), \tag{12.3}$$

where we have assumed that the two atomic orbitals χ_a and χ_b are normalized and real, and introduced

$$S = \langle \chi_a | \chi_b \rangle = \langle \chi_b | \chi_a \rangle. \tag{12.4}$$

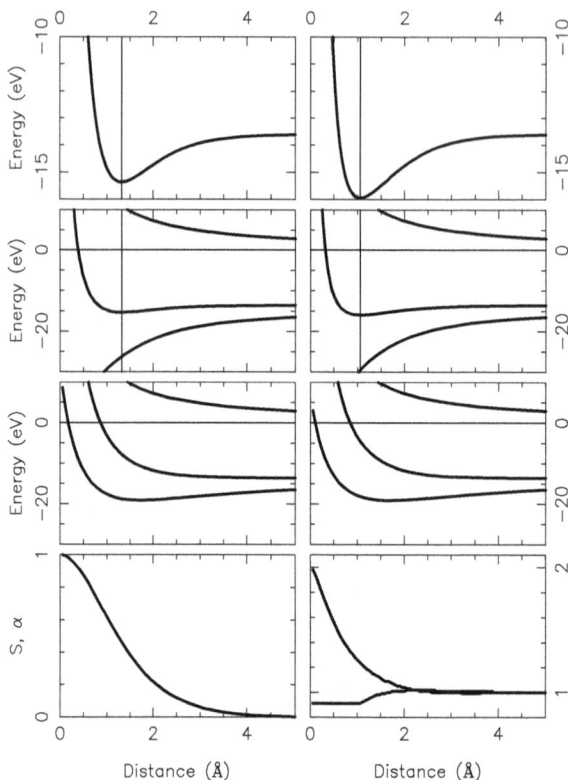

Figure 12.3: The total energy as a function of the distance between the two nuclei for the hydrogen molecular ion. The electronic energy is calculated using the bonding and antibonding wave functions of equations (12.2) and (12.6), respectively, where in the left part atomic orbitals of the type $\chi(\vec{r}) = \frac{\alpha^{3/2}}{a_0^{3/2}\sqrt{\pi}}e^{-\alpha|\vec{r}-\vec{R}|/a_0}$ with $\alpha = 1$ are used, while in the right-half α is determined so that the electronic energy is minimized. In the second and third rows of diagrams, the top curve shows the energy of the nucleus-nucleus interaction, the bottom curve the electronic energy, and the middle curve the sum of the two. In the second row of the panels, it is assumed that the binding orbital is occupied while the antibonding orbital is occupied in the third row of panels. In the bottom panels (left) S from equation (12.4), and (right) the optimized values of α, where the lower curve shows the value for the antibonding and the upper curve the value for the binding orbital. Finally, the thin vertical lines in the top diagrams show the position of the minimum of the total energy curve. In the top diagrams, a finer energy scale has been used than in the panels of the second row.

From this,

$$N_b = \frac{1}{\sqrt{2 + 2S}}. \tag{12.5}$$

We could also have produced an antibonding wave function from the 1s functions of the two hydrogen atoms,

$$\psi_a(\vec{r}) = N_a \cdot [\chi_a(\vec{r}) - \chi_b(\vec{r})]. \tag{12.6}$$

However, this leads to a reduction of the electron density between the nuclei. Furthermore, this antibonding orbital would have nodes, so that the spatial changes of the wave function in position space increase, whereby also the kinetic energy increases. If this orbital were occupied, we would have no binding interaction between the two nuclei, and the structure of the lowest total energy would be that at which both nuclei are infinitely far apart (see Figure 12.3). In this case, the normalization constant is given by

$$N_a = \frac{1}{\sqrt{2 - 2S}}.$$
(12.7)

In our case, $S > 0$ (see Figures 12.2 and 12.3), so that $N_a > N_b$. As a result, the bonding of ψ_b is weaker than the antibonding of ψ_a. We will explain this in more detail in Chapter 13 and also use it there.

Overall, we have seen through this example how we can generate molecular orbitals from atomic orbitals that can lead to a bonding or antibonding interaction between the nuclei (see Figure 12.4). Further, we see that, as with atoms, there are many orbitals for molecules of which not all may be occupied. In the ground state, the energetically lowest orbitals are occupied.

When creating the molecular orbitals, we started with the atomic orbitals of the well-separated atoms. Conversely, we could have started with the so-called united-atom limit (which is rarely done), which in our case would be the He⁺ Ion. Also for this there are atomic orbitals, which are very similar to those of the hydrogen atom, but decay faster as a function of the distance to the nucleus. So for the 1s orbital we have e^{-2r/a_0} instead of e^{-r/a_0} for the hydrogen atom (here a_0 is the Bohr radius). More generally, it is useful to use e^{-ar/a_0} and determine α so that the total energy becomes lowest. The results that would be obtained by this are also shown in Figure 12.3, and indeed this gives a lowest total energy at a H–H distance of less than 1 Å, which agrees very well with experimental results. As can also be seen in the figure, the optimized value of α at this bond length is slightly larger than 1 (rather 1.25), which is consistent with our simplified considerations that the electron experiences a potential that has contributions from both nuclei.

12.3 HeH²⁺

Certainly, HeH²⁺ is not the most important system of chemistry, but can be used to explain a couple of additional aspects.

From the atomic 1s orbitals of the He and H atoms (χ_a and χ_b), we again form a bonding and an antibonding combination,

$$\psi_b(\vec{r}) = c_{ba}\chi_a(\vec{r}) + c_{bb}\chi_b(\vec{r})$$
$$\psi_a(\vec{r}) = c_{aa}\chi_a(\vec{r}) - c_{ab}\chi_b(\vec{r}),$$
(12.8)

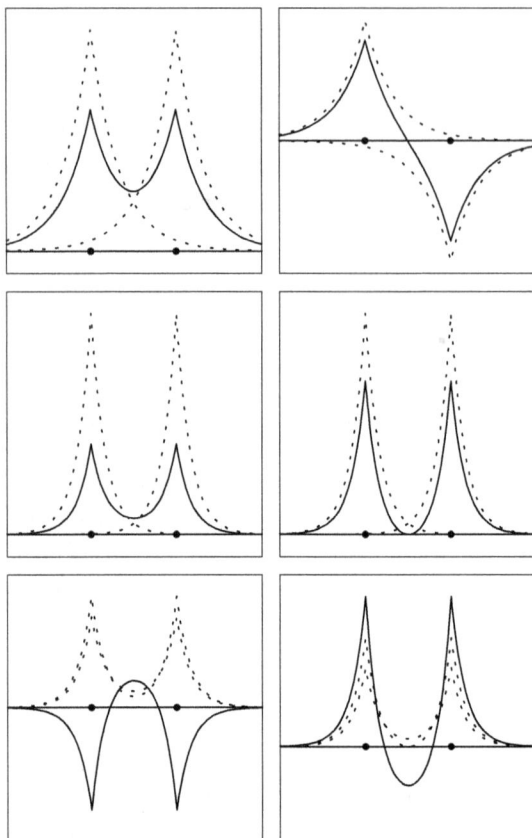

Figure 12.4: The bonding and antibonding orbitals, which can be generated from two $1s$ orbitals, χ_a and χ_b. On the left the results for the bonding (ψ_b) orbital are shown, on the right the ones for the antibonding (ψ_a) orbital. In the top panels, the wave functions (solid curves) are compared to the two atom-centered $1s$ orbitals (dotted curves), while in the middle these functions are squared (i. e., giving the electron densities). Finally, the bottom panels show the difference densities $|\psi_b|^2 - \frac{1}{2}(|\chi_a|^2 + |\chi_b|^2)$ and $|\psi_a|^2 - \frac{1}{2}(|\chi_a|^2 + |\chi_b|^2)$ (the solid curves that have been multiplied by 5) in comparison to the individual contributions $[|\psi_b|^2$, or $|\psi_a|^2$, and $\frac{1}{2}(|\chi_a|^2 + |\chi_b|^2)]$ (dotted curves).

but in this case there is no symmetry that dictates that $|c_{ba}| = |c_{bb}|$ or $|c_{aa}| = |c_{ab}|$ must be fulfilled. This means that the two orbitals are distributed asymmetrically among the two atoms. The only thing that is known is that the orbitals must be orthonormal,

$$c_{ba}^2 + c_{bb}^2 + 2Sc_{ba}c_{bb} = 1$$
$$c_{aa}^2 + c_{ab}^2 - 2Sc_{aa}c_{ab} = 1$$
$$c_{ba}c_{aa} - c_{bb}c_{ab} + S(c_{bb}c_{aa} - c_{ab}c_{ba}) = 0, \qquad (12.9)$$

when assuming that the atomic orbitals are real and positive.

But again, we have created two new molecular orbitals from the two atomic orbitals, one of which is a bonding orbital and the other an antibonding orbital. And again, the bonding orbital will have a lower orbital energy and the antibonding one a higher energy. Here, further details are not to be discussed—this will be done in Section 13.1.

12.4 H_2: the LCAO-MO picture

For the H_2 molecule, as for the H_2^+ molecular ion, we will use the Born–Oppenheimer approximation and, therefore, determine the electronic wave function for the two electrons as they move in the field of the two nuclei, that are kept at fixed positions, and of each other. For this electronic wave function, we will in this chapter use the Hartree–Fock approximation, and write accordingly

$$\Psi_e(\vec{x}_1, \vec{x}_2) = \frac{1}{\sqrt{2}} \begin{vmatrix} \psi_1(\vec{x}_1) & \psi_2(\vec{x}_1) \\ \psi_1(\vec{x}_2) & \psi_2(\vec{x}_2) \end{vmatrix}. \tag{12.10}$$

The molecular orbitals $\{\psi_i(\vec{x})\}$ are distributed more or less over the whole H_2 molecule. As in the simpler example of H_2^+, we will write them in terms of atomic orbitals:

$$\psi_i(\vec{x}) = \sum_X \sum_j \chi_{Xj} c_{Xji}. \tag{12.11}$$

Here, χ_{Xj} is the jth atomic orbital of atom X. In our case, these would be the two $1s$ orbitals centered at the two hydrogen atoms. The coefficients $\{c_{Xji}\}$ are the ones we will determine by solving the Hartree–Fock–Roothaan equations. This method, i. e., to use the Hartree–Fock approximation for the electronic wavefunction, and then to write the single one-electron wavefunctions as a linear combination of atom-centered functions, is called the LCAO-MO method. LCAO means Linear Combination of Atomic Orbital, while MO means Molecular Orbital.

For the H_2 molecule, the LCAO-MO approach becomes particularly simple. The two orbitals ψ_1 and ψ_2 will have the same spatial dependence, but different spin dependencies,

$$\psi_1(\vec{x}) = \psi(\vec{r})\alpha$$
$$\psi_2(\vec{x}) = \psi(\vec{r})\beta. \tag{12.12}$$

$\psi(\vec{r})$, on the other hand, is written as linear combinations of the $1s$ functions of the hydrogen atoms, and because of the symmetry of the molecule, the bonding and antibonding molecular orbitals obtained from equations (12.2) and (12.6) will result. Of these two, the bonding orbital leads to the lowest total energy and $\psi(\vec{r})$, therefore becomes equal to $\psi_b(\vec{r})$ of equation (12.2),

$$\psi(\vec{r}) = N_b \cdot [\chi_{A,1s}(\vec{r}) + \chi_{B,1s}(\vec{r})]. \tag{12.13}$$

A and B denote the two H atoms. To simplify the notation, we may use the same notation as for H_2^+,

$$\chi_a(\vec{r}) = \chi_{A,1s}(\vec{r})$$
$$\chi_b(\vec{r}) = \chi_{B,1s}(\vec{r}). \tag{12.14}$$

The calculation of the total energy is now more complex because we have to take into account that the two electrons interact with each other. Therefore, the total energy is not only the sum of the energies of the individual electrons in their orbitals, although this sum already makes a significant contribution.

It also turns out that the description of the total energy as a function of the distance between the two cores is relatively accurate if the distance is not too large, but for larger distances there are significant deviations (see Figure 12.5). The cause of this behavior is discussed in Section 12.6.

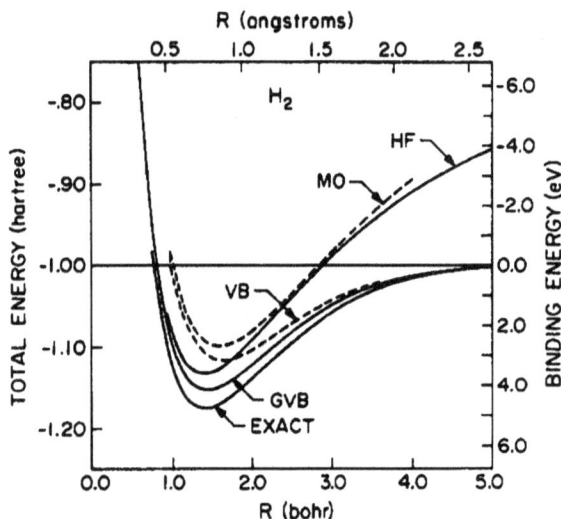

Figure 12.5: Total energy as a function of the distance between the two nuclei for the hydrogen molecule. MO and HF show results found with the Hartree–Fock approximation. The H 1s atomic orbitals were used in the construction of the molecular orbitals for the MO results, while the molecular orbitals were optimized for the HF results. Similarly, GVB and VB show results obtained with the valence bond theory using the H 1s atomic orbitals for the VB results, while the orbitals in the GVB results were optimized. This means, as in Figure 12.3, that α has been optimized in the exponential functions for the HF and GVB results, whereas this is not the case for the MO and VB results. Finally, the results marked out by "Exact" represent very accurate, numerical results. Reproduced with permission from American Chemical Society from William A. Goddard III, Thom H. Dunning, Jr., William J. Hunt and P. Jeffrey Hay: *Generalized Valence Bond Description of Bonding in Low-Lying States of Molecules*, Acc. Chem. Res. **6**, 368–376 (1973).

12.5 H$_2$: the VB picture

The wave function in equation (12.10) is an approximation to the exact electronic wave function of the H$_2$ molecule, but not necessarily the only or even the best possible one. But it is an example of the most commonly used electronic wave functions for (smaller and larger) molecules. However, here we will discuss another approximation, although it is not used so often. The approximation is the so-called valence bond model. This model is also called the Heitler–London model after Walter Heitler and Fritz London, who already in 1927 presented this as an approximate treatment of H$_2$ and thereby presented the first attempt to use quantum theory to describe a chemical bond.

In Figure 12.5, we see several curves that are supposed to approximate the total energy of the H$_2$ molecule as a function of the bond length. The exact curve is the lowest. That this is so is a consequence of the variational principle. The total energy is a sum of the electronic energy and the repulsive energy of the two nuclei. The latter is treated exactly in all methods, whereas the electronic energy is approximated, which means that the approximated electronic energy is never below the exact electronic energy. In Figure 12.5, we see that the Hartree–Fock approximation leads to the highest-lying curves—thus, it is the least accurate approximation. The other approximate curves originate from the valence bond model, which means that this model is not at all that bad for this system.

According to the valence bond model, electron pairs are occupying bonding orbitals between two atoms. Specifically, this means that for H$_2$ the electronic wave function is approximated according to

$$\Psi_e(\vec{x}_1, \vec{x}_2) = \frac{1}{\sqrt{2 + 2S^2}} \left[\chi_a(\vec{r}_1)\chi_b(\vec{r}_2) + \chi_b(\vec{r}_1)\chi_a(\vec{r}_2) \right] \cdot \frac{1}{\sqrt{2}} \left[\alpha(1)\beta(2) - \beta(1)\alpha(2) \right]. \quad (12.15)$$

As above,

$$S = \langle \chi_a | \chi_b \rangle. \quad (12.16)$$

As we will see below, this wave function can not be expressed as a single Slater determinant. Therefore, it does not match the Hartree–Fock approximation, or the orbital picture. Furthermore, the VB image is less accurate when describing bonds that contain a larger ionic (rather than covalent) character.

12.6 H$_2$: correlation

The orbital model corresponds to employing the Hartree–Fock approximation. That is, we solve the Hartree–Fock equations (or rather the Hartree–Fock–Roothaan equations) assuming that the N energetic lowest orbitals are occupied by electrons (N is the total number of electrons). The approximate electronic wave function is thereby written as a single Slater determinant, and this Slater determinant contains the N energetically lowest orbitals.

In principle, we would not necessarily have to distribute the N electrons among the N energetically lowest orbitals, but could also form Slater determinants from any set of N orbitals. Each of these Slater determinants represents a so-called configuration. The Slater determinant with the N energetically lowest orbitals is then called the ground-state electron configuration. This means that the orbital model, or the Hartree–Fock approximation, can also be referred to as a single-configuration approximation.

The Hartree–Fock approximation, or the single-configuration approximation, or the orbital model is, like any approximation, rarely exact. Often the errors are acceptably small, but there are also situations where this is not the case. We have seen this a few times: for the singlet and triplet states of the excited He atom, as well as in the Hartree–Fock approximation for the H_2 molecule for larger interatomic distances.

The difference between the exact wave function and the best possible Hartree–Fock wave function (that which one obtains when solving the Hartree–Fock equations as exactly as possible) is called correlation. The fact that correlation can be important is also seen in Figure 12.5 above all for larger interatomic distances. We will use this molecule here to discuss correlation a little further.

The fact that there are correlation effects in the general case can also be understood in the following way. When ignoring correlation effects, the Hartree–Fock equations are solved. This method determines the wave functions of the individual orbitals. For each electron, one considers the potential created by the electrons in the other orbitals. This means that each electron is experiencing an average potential from the other electrons. The effect that the electrons actually repel each other (because of the same charge), and therefore at any point in time try to avoid each other, is not taken into account here. In reality, the electrons do not move independently, but the movements of the electrons are correlated: when one electron comes near another electron, the other electron tries to avoid it.

These effects can be described by an improved electronic wave function. In doing so, one will have to leave the one-configuration approximation, and accordingly write the electronic wave function by means of more Slater determinants (or configurations). If this is the case, it is no longer possible to identify orbitals of the individual electrons: the orbital picture fails.

This problem will now be discussed with the help of the H_2 molecule; cf. Figure 12.5. With the Hartree–Fock approximation, the electronic wave function becomes

$$\Psi_e(\vec{x}_1, \vec{x}_2) = \frac{1}{\sqrt{2}}[\psi_b(\vec{r}_1)\alpha(1)\psi_b(\vec{r}_2)\beta(2) - \psi_b(\vec{r}_2)\alpha(2)\psi_b(\vec{r}_1)\beta(1)]$$

$$= \frac{1}{\sqrt{2}}\psi_b(\vec{r}_1)\psi_b(\vec{r}_2)[\alpha(1)\beta(2) - \alpha(2)\beta(1)]. \tag{12.17}$$

ψ_b is given by equation (12.13),

$$\psi_b = (2 + 2S)^{-1/2}(\chi_a + \chi_b).$$ (12.18)

By inserting, we get from equation (12.17),

$$\begin{aligned}
\Psi_e &= \frac{1}{2 + 2S}[\chi_a(\vec{r}_1) + \chi_b(\vec{r}_1)][\chi_a(\vec{r}_2) + \chi_b(\vec{r}_2)]\frac{1}{\sqrt{2}}[\alpha(1)\beta(2) - \alpha(2)\beta(1)] \\
&= \frac{1}{2 + 2S}[\chi_a(\vec{r}_1)\chi_a(\vec{r}_2) + \chi_b(\vec{r}_1)\chi_b(\vec{r}_2) + \chi_b(\vec{r}_1)\chi_a(\vec{r}_2) + \chi_a(\vec{r}_1)\chi_b(\vec{r}_2)] \\
&\quad \cdot \frac{1}{\sqrt{2}}[\alpha(1)\beta(2) - \alpha(2)\beta(1)] \\
&= \frac{1}{2 + 2S}\{[\chi_a(\vec{r}_1)\chi_a(\vec{r}_2) + \chi_b(\vec{r}_1)\chi_b(\vec{r}_2)] + [\chi_b(\vec{r}_1)\chi_a(\vec{r}_2) + \chi_a(\vec{r}_1)\chi_b(\vec{r}_2)]\} \\
&\quad \cdot \frac{1}{\sqrt{2}}[\alpha(1)\beta(2) - \alpha(2)\beta(1)].
\end{aligned}$$ (12.19)

The first term inside the curly brackets describes an electron distribution in which both electrons are located on the same atom, and the second term describes an electron distribution with one electron on each atom. Thus, the first term corresponds to H$^+$-H$^-$ and H$^-$-H$^+$ situations and is therefore called an ionic term. The other term corresponds to H-H situations and is therefore called a covalent term.

This separation is completely independent of the distance R between the two atomic nuclei. For small distances, the result may be useful, but for larger distances one would intuitively expect that the system consists of two isolated, neutral hydrogen atoms, and that therefore the description by means of the Hartree–Fock approximation or the orbital picture cannot be very accurate. This then explains why the Hartree–Fock results in Figure 12.5 are not so accurate.

It is possible to modify the wave function in equation (12.19) so that the two terms are weighted differently (and become R-dependent). This can be done, e. g., by using the wave function

$$\Psi_e'' = C\{c[\chi_a(\vec{r}_1)\chi_a(\vec{r}_2) + \chi_b(\vec{r}_1)\chi_b(\vec{r}_2)] + [\chi_b(\vec{r}_1)\chi_a(\vec{r}_2) + \chi_a(\vec{r}_1)\chi_b(\vec{r}_2)]\} \\ \cdot [\alpha(1)\beta(2) - \alpha(2)\beta(1)].$$ (12.20)

With $c = 0$, we have no ionic contributions, while $c = 1$ results in the wave function of equation (12.19). With the wave function of equation (12.20), the parameter c may depend on R. Furthermore, C is a constant which ensures that the wave function is normalized.

Alternatively, instead of the bonding orbital, ψ_b from equation (12.18), we might use the antibonding orbital,

$$\psi_a = (2 - 2S)^{-1/2}(\chi_a - \chi_b)$$ (12.21)

and from this create a Slater determinant for an excited configuration

$$
\begin{aligned}
\Psi'_e &= \frac{1}{2-2S}[\chi_a(\vec{r}_1) - \chi_b(\vec{r}_1)][\chi_a(\vec{r}_2) - \chi_b(\vec{r}_2)]\frac{1}{\sqrt{2}}[\alpha(1)\beta(2) - \alpha(2)\beta(1)] \\
&= \frac{1}{2-2S}[\chi_a(\vec{r}_1)\chi_a(\vec{r}_2) + \chi_b(\vec{r}_1)\chi_b(\vec{r}_2) - \chi_b(\vec{r}_1)\chi_a(\vec{r}_2) - \chi_a(\vec{r}_1)\chi_b(2)] \\
&\quad \cdot \frac{1}{\sqrt{2}}[\alpha(1)\beta(2) - \alpha(2)\beta(1)] \\
&= \frac{1}{2-2S}\{[\chi_a(\vec{r}_1)\chi_a(\vec{r}_2) + \chi_b(\vec{r}_1)\chi_b(\vec{r}_2)] - [\chi_b(\vec{r}_1)\chi_a(\vec{r}_2) + \chi_a(\vec{r}_1)\chi_b(\vec{r}_2)]\} \\
&\quad \cdot \frac{1}{\sqrt{2}}[\alpha(1)\beta(2) - \alpha(2)\beta(1)].
\end{aligned}
\tag{12.22}
$$

By comparing with equation (12.19), we see that the wave function in equation (12.20) can also be written with the help of the two Slater determinants:

$$
\begin{aligned}
\Psi''_e &= c_1\Psi_e + c_2\Psi'_e \\
&= c_1\frac{1}{\sqrt{2}}\begin{vmatrix} \psi_b(\vec{r}_1)\alpha(1) & \psi_b(\vec{r}_1)\beta(1) \\ \psi_b(\vec{r}_2)\alpha(2) & \psi_b(\vec{r}_2)\beta(2) \end{vmatrix} + c_2\frac{1}{\sqrt{2}}\begin{vmatrix} \psi_a(\vec{r}_1)\alpha(1) & \psi_a(\vec{r}_1)\beta(1) \\ \psi_a(\vec{r}_2)\alpha(2) & \psi_a(\vec{r}_2)\beta(2) \end{vmatrix},
\end{aligned}
\tag{12.23}
$$

which can give a relation between c_1 and c_2 on the one hand and c and C of equation (12.20) on the other hand,

$$
\begin{aligned}
C \cdot c &= \frac{c_1}{2+2S} + \frac{c_2}{2-2S} \\
C &= \frac{c_1}{2+2S} - \frac{c_2}{2-2S}.
\end{aligned}
\tag{12.24}
$$

Equation (12.23) immediately shows that Ψ''_e is written as the sum of two configurations. Or: that the Hartree–Fock approximation is no longer employed, which means that the orbital picture must be abandoned. Furthermore, we recognize that the electronic wave function of the valence bond model, Equation (12.15), can also be written as the sum of two Slater determinants by choosing $c = 0$.

12.7 H_2: energy contributions

Finally, we will discuss the different contributions to the total energy of the H_2 molecule at the equilibrium distance. For this, we refer to Figure 12.6.

In the ground state, the lowest total energy of the molecule according to the Born–Oppenheimer approximation is −31.957 eV. But as we discussed in Chapter 6, the atomic nuclei are not at fixed positions, but have a zero-point energy. We determine this by approximating the total energy curve as a function of bond length through a harmonic potential. Using the methods of Chapter 6, we then find that the zero point energy equals 0.271 eV.

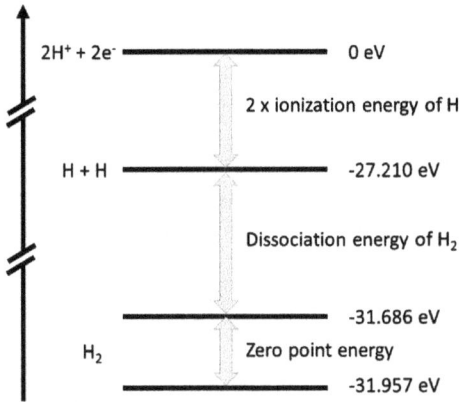

Figure 12.6: The different energy contributions to the total energy of the H_2 molecule.

To bring the two atoms infinitely far apart, we have to provide an additional 4.476 eV. This corresponds to the binding energy and thus takes into account the zero point energy. The two isolated, noninteracting hydrogen atoms then have a total energy of -27.210 eV. This energy is relative to the zero of the potential (i. e., for particles at rest infinitely far away from the system of interest) at which the two electrons of the two atoms are infinitely distant from the nuclei. The energy zero then corresponds to $H^+ + H^+ + e^- + e^-$ in our case.

12.8 Problems

1. Construct the ψ_b and ψ_a functions for H_2^+. Why are the absolute values of the coefficients the same, $|c_a| = |c_b|$? Sketch these functions. What function describes the bonding orbital? Which is the antibonding orbital? Justify the answers.
2. Compare the molecular orbital description and the valence bond model for H_2.
3. When, how, and why does the LCAO-MO description for H_2 fail?
4. Sketch the total energy as a function of the interatomic distance for H_2^+, if
   (a) the bonding orbital is occupied, or if
   (b) the antibonding orbital is occupied.
5. Explain the term LCAO-MO.
6. Explain the role of correlation in describing the energy of the H_2 system as a function of H–H distance.

13 Other diatomic molecules

13.1 A simple model for an AB molecule

In the last chapter, Chapter 12, we discussed three examples, the H_2^+ molecular ion, the H_2 molecule, and the rather exotic HeH^{2+} molecular ion, with the special emphasis on how we can understand the formation of chemical bonds within the orbital picture. Although the orbital picture is an approximation, it can be used to rationalize the bonding properties of very many compounds, including the changes that occur in chemical reactions.

According to the orbital picture, chemical bonds are formed by joining two fragments (e. g., atoms), and then the orbitals of the individual fragments form bonding and antibonding molecular orbitals. Because this is so fundamental, here we will discuss the mathematical details for a simple model, and we will thereby also learn more about other properties of the molecular orbitals.

The model consists of a diatomic AB molecule, for which it can be assumed that we can describe the molecular orbitals as linear combinations of only two atomic orbitals,

$$\psi_i(\vec{r}) = c_{A,i}\chi_A(\vec{r}) + c_{B,i}\chi_B(\vec{r}). \tag{13.1}$$

The index i describes the (two) different molecular orbitals, while χ_A and χ_B are the atomic orbitals on atom A and atom B, respectively. We mention that we will not consider spin effects here

We assume that we have solved the Hartree–Fock–Roothaan equations for this system,

$$\begin{pmatrix} F_{AA} & F_{AB} \\ F_{BA} & F_{BB} \end{pmatrix} \cdot \begin{pmatrix} c_{A,i} \\ c_{B,i} \end{pmatrix} = \epsilon_i \cdot \begin{pmatrix} S_{AA} & S_{AB} \\ S_{BA} & S_{BB} \end{pmatrix} \cdot \begin{pmatrix} c_{A,i} \\ c_{B,i} \end{pmatrix}. \tag{13.2}$$

Here,

$$F_{XY} = \langle \chi_X | \hat{F} | \chi_Y \rangle$$
$$S_{XY} = \langle \chi_X | \chi_Y \rangle \tag{13.3}$$

are the elements of the Fock and overlap matrix, respectively, and $\hat{F}$ is the Fock operator. X and Y are A and/or B.

The Hartree–Fock–Roothaan equations (13.2) can be written as a set of two coupled, linear equations:

$$\begin{pmatrix} F_{AA} - \epsilon_i S_{AA} & F_{AB} - \epsilon_i S_{AB} \\ F_{BA} - \epsilon_i S_{BA} & F_{BB} - \epsilon_i S_{BB} \end{pmatrix} \cdot \begin{pmatrix} c_{A,i} \\ c_{B,i} \end{pmatrix} = \begin{pmatrix} 0 \\ 0 \end{pmatrix}. \tag{13.4}$$

Linear equations with only 0 on the right-hand side are called homogeneous equations, and they always have the trivial solution

$$c_{A,i} = c_{B,i} = 0. \tag{13.5}$$

https://doi.org/10.1515/9783110742206-013

But here this solution is not of interest or relevance. It corresponds to the case that the molecular orbitals disappear.

Instead, we investigate if it is possible to identify other solutions. We recognize that the orbital energy ϵ_i is not fixed either, and we therefore ask ourselves if there are cases (i. e., values of ϵ_i) for which equation (13.4) has other solutions than the trivial one. This is the case if the two equations are linearly dependent (see the discussion in Section 9.5), i. e., if

$$\begin{vmatrix} F_{AA} - \epsilon_i S_{AA} & F_{AB} - \epsilon_i S_{AB} \\ F_{BA} - \epsilon_i S_{BA} & F_{BB} - \epsilon_i S_{BB} \end{vmatrix} = 0, \tag{13.6}$$

or, in other words,

$$\begin{aligned} 0 &= (F_{AA} - \epsilon_i S_{AA})(F_{BB} - \epsilon_i S_{BB}) - (F_{AB} - \epsilon_i S_{AB})(F_{BA} - \epsilon_i S_{BA}) \\ &= (S_{AA}S_{BB} - S_{AB}S_{BA})\epsilon_i^2 + (-F_{AA}S_{BB} - F_{BB}S_{AA} + F_{AB}S_{BA} + F_{BA}S_{AB})\epsilon_i \\ &\quad + (F_{AA}F_{BB} - F_{AB}F_{BA}) \equiv A\epsilon_i^2 + B\epsilon_i + C. \end{aligned} \tag{13.7}$$

It is well known that this equation has two solutions,

$$\epsilon_i = -\frac{B}{2A} \pm \sqrt{\frac{B^2}{4A^2} - \frac{C}{A}}. \tag{13.8}$$

We will now assume that the atomic orbitals are real and normalized. Thus,

$$S_{AA} = S_{BB} = 1$$
$$S_{AB} = S_{BA} = S$$
$$F_{AB} = F_{BA} \tag{13.9}$$

with S equal to a real number. When the atomic orbitals are real, it is usually true that $F_{AB} = F_{BA}$ and $S_{AB} = S_{BA}$ have opposite signs. For example, for the 1s orbitals of the H atoms in the H_2 molecule

$$S_{AB} = S_{BA} > 0$$
$$F_{AB} = F_{BA} < 0. \tag{13.10}$$

In equation (13.7), we obtain by using equation (13.9),

$$A = 1 - S^2$$
$$B = -(F_{AA} + F_{BB} - 2SF_{AB})$$
$$C = F_{AA}F_{BB} - F_{AB}^2. \tag{13.11}$$

Inserting this in equation (13.8), we obtain

$$\begin{aligned} \epsilon_i &= \frac{1}{2 - 2S^2}\{F_{AA} + F_{BB} - 2SF_{AB} \\ &\quad \pm [(F_{AA} + F_{BB} - 2SF_{AB})^2 - 4(1 - S^2)(F_{AA}F_{BB} - F_{AB}^2)]^{1/2}\}. \end{aligned} \tag{13.12}$$

To obtain the corresponding wave functions, we insert the orbital energies from equation (13.12) into one of the equations in equation (13.4). We know that with the values of ϵ_i from equation (13.12) the two equations are identical, and we therefore only have to consider one of them. We then get, e. g.,

$$(F_{AA} - \epsilon_i)c_{A,i} + (F_{AB} - \epsilon_i S)c_{B,i} = 0, \tag{13.13}$$

or

$$c_{B,i} = -\frac{\epsilon_i - F_{AA}}{S\epsilon_i - F_{AB}}c_{A,i}. \tag{13.14}$$

Because the orbital must be normalized, we also have (assuming that the coefficients $c_{A,i}$ and $c_{B,i}$ are real)

$$1 = c_{A,i}^2 + c_{B,i}^2 + 2Sc_{A,i}c_{B,i} = c_{A,i}^2\left(1 + \frac{(\epsilon_i - F_{AA})^2}{(S\epsilon_i - F_{AB})^2} - 2S\frac{\epsilon_i - F_{AA}}{S\epsilon_i - F_{AB}}\right). \tag{13.15}$$

From this, first $c_{A,i}$ and then $c_{B,i}$ with the help of equation (13.14) can be determined.

Through a detailed analysis of the mathematical results the following findings are observed:

- The energies F_{AA} and F_{BB} are not the orbital energies of the atomic orbitals (ϵ_A and ϵ_B) for the noninteracting atoms. This is because $\hat{F}$ contains the potential of both nuclei and of all electrons. As a first (often good) approximation you can, however, ignore this difference.
- Depending on the system, $\epsilon_A - F_{AA}$ (or $\epsilon_B - F_{BB}$) can be both positive and negative. So, general statements about the relative ordering of ϵ_A and F_{AA} (or ϵ_B and F_{BB}) are not possible.
- The two orbital energies in equation (13.12) are not symmetrically placed around F_{AA} and F_{BB}. That would be the case if $S = 0$.
- In fact, however, for $S \neq 0$, the higher orbital energy is more distant from $\frac{1}{2}(F_{AA} + F_{BB})$ than the lower orbital energy. In other words, this means that the antibonding orbital is more antibonding than the bonding orbital is bonding.
- For example, suppose that $F_{AA} < F_{BB}$, then the bonding orbital (with the lowest orbital energy) has the larger contributions from χ_A and the smaller ones from χ_B (i. e., for this orbital, $|c_{A,i}| > |c_{B,i}|$). The reverse is true for the antibonding orbital.
- The energy of the bonding orbital is lower than the smallest value of F_{AA} and F_{BB}, while the energy of the antibonding orbital is higher than the highest value of F_{AA} and F_{BB}.
- If we assume that

$$F_{AA} = F_{BB}, \tag{13.16}$$

(i. e., we have a homonuclear, diatomic molecule), the formulas become particularly simple. Then

$$\epsilon_i = \frac{F_{AA} - SF_{AB}}{1 - S^2} \pm \frac{1}{1 - S^2}[(F_{AA} - SF_{AB})^2 - (1 - S^2)(F_{AA}^2 - F_{AB}^2)]^{1/2}$$

$$= \frac{F_{AA} - SF_{AB}}{1 - S^2} \pm \frac{F_{AB} - SF_{AA}}{1 - S^2} = \frac{F_{AA} \pm SF_{AB}}{1 \pm S}. \tag{13.17}$$

From this one can clearly see that the orbital energies are not symmetric about $F_{AA} = F_{BB}$.

- In the case of equation (13.16), it is easy to show that

$$|c_{A,i}| = |c_{B,i}|. \tag{13.18}$$

- On the other hand, if

$$S = F_{AB} = 0, \tag{13.19}$$

the two molecular orbitals become identical to the atomic orbitals.
- In this case, however, as mentioned above, the energies F_{AA} and F_{BB} are not the orbital energies of the isolated atoms. Instead, F_{AA}, for example, is the energy that an electron would have if it were in the orbital χ_A, but would experience the potential of all nuclei and electrons.
- Energetically low orbitals may be due to low atomic orbital energies (i. e., F_{AA} and/or F_{BB} is low), and/or that the atomic orbitals have strong interactions (i. e., $|F_{AB}|$ and $|S|$ are large).
- Similarly, energetically high orbitals are obtained from high-energy atomic orbitals (i. e., F_{AA} and/or F_{BB} is high), and/or that the atomic orbitals have strong interactions (i. e., $|F_{AB}|$ and $|S|$ are large)
- With reference to Figure 13.1, $\Delta_1 \leq \Delta_2$ in the general case.

Some of these aspects are shown qualitatively and semiquantitatively in Figure 13.1. Further aspects can be seen in Figure 13.2. The left part of Figure 13.2 shows the orbitals and their energies for the molecular orbitals formed from 1s orbitals on two H atoms. The asymmetry of the energies of the bonding and antibonding orbitals can be identified, as well as the symmetric spatial distribution of the orbitals on the two atoms. In the right part, an example is shown where two different atomic orbitals interact with each other. As an example, we consider the HF molecule placed along the z axis. For this, a $2p_z$ function on the F atom is shown together with an 1s orbital on the H atom. Here, one can see how the energetic lowest (bonding) orbital has the largest contributions from the atomic F $2p_z$ orbital, while the energetic highest (antibonding) orbital has the largest contributions from the atomic H 1s orbital.

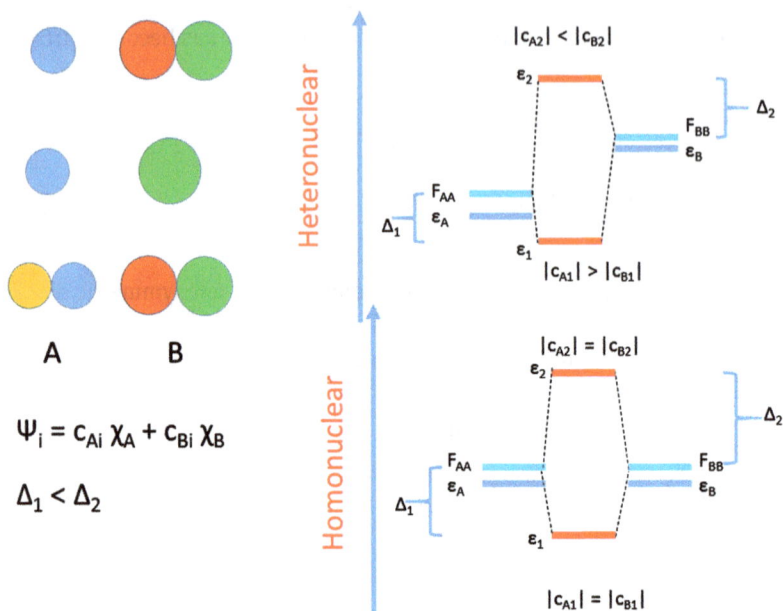

$$\Psi_i = c_{Ai}\, \chi_A + c_{Bi}\, \chi_B$$

$$\Delta_1 < \Delta_2$$

Figure 13.1: Schematic presentation of the results of Section 13.1. In the top left corner are shown three examples of pairs of atom-centered orbitals that can interact with each other: (s,p), (s,s), and (p,p).

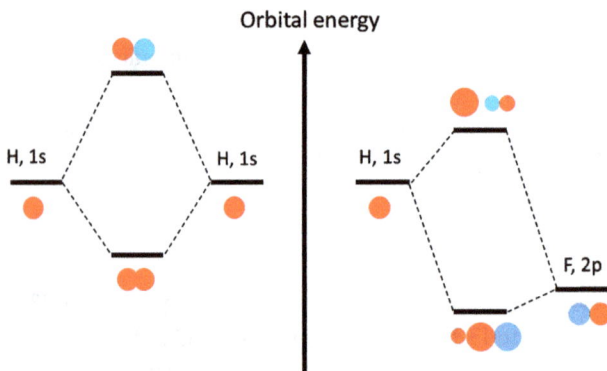

Figure 13.2: Orbital picture for (left) H_2 and (right) HF.

The interactions (i. e., F_{AB} and S) of the two atomic orbitals depend both on the distance between the two atoms and on the type of the two atomic orbitals. This is shown in Figure 13.3 for the example of the overlap between s and p functions. Finally, there are cases where symmetry causes there to be no interaction between the two atomic orbitals. The example in the lowest part of Figure 13.4 illustrates this.

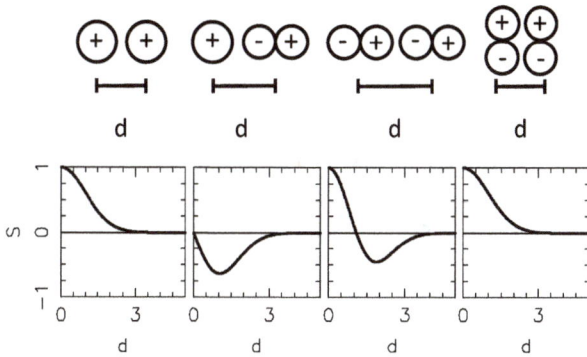

Figure 13.3: The variation of the overlap S between two s and/or p atomic orbitals as a function of the distance d between the two atoms for four different cases. From left to right, the examples show (s, s), (s, p_z), (p_z, p_z), (p_y, p_y), where it is assumed that the z axis forms the molecular axis.

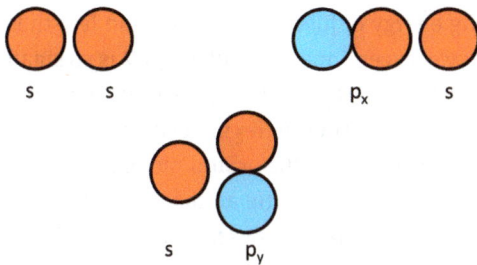

Figure 13.4: Schematic representation of the interaction between two atomic orbitals for three different cases. In the lowest case, the interactions disappear because of different symmetry properties of the two atomic orbitals.

13.2 He₂

We can now use the results from the previous subchapter to explain the differences between H_2 and He_2. From the two 1s orbitals, we can form a bonding and an antibonding combination. But for He_2, unlike H_2, we must fill both orbitals, each with two electrons (see Figure 13.5). In our arguments here, we shall neglect the differences between F_{AA} and ϵ_A. The bonding combination is then always lower in energy than each of the two atomic orbitals, while the antibonding combination is higher. Because the energy of the antibonding combination differs more from the energy of the atomic orbitals than the energy of the bonding combination does [this can be seen from the discussion above, including equation (13.17), if we set $\epsilon_A = F_{AA}$ and $\epsilon_B = F_{BB}$ with the notation from Section 13.1], the total interaction between the two atoms in He_2 will be slightly more antibonding than bonding: The molecule is unstable, and the two atoms will separate.

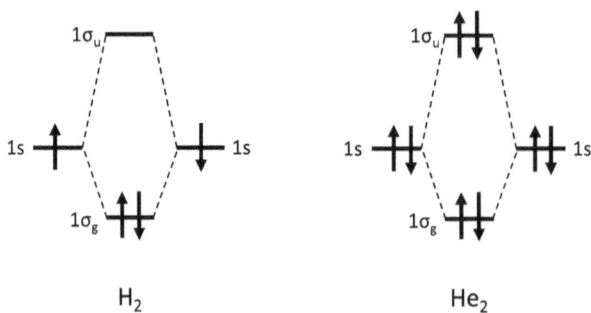

Figure 13.5: Orbital picture for (left) H_2 and (right) He_2.

From these results, we obtain that He_2 does not exist, which is largely true. But there are a few additional comments that are not completely unimportant.

First, He_2 is not completely unstable. In fact, two He atoms bind to each other with a bond length of about 52 Å and a binding energy of just $0.1\,\mu eV$. The bond length is exceptionally large and the binding energy is exceptionally small. $0.1\,\mu eV$ is so small that the two atoms separate at room temperature, so He_2 is only stable at very low temperatures (below 1.3 mK). But in general, the stability of the He_2 molecule is in contradiction to what we have learnt so far. The reason is that the orbital picture is unable to describe the small binding interaction between the two He atoms. Ultimately, the orbital picture, or the Hartree–Fock approximation, fails to describe this weak interaction.

Another way to stabilize He_2 is to excite one electron from the antibonding orbital that we have formed from the atomic 1s orbitals into a bonding orbital from the atomic 2s orbitals. Thus, the He–He interaction is binding, although the molecule is excited. Eventually, the excited electron falls back into the antibonding orbital, and the molecule breaks. Such a system, which is stable only in the excited state, is called an excimer (**Exci**ted Di**mer**).

13.3 HeH

In Section 12.3, we discussed the system HeH^{2+}. As a continuation of this discussion we will briefly discuss HeH. In this case, we have three electrons, and in the simplest case, two of them occupy a bonding orbital, while only one occupies an antibonding orbital. Also, in the simplest case, both orbitals are formed from the 1s orbitals of H and He, with the bonding orbital having a greater contribution from the He 1s function, and the antibonding orbital having a greater contribution from the H 1s function. Because only one electron is in the antibonding orbital, while two electrons occupy the bonding orbital, one might think that HeH is stable. In fact, HeH is unstable in the ground state, which means that the lowest total energy in the ground state is found for an infinitely large distance between the two nuclei. On the other hand, there are electronically excited states of the HeH molecule with lowest energy for finite distances between the nuclei; so also for HeH the orbital picture has problems to describe the bonding properties correctly.

13.4 Other diatomic molecules

Not just for diatomic molecules, but in the general case we proceed as we have demonstrated for H_2, He_2, and HeH. Thus, the molecular orbitals are written as linear combinations of atomic centered orbitals (LCAO),

$$\psi_k(\vec{x}) = \sum_{\vec{R},(lm),\alpha} c_{k,\vec{R},(lm),\alpha} \chi_{\vec{R},(lm),\alpha}(\vec{x}). \tag{13.20}$$

Here, $\vec{R}$ describes the position of the atom at which the function is centered, (lm) the angular dependence of the function, and α other dependencies (e. g., principal quantum number, spatial range of the function, and spin dependence). For example, with the help of computer calculations one can determine the coefficients $c_{k,\vec{R},(lm),\alpha}$.

For larger molecules, the orbitals are rarely so localized that they can be interpreted as bonding or antibonding orbitals between two neighboring atoms.

For diatomic homonuclear molecules, all orbitals share the same contribution from both atoms, and there are always pairs of bonding and antibonding orbitals. In that case, both atoms have the same number of electrons, and the bond is purely covalent. Most of this applies also to diatomic, heteroatomic molecules with the difference that the individual orbitals no longer have the same contribution from the two atoms. This further implies that we have unequally many electrons on the two atoms, and the bond becomes at least partially ionic.

All atomic orbitals that contribute to the same molecular orbital must have the same symmetry properties (this is a consequence of the mathematical discipline called group theory, that shall not be discussed further here), which means that a molecular orbital can be classified according to its symmetry properties. For the two-atomic molecules, the molecular orbitals can be separated into those that are fully rotationally symmetric about the molecular axis (i. e., σ orbitals), those that have one nodal plane that contains the molecular axis (π orbitals), those that have two nodal planes that contain the molecular axis (δ orbitals), etc. It then holds that atomic s orbitals can form only σ orbitals, whereas atomic p orbitals can participate in the formation of σ orbitals as well as in the formation of π orbitals. Atomic d orbitals can participate in the formation of molecular σ, π, and δ orbitals.

Homonuclear, diatomic molecules possess another element of symmetry, namely inversion at a center, which lies in the middle between the two atomic nuclei. The orbitals can then be separated into g and u orbitals. g (even) orbitals are those that are symmetric with respect to inversion, whereas u (odd) orbitals are antisymmetric.

Figure 13.6 shows examples of such orbitals for the H_2^+ molecular ion. Here, it should be emphasized, what holds true for all systems: there are infinitely many orbitals for an atom, molecule, ..., even if not all are occupied.

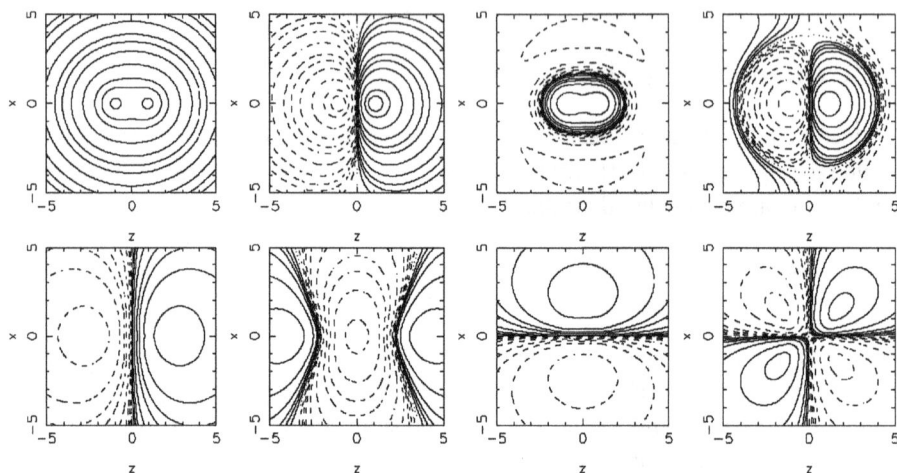

Figure 13.6: Different orbitals for the H_2^+ molecular ion. The two nuclei are placed at $z = \pm 1$ a. u. and the orbitals are $1\sigma_g$, $1\sigma_u$, $2\sigma_g$, $2\sigma_u$ in the upper row, and $3\sigma_g$, $3\sigma_u$, $1\pi_g$, $1\pi_u$ in the lower row. Solid and dashed curves indicate positive and negative values, while zero is indicated by dotted curves.

Figure 13.7 shows a different representation of the orbitals of the H_2^+ molecular ion, and here one can clearly see the differences between bonding and antibonding orbitals. The orbitals are drawn along a diagonal in Figure 13.6. Except for σ orbitals, it is not useful to draw the orbitals along the bond axis, because $\pi, \delta, \ldots$ orbital become identically zero there. Therefore, they are drawn along another straight line in Figure 13.7.

Finally, the orbitals can be separated into bonding and antibonding. The latter are indicated by an asterisk, $*$.

After having determined all (relevant) orbitals of a molecule, diagrams such as those in Figure 13.8 for N_2, Figure 13.9 for F_2, and Figure 13.10 for HF can be constructed. The occupation of the orbitals is obtained in the same way as for the isolated atoms. That is, the orbitals are filled from below with increasing energy, and never more than one electron (considering the spin) is placed in one orbital. For energetically degenerate orbitals, spatially distinct orbitals are first occupied by electrons with parallel spins (Hund's rules).

For the whole series $Li_2 - F_2$ one can then occupy the orbitals as shown schematically in Figure 13.11. There are two special features that deserve to be emphasized. First, it can be seen how the relative order of the orbitals changes (between N_2 and O_2). Furthermore, for both B_2 and O_2 π orbitals are only half-filled with electrons with

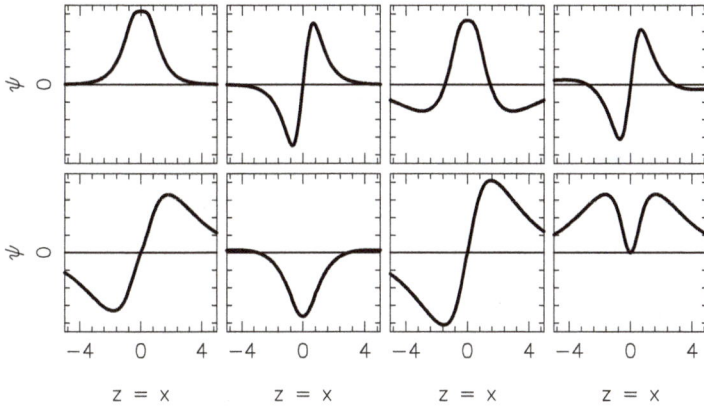

Figure 13.7: As Figure 13.6 but along the line $z = x$. Among other things, this illustrates the separation into g and u orbitals.

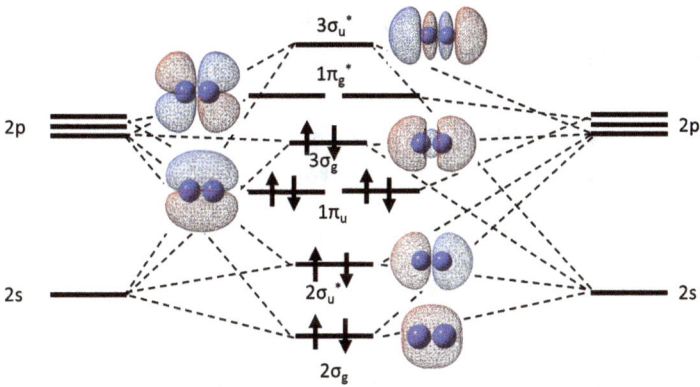

Figure 13.8: Orbital picture for N_2.

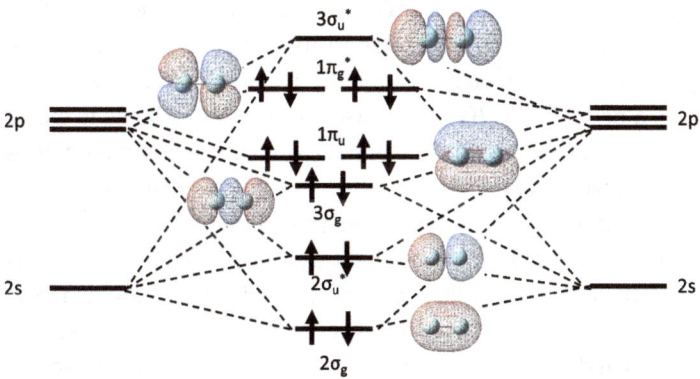

Figure 13.9: Orbital picture for F_2.

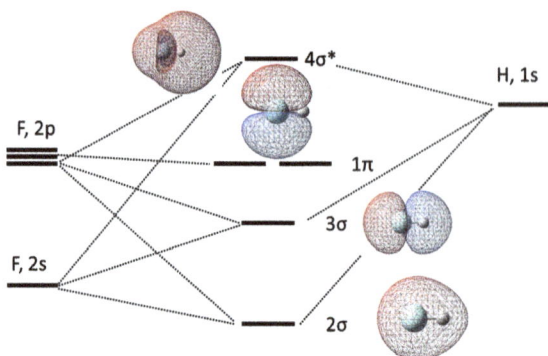

Figure 13.10: Orbital picture for HF.

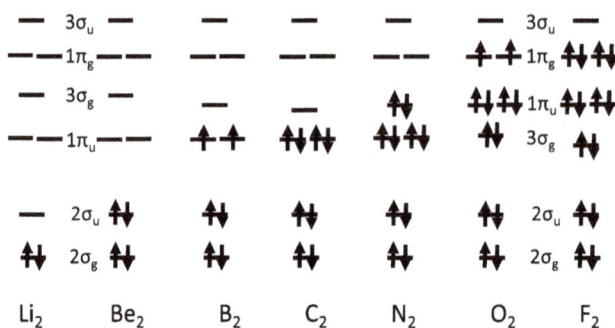

Figure 13.11: Schematic orbital diagram for Li_2, Be_2, B_2, C_2, N_2, O_2, and F_2.

parallel spins. This creates triplet states. That O_2 has a triplet state should be well known.

The first observation can be explained as follows. For the sake of simplicity, let us assume that the z axis is the molecular axis. Then we can separate the orbitals according to how they behave with respect to the rotation about the axis of the molecule. The atomic s and p_z orbitals are completely rotationally symmetric, while the atomic p_x and p_y orbitals have a nodal plane. If we then use these atomic orbitals to create the molecular orbitals, the molecular orbitals will have similar symmetry properties and each molecular orbital will have the same symmetry properties. Completely rotationally symmetric σ orbitals are then generated from s and p_z atomic orbitals, while π orbitals with one nodal plane are formed by atomic p_x and p_y orbitals. If we focus on the valence orbitals, we have two σ orbitals per atom (s and p_z) and can then form a total of four molecular orbitals with σ symmetry. These have different contributions from the s and p_z orbitals and, therefore, their energy as a function of the molecule changes in a different way than what is the case for the two energetically degenerate π orbitals. These are formed from p_x and p_y atomic orbitals and have no contributions from the s atomic orbitals. This difference can be clearly seen in Figures 13.8 and 13.9 for the $3\sigma_g$ orbitals.

While Figure 13.11 represents a schematic representation, in Figure 13.12 we show the calculated energies of the occupied orbitals for some of the molecules of Figure 13.11. The fact that the energies of the occupied (spin-up) and unoccupied (spin-down) π orbitals are identical for B_2 and O_2 is an approximation: because we have a different number of spin-up and spin-down electrons, exchange interactions could make the orbital energies different (see Section 10.8).

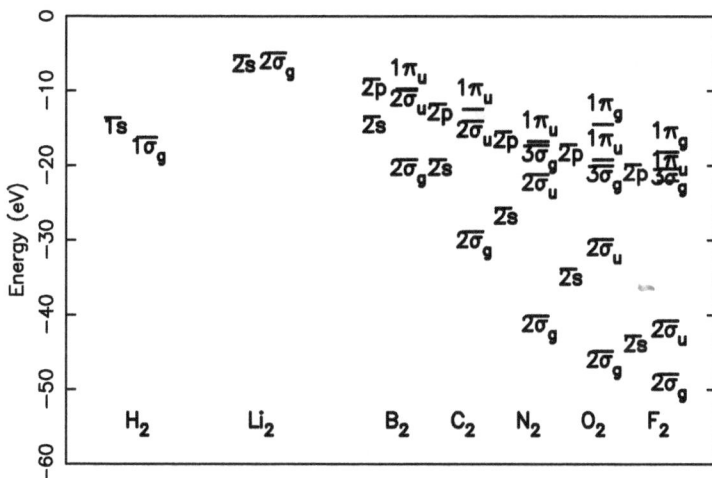

Figure 13.12: Energies of the occupied valence orbitals for H_2, Li_2, B_2, C_2, N_2, O_2, and F_2 in comparison to the energies of the atomic orbitals of the isolated atoms. For each element, the horizontal lines to the left mark the energies for the isolated atoms, whereas those to the right mark those for the diatomic molecules.

If one calculates in Figure 13.11 how many bonding orbitals are doubly occupied, and subtracts from this the number of doubly occupied antibonding orbitals, one obtains the so-called bond order. This is shown in Figure 13.13. Here, one sees how a large bond order correlates at least reasonably well with a small bond length, a high binding energy, and a high vibrational frequency.

For some atoms, molecular orbitals are created that consist of functions with several different (lm) from the same atom. These so-called hybrid orbitals are very important for generating directional bonds in molecules. In Figures 13.14 and 13.15, we show what lies behind this concept. We have seen above that the formation of an increased electron density between the atoms leads to a particularly strong bond. We consider the atom A in the upper part of Figure 13.14, on which an s and a p function are centered. If we try to create a bond to atom B, the s function is not optimal for increasing the electron density between the two atoms, even if the energy of this orbital is lower

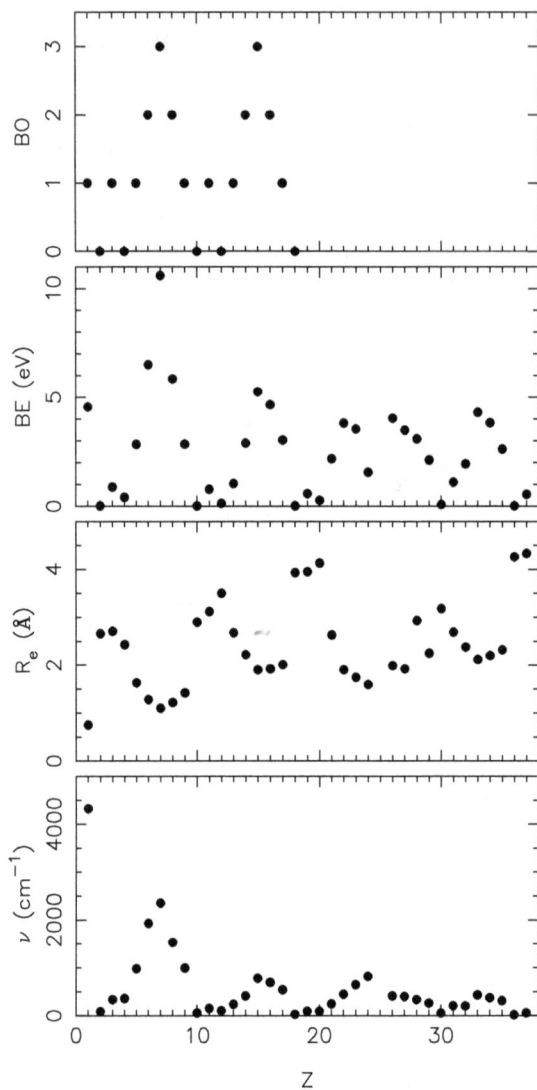

Figure 13.13: Calculated values of different properties of homonuclear, diatomic molecules as a function of atomic number of the atoms. BO is the bond order, BE the binding energy, R_e the bond length, and v the vibrational frequency.

than that of the p orbital. On the other hand, the p function allows better for an increased electron density between the two atoms, but, as mentioned above, electrons in such orbitals have a higher energy. Although the p function is more directed toward the B atom, so that a greater decrease in the energy of the molecular orbital is to be expected, it is not clear which of the two scenarios in Figure 13.15 will happen and both scenarios are in principle possible. Ultimately, what exactly will happen depends on

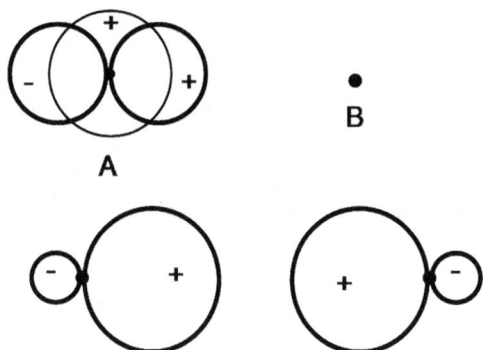

Figure 13.14: The formation of *sp* hybrid orbitals.

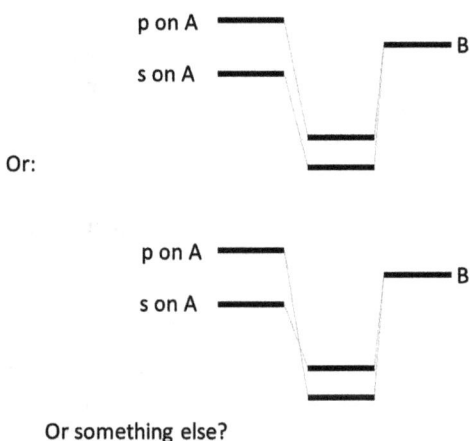

Or:

Or something else?

Figure 13.15: Possible scenarios for the energies of molecular orbitals that arise through the bond between an *s* or *p* function of atom A with some function on atom B.

the atomic type. For some atoms, however, two new orbitals may be generated,

$$\chi_1 = \frac{1}{\sqrt{2}}(\chi_s + \chi_p)$$

$$\chi_2 = \frac{1}{\sqrt{2}}(\chi_s - \chi_p), \tag{13.21}$$

which are shown in the lower part of Figure 13.14. We see that, above all, the first (χ_1) orbital is well directed toward atom B and, therefore, is optimally suited to lead to an increase in the electron density between the two atoms. If only one orbital, but not the other, is occupied, we can thereby create a stable bond at relatively low energy costs.

Through this formation of so-called hybrid orbitals, we have transferred a fraction of the electrons from the energetically lower *s* level to the energetically higher *p* level (this is called promotion), which costs energy, but this is more than compensated by the stronger bond that thereby can be created.

In particular, carbon is very good at producing different types of directed hybrid orbitals (see Figure 13.16). This explains why carbon chemistry (organic chemistry) is so rich. Also for compounds with gold, there are often hybrid orbitals on the gold atoms. Then not atomic s and p, but atomic s and d orbitals are hybridized.

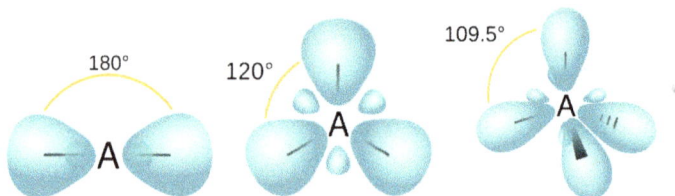

Figure 13.16: *sp*, *sp*^{2}, and *sp*^{3} hybrid orbitals for carbon. Copied on 03.02.17 from https://commons.wikimedia.org/wiki/File_AAE4h.svg, https://commons.wikimedia.org/wiki/File_AAE3h.svg and https://commons.wikimedia.org/wiki/File_AAE2h.svg. By Jfmelero (Own work) [CC BY-SA 3.0 (http://creativecommons.org/licenses/by-sa/3.0)], via Wikimedia Commons.

Finally, we compare hetero and homonuclear, diatomic molecules through an example. For molecules such as N_2 (see Figure 13.8) and CO (see Figure 13.17), we obtain orbital diagrams as shown in Figures 13.8 and 13.17. We see that for CO for each pair of bonding and antibonding orbitals, the first one has the largest weight on O, while the last one has the largest weight on C. If one adds the electron distributions of all occupied orbitals one obtains thereby more electrons on O than on C. Although also the nucleus of O has a larger charge than that of C, the final result is that the molecule is polar.

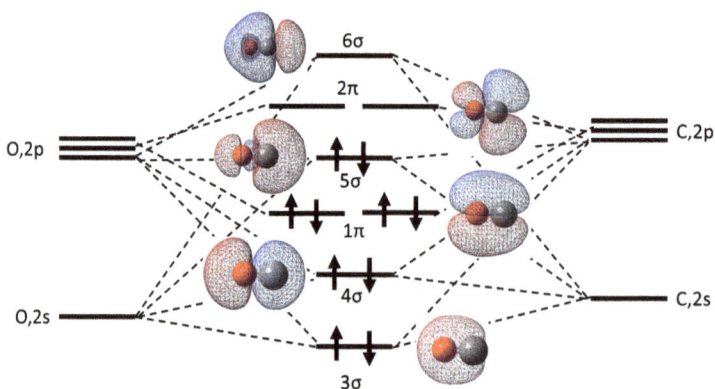

Figure 13.17: The formation of molecular orbitals from the atomic orbitals for CO.

13.5 Problems with answers

1. **Problem:** Consider a two-atomic, heteroatomic molecule placed along the z axis and use the Hartree–Fock–Roothaan approximation. The molecular orbitals are described by means of one s and three p functions on each atom. Set up the secular equation for this system and explain which matrix elements that occur in it must vanish.

Answer: We denote the two atoms with A and B and the s, p_x, p_y, and p_z functions on atom X with $\chi_{X,s}$, $\chi_{X,x}$, $\chi_{X,y}$, and $\chi_{X,z}$, e. g.,

$$\langle \chi_{X,u} | \hat{F} | \chi_{Y,v} \rangle = F_{Xu,Yv}$$

$$\langle \chi_{X,u} | \chi_{Y,v} \rangle = O_{Xu,Yv}. \tag{13.22}$$

Then we use the 8×8 secular equation:

$$
\begin{pmatrix}
F_{As,As} & F_{As,Ax} & F_{As,Ay} & F_{As,Az} & F_{As,Bs} & F_{As,Bx} & F_{As,By} & F_{As,Bz} \\
F_{Ax,As} & F_{Ax,Ax} & F_{Ax,Ay} & F_{Ax,Az} & F_{Ax,Bs} & F_{Ax,Bx} & F_{Ax,By} & F_{Ax,Bz} \\
F_{Ay,As} & F_{Ay,Ax} & F_{Ay,Ay} & F_{Ay,Az} & F_{Ay,Bs} & F_{Ay,Bx} & F_{Ay,By} & F_{Ay,Bz} \\
F_{Az,As} & F_{Az,Ax} & F_{Az,Ay} & F_{Az,Az} & F_{Az,Bs} & F_{Az,Bx} & F_{Az,By} & F_{Az,Bz} \\
F_{Bs,As} & F_{Bs,Ax} & F_{Bs,Ay} & F_{Bs,Az} & F_{Bs,Bs} & F_{Bs,Bx} & F_{Bs,By} & F_{Bs,Bz} \\
F_{Bx,As} & F_{Bx,Ax} & F_{Bx,Ay} & F_{Bx,Az} & F_{Bx,Bs} & F_{Bx,Bx} & F_{Bx,By} & F_{Bx,Bz} \\
F_{By,As} & F_{By,Ax} & F_{By,Ay} & F_{By,Az} & F_{By,Bs} & F_{By,Bx} & F_{By,By} & F_{By,Bz} \\
F_{Bz,As} & F_{Bz,Ax} & F_{Bz,Ay} & F_{Bz,Az} & F_{Bz,Bs} & F_{Bz,Bx} & F_{Bz,By} & F_{Bz,Bz}
\end{pmatrix}
$$

$$
\times
\begin{pmatrix}
c_{A,s} \\
c_{A,x} \\
c_{A,y} \\
c_{A,z} \\
c_{B,s} \\
c_{B,x} \\
c_{B,y} \\
c_{B,z}
\end{pmatrix}
$$

$$
= \epsilon
\begin{pmatrix}
O_{As,As} & O_{As,Ax} & O_{As,Ay} & O_{As,Az} & O_{As,Bs} & O_{As,Bx} & O_{As,By} & O_{As,Bz} \\
O_{Ax,As} & O_{Ax,Ax} & O_{Ax,Ay} & O_{Ax,Az} & O_{Ax,Bs} & O_{Ax,Bx} & O_{Ax,By} & O_{Ax,Bz} \\
O_{Ay,As} & O_{Ay,Ax} & O_{Ay,Ay} & O_{Ay,Az} & O_{Ay,Bs} & O_{Ay,Bx} & O_{Ay,By} & O_{Ay,Bz} \\
O_{Az,As} & O_{Az,Ax} & O_{Az,Ay} & O_{Az,Az} & O_{Az,Bs} & O_{Az,Bx} & O_{Az,By} & O_{Az,Bz} \\
O_{Bs,As} & O_{Bs,Ax} & O_{Bs,Ay} & O_{Bs,Az} & O_{Bs,Bs} & O_{Bs,Bx} & O_{Bs,By} & O_{Bs,Bz} \\
O_{Bx,As} & O_{Bx,Ax} & O_{Bx,Ay} & O_{Bx,Az} & O_{Bx,Bs} & O_{Bx,Bx} & O_{Bx,By} & O_{Bx,Bz} \\
O_{By,As} & O_{By,Ax} & O_{By,Ay} & O_{By,Az} & O_{By,Bs} & O_{By,Bx} & O_{By,By} & O_{By,Bz} \\
O_{Bz,As} & O_{Bz,Ax} & O_{Bz,Ay} & O_{Bz,Az} & O_{Bz,Bs} & O_{Bz,Bx} & O_{Bz,By} & O_{Bz,Bz}
\end{pmatrix}
$$

$$
\times
\begin{pmatrix}
c_{A,s} \\
c_{A,x} \\
c_{A,y} \\
c_{A,z} \\
c_{B,s} \\
c_{B,x} \\
c_{B,y} \\
c_{B,z}
\end{pmatrix}.
\tag{13.23}
$$

The matrix elements $\langle \chi_1 | \chi_2 \rangle$ and $\langle \chi_1 | \hat{F} | \chi_2 \rangle$ are zero if χ_1 and χ_2 possess different symmetry properties. In our case, the rotational symmetry around the molecular axis is relevant, and for these the s and p_z functions are totally rotationally symmetric, while the p_x and p_y functions have one node. This means that all matrix elements vanish between (s, p_x), (s, p_y), (p_z, p_x), (p_z, p_y), and (p_x, p_y). Accordingly, the secular equation simplifies, too:

$$
\begin{pmatrix}
F_{As,As} & 0 & 0 & F_{As,Az} & F_{As,Bs} & 0 & 0 & F_{As,Bz} \\
0 & F_{Ax,Ax} & 0 & 0 & 0 & F_{Ax,Bx} & 0 & 0 \\
0 & 0 & F_{Ay,Ay} & 0 & 0 & 0 & F_{Ay,By} & 0 \\
F_{Az,As} & 0 & 0 & F_{Az,Az} & F_{Az,Bs} & 0 & 0 & F_{Az,Bz} \\
F_{Bs,As} & 0 & 0 & F_{Bs,Az} & F_{Bs,Bs} & 0 & 0 & F_{Bs,Bz} \\
0 & F_{Bx,Ax} & 0 & 0 & 0 & F_{Bx,Bx} & 0 & 0 \\
0 & 0 & F_{By,Ay} & 0 & 0 & 0 & F_{By,By} & 0 \\
F_{Bz,As} & 0 & 0 & F_{Bz,Az} & F_{Bz,Bs} & 0 & 0 & F_{Bz,Bz}
\end{pmatrix}
$$

$$
\times
\begin{pmatrix}
c_{A,s} \\
c_{A,x} \\
c_{A,y} \\
c_{A,z} \\
c_{B,s} \\
c_{B,x} \\
c_{B,y} \\
c_{B,z}
\end{pmatrix}
$$

$$
= \epsilon
\begin{pmatrix}
O_{As,As} & 0 & 0 & O_{As,Az} & O_{As,Bs} & 0 & 0 & O_{As,Bz} \\
0 & O_{Ax,Ax} & 0 & 0 & 0 & O_{Ax,Bx} & 0 & 0 \\
0 & 0 & O_{Ay,Ay} & 0 & 0 & 0 & O_{Ay,By} & 0 \\
O_{Az,As} & 0 & 0 & O_{Az,Az} & O_{Az,Bs} & 0 & 0 & O_{Az,Bz} \\
O_{Bs,As} & 0 & 0 & O_{Bs,Az} & O_{Bs,Bs} & 0 & 0 & O_{Bs,Bz} \\
0 & O_{Bx,Ax} & 0 & 0 & 0 & O_{Bx,Bx} & 0 & 0 \\
0 & 0 & O_{By,Ay} & 0 & 0 & 0 & O_{By,By} & 0 \\
O_{Bz,As} & 0 & 0 & O_{Bz,Az} & O_{Bz,Bs} & 0 & 0 & O_{Bz,Bz}
\end{pmatrix}
$$

$$\times \begin{pmatrix} c_{A,s} \\ c_{A,x} \\ c_{A,y} \\ c_{A,z} \\ c_{B,s} \\ c_{B,x} \\ c_{B,y} \\ c_{B,z} \end{pmatrix}. \tag{13.24}$$

13.6 Problems

1. With the LCAO-MO method, the bonding and antibonding orbitals for H_2 can be written as $N_b \cdot [\chi_a(\vec{r}) + \chi_b(\vec{r})]$ and $N_a \cdot [\chi_a(\vec{r}) - \chi_b(\vec{r})]$, respectively. Explain why $|N_a| > |N_b|$, and what consequences this has for the splitting of the orbital energies.
2. Compare the molecular orbitals of homo- and heteronuclear diatomic molecules.
3. Sketch the molecular orbitals and their orbital energies for the CN^- radical.
4. Describe how the orbital energies change qualitatively when considering the series of two-atomic homonuclear molecules $Li_2 \rightarrow F_2$.
5. Sketch the orbitals and their energies for HHe.
6. Explain the term "bond order."
7. Explain through a few examples molecular orbital energy diagram.
8. Explain the term "hybridization."
9. Consider a two-atomic homoatomic molecule with the z axis along the bond and use the Hartree–Fock–Roothaan approximation. The molecular orbitals are described by means of one s and three p functions on each atom. Set up the secular equation for this system and explain which matrix elements must vanish, and which must be identical.
10. Describe how the orbital picture can explain that the HF molecule can be understood as $H^{+q}F^{-q}$.
11. How does the overlap integral depend on the interatomic distance for (i) two s, (ii) two p, and (iii) one p and one s function (note: there are several subcases)?
12. Consider the two-atomic LiH molecule with the bond along the z axis, and use the Hartree–Fock–Roothaan approximation. The molecular orbitals are described through the $1s$ and $2s$ functions on Li and the $1s$ function on H. Set up the secular equation for this system and explain which matrix elements must vanish, and which must be identical.
13. Consider the two-atomic HF molecule lying along the z axis, and use the Hartree–Fock–Roothaan approximation. The molecular orbitals are described by the $1s$, $2s$, and $2p$ functions on F and the $1s$ function on H. Set up the secular equation for this system and explain which matrix elements must vanish, and which must be identical.

14 Larger systems: methods

14.1 Hartree–Fock–Roothaan method

Since we in this chapter will need some results from previous chapters, we will briefly review those here.

With the Hartree–Fock approximation (or the orbital picture), the exact solution to the electronic Schrödinger equation,

$$\hat{H}_e \Psi_e = E_e \Psi_e, \tag{14.1}$$

is approximated through a Slater determinant,

$$\Psi_e(\tilde{x}_1, \tilde{x}_2, \ldots, \tilde{x}_N) \simeq \frac{1}{\sqrt{N!}} \begin{vmatrix} \psi_1(\tilde{x}_1) & \psi_2(\tilde{x}_1) & \cdots & \psi_N(\tilde{x}_1) \\ \psi_1(\tilde{x}_2) & \psi_2(\tilde{x}_2) & \cdots & \psi_N(\tilde{x}_2) \\ \vdots & \vdots & \ddots & \vdots \\ \psi_1(\tilde{x}_N) & \psi_2(\tilde{x}_N) & \cdots & \psi_N(\tilde{x}_N) \end{vmatrix}. \tag{14.2}$$

By applying the variational principle, we find that the single-electron wave functions $\{\psi_k\}$ must satisfy the Hartree–Fock equations,

$$\hat{F}\psi_k = \epsilon_k \psi_k. \tag{14.3}$$

$\hat{F}$ is the Fock operator,

$$\hat{F} = \hat{h}_1 + \sum_{i=1}^{N} (\hat{J}_i - \hat{K}_i), \tag{14.4}$$

where $\hat{h}_1$ is the operator of the kinetic energy and the potential energy due to the nuclei

$$\hat{h}_1 = -\frac{\hbar^2}{2m}\nabla^2 - \sum_k \frac{Z_k e^2}{4\pi\epsilon_0 |\vec{R}_k - \vec{r}|} \tag{14.5}$$

and the $\hat{J}_i$ and $\hat{K}_i$ operators are derived from the interactions among the electrons,

$$\hat{J}_i \psi_k(\tilde{x}_1) = \int \psi_i^*(\tilde{x}_2) \hat{h}_2 \psi_i(\tilde{x}_2) \psi_k(\tilde{x}_1)\, d\tilde{x}_2 = \frac{e^2}{4\pi\epsilon_0} \int \frac{|\psi_i(\tilde{x}_2)|^2 \psi_k(\tilde{x}_1)}{|\vec{r}_2 - \vec{r}_1|}\, d\tilde{x}_2$$

$$= \frac{e^2}{4\pi\epsilon_0} \int \frac{|\psi_i(\tilde{x}_2)|^2}{|\vec{r}_2 - \vec{r}_1|}\, d\tilde{x}_2 \psi_k(\tilde{x}_1) \tag{14.6}$$

https://doi.org/10.1515/9783110742206-014

and

$$\hat{K}_i \psi_k(\vec{x}_1) = \int \psi_i^*(\vec{x}_2) \hat{h}_2 \psi_i(\vec{x}_1) \psi_k(\vec{x}_2) \, d\vec{x}_2 = \frac{e^2}{4\pi\epsilon_0} \int \frac{\psi_i^*(\vec{x}_2) \psi_k(\vec{x}_2) \psi_i(\vec{x}_1)}{|\vec{r}_2 - \vec{r}_1|} \, d\vec{x}_2$$

$$= \frac{e^2}{4\pi\epsilon_0} \int \frac{\psi_i^*(\vec{x}_2) \psi_k(\vec{x}_2)}{|\vec{r}_2 - \vec{r}_1|} \, d\vec{x}_2 \psi_i(\vec{x}_1). \tag{14.7}$$

Here, $\vec{x}$ is a combined position and spin coordinate,

$$\vec{x} = (\vec{r}, \sigma) \tag{14.8}$$

where the spin variable σ can take only two values, α or β. Therefore, the integration over $\vec{x}$ is not to be taken quite literal,

$$\int \cdots d\vec{x} \equiv \sum_{\sigma=\alpha,\beta} \int \cdots d\vec{r}. \tag{14.9}$$

As mentioned earlier, the $\hat{J}_i$ operators are called Coulomb operators, while the $\hat{K}_i$ operators are called exchange operators.

From equation (14.6), we see that

$$\sum_i \hat{J}_i \psi_k(\vec{x}_1) = \frac{e^2}{4\pi\epsilon_0} \int \frac{\sum_i |\psi_i(\vec{x}_2)|^2}{|\vec{r}_2 - \vec{r}_1|} \, d\vec{x}_2 \psi_k(\vec{x}_1)$$

$$= \frac{e^2}{4\pi\epsilon_0} \int \frac{\rho(\vec{r}_2)}{|\vec{r}_2 - \vec{r}_1|} \, d\vec{r}_2 \psi_k(\vec{x}_1) \equiv V_C(\vec{r}_1) \psi_k(\vec{x}_1). \tag{14.10}$$

Here, $V_C(\vec{r}_1)$ is the electrostatic potential that is produced by the total electron density $\rho(\vec{r})$ at the point $\vec{r}_1$.

We will separate the orbitals into the two spin parts, $\sigma = \alpha$ and $\sigma = \beta$,

$$\psi_i(\vec{x}) = \psi_{i\alpha}(\vec{r})\alpha + \psi_{i\beta}(\vec{r})\beta, \tag{14.11}$$

where the two indices α and β are only used to distinguish the two functions $\psi_{i\alpha}(\vec{r})$ and $\psi_{i\beta}(\vec{r})$.

Often a particular orbital has either an α or β spin, so that one of the two functions $\psi_{i\alpha}(\vec{r})$ and $\psi_{i\beta}(\vec{r})$ becomes identically zero.

It is convenient to use a vector notation for the spin components, so that

$$\alpha = \begin{pmatrix} 1 \\ 0 \end{pmatrix}$$

$$\beta = \begin{pmatrix} 0 \\ 1 \end{pmatrix}, \tag{14.12}$$

whereby

$$\psi_i(\vec{x}) \rightarrow \vec{\psi}_i(\vec{x}) = \begin{pmatrix} \psi_{i\alpha}(\vec{r}) \\ \psi_{i\beta}(\vec{r}) \end{pmatrix}. \tag{14.13}$$

The electron density in position space for this orbital is then

$$\rho_i(\vec{r}) = [\vec{\psi}_i^*(\vec{r})]^T \cdot \vec{\psi}_i(\vec{r}) = \begin{pmatrix} \psi_{i\alpha}^*(\vec{r}) & \psi_{i\beta}^*(\vec{r}) \end{pmatrix} \cdot \begin{pmatrix} \psi_{i\alpha}(\vec{r}) \\ \psi_{i\beta}(\vec{r}) \end{pmatrix}$$

$$= |\psi_{i\alpha}(\vec{r})|^2 + |\psi_{i\beta}(\vec{r})|^2 \equiv \rho_{i\alpha}(\vec{r}) + \rho_{i\beta}(\vec{r}). \tag{14.14}$$

By summation over all orbitals, one gets then the total electron density,

$$\rho(\vec{r}) = \sum_i \rho_i(\vec{r}). \tag{14.15}$$

In a practical calculation, it is necessary to approximate the one-electron wave functions. Thus, we will use the Hartree–Fock–Roothaan approach and accordingly expand the one-electron wave functions in a preselected set of basis functions,

$$\psi_l(\vec{x}) = \sum_{p=1}^{N_b} \chi_p(\vec{x}) c_{pl}. \tag{14.16}$$

The basis functions $\{\chi_p\}$ are preselected and the calculation will "only" provide the expansion coefficients $\{c_{pl}\}$.

The equation whereby the coefficients are to be determined (the secular equation) is in matrix form

$$\underline{\underline{F}} \cdot \underline{c}_l = \epsilon_l \cdot \underline{\underline{O}} \cdot \underline{c}_l, \tag{14.17}$$

where $\underline{\underline{F}}$ is the Fock matrix with the elements

$$F_{pm} = \langle \chi_p | \hat{h}_1 | \chi_m \rangle + \sum_{i=1}^{N} \sum_{n,q=1}^{N_b} c_{ni} c_{qi}^* [\langle \chi_p \chi_q | \hat{h}_2 | \chi_m \chi_n \rangle - \langle \chi_q \chi_p | \hat{h}_2 | \chi_m \chi_n \rangle] \tag{14.18}$$

and $\underline{\underline{O}}$ is the overlap matrix with the elements

$$O_{pm} = \langle \chi_p | \chi_m \rangle. \tag{14.19}$$

Finally,

$$\langle \chi_p | \hat{h}_1 | \chi_m \rangle = \int \chi_p^*(\vec{x}) \hat{h}_1 \chi_m(\vec{x}) \, d\vec{x}$$

$$= \int \chi_p^*(\vec{x}) \left[-\frac{\hbar^2}{2m} \nabla^2 - \sum_k \frac{Z_k e^2}{4\pi\epsilon_0 |\vec{R}_k - \vec{r}|} \right] \chi_m(\vec{x}) \, d\vec{x}$$

$$\langle \chi_p \chi_q | \hat{h}_2 | \chi_m \chi_n \rangle = \int \int \chi_p^*(\vec{x}_1) \chi_q^*(\vec{x}_2) \hat{h}_2 \chi_m(\vec{x}_1) \chi_n(\vec{x}_2) \, d\vec{x}_1 \, d\vec{x}_2$$

$$= \int \int \chi_p^*(\vec{x}_1) \chi_q^*(\vec{x}_2) \frac{e^2}{4\pi\epsilon_0 |\vec{r}_1 - \vec{r}_2|} \chi_m(\vec{x}_1) \chi_n(\vec{x}_2) \, d\vec{x}_1 \, d\vec{x}_2. \tag{14.20}$$

$\underline{c}_l$ contains the coefficients of the lth orbital.

Explicitly, equation (14.17) is

$$\sum_{m=1}^{N_b}\left\{\langle\chi_p|\hat{h}_1|\chi_m\rangle + \sum_{i=1}^{N}\sum_{n,q=1}^{N_b} c_{ni}c_{qi}^*\left[\langle\chi_p\chi_q|\hat{h}_2|\chi_m\chi_n\rangle - \langle\chi_q\chi_p|\hat{h}_2|\chi_m\chi_n\rangle\right]\right\}c_{ml}$$

$$= \epsilon_l\sum_{m=1}^{N_b}\langle\chi_p|\chi_m\rangle c_{ml}. \tag{14.21}$$

By solving this equation, one obtains as many solutions as there are basis functions, i. e., N_b. These solutions have specific orbital energies ϵ_l and coefficients, $\{c_{pl}\}$. For the ground state configuration, those N orbitals with the lowest orbital energies are filled, while the others are unoccupied. This way of occupying the orbitals is indirectly assumed above in equation (14.21) so that the i summation runs over exactly the occupied orbitals. Accordingly, we will also assume that

$$\epsilon_1 \le \epsilon_2 \le \epsilon_3 \le \cdots \le \epsilon_{N_b}. \tag{14.22}$$

14.2 Basis sets

So far, we have not discussed how to choose the basis functions $\{\chi_p\}$. There are three important criteria for choosing them that, however, cannot all be met optimally at the same time:

- The basic functions should be chosen so that the approximation equation (14.16) is good.
- The basic functions should be chosen so that all integrals that are given in equations (14.20) and (14.21) can be calculated quickly, accurately, and at best analytically.
- In principle, any accuracy can be achieved by choosing a sufficiently large number (which may be very many) of basic functions. But because the time required for the numerical treatment of equation (14.17) scales with the number of basis functions N_b to the third power, N_b should not be too large.

For the hydrogen atom and other single-electron atoms, we have seen that the electronic wave functions can be written by means of exponential functions. It is to be expected that such functions are also found for multielectron atoms. Therefore, basis functions of the type

$$\chi(\vec{r}) = \chi_{\vec{R},\zeta,n,l,m}(\vec{r}) = \frac{(2\zeta)^{n+1/2}}{(2n!)^{1/2}}r_R^{n-1}e^{-\zeta r_R}Y_{lm}(\theta_R,\phi_R), \tag{14.23}$$

that need to be multiplied by a spin function form a set of functions that are likely to perform well on the first and third condition. These functions are called Slater-Type Orbitals (STOs). Unfortunately, the second condition is not fulfilled well and, therefore, such functions are hardly used. In equation (14.23), (r_R, θ_R, ϕ_R) are the polar coordinates of $\vec{r}$ relative to the point $\vec{R}$.

Instead, it has been found that Gauss functions (also called Gaussians or GTOs, named after Carl Friedrich Gauss),

$$\chi(\vec{r}) = \chi_{\vec{R},\alpha,n,l,m}(\vec{r}) = 2^{n+1} \frac{\alpha^{(2n+1)/4}}{[(2n-1)!!]^{1/2}(2\pi)^{1/4}} r_R^{n-1} e^{-\alpha r_R^2} Y_{lm}(\theta_R, \phi_R) \tag{14.24}$$

(again multiplied by a spin function and again assuming that the atom on which the function is placed at the origin of the coordinate system), provide a good compromise for the three conditions. In equation (14.24),

$$(2n-1)!! = (2n-1) \cdot (2n-3) \cdots 1. \tag{14.25}$$

We can estimate how good these basis functions are by comparing them with the exact wave function for the 1s electron of the hydrogen atom. This is shown in Figures 14.1 and 14.2. In Figure 14.1, we have used only a single Gaussian function with $(n, l, m) = (1, 0, 0)$ and optimized α (see problem 2 in Section 9.9). It can be seen that the approximation is not optimal, but not bad either. However, especially in the area closest to the nucleus, the approximation is inaccurate. Here, the correct function is nondifferentiable (has a cusp), but the Gauss function is differentiable at this point. This is the so-called cusp problem. Even using several Gaussian functions, as in Figure 14.2, this problem cannot be removed, although the resulting, approximate function better describes the correct function.

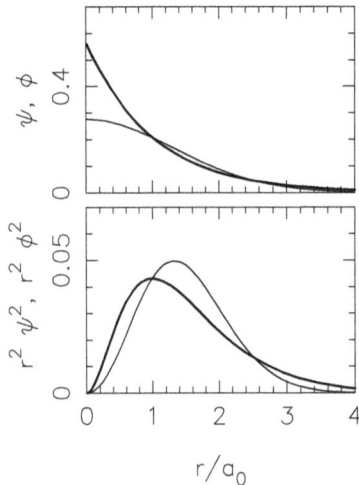

Figure 14.1: The exact 1s wave function of the hydrogen atom (thick curve) compared to an approximated wave function consisting of a single Gaussian function (thin curve).

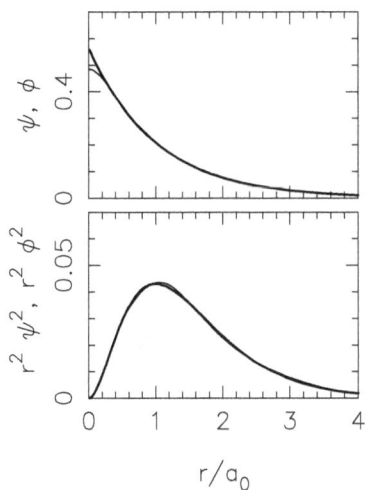

Figure 14.2: The exact $1s$ wave function of the hydrogen atom (thick curve) compared to an approximated wave function consisting of a linear combination of three Gaussian functions (thin curve), all with $(n, l, m) = (1, 0, 0)$ but different α.

Because of these and similar results, one normally uses more Gauss functions than one has (occupied) atomic orbitals. These will then differ in the decay constants α. If α is very small, the Gaussian function is very delocalized, and then one speaks of a diffuse basis function. Furthermore, it turns out that it may be beneficial when studying molecules to use functions with higher l than one has for the isolated atoms. This could then be, e. g., p functions for hydrogen atoms and d functions for carbon atoms. Such additional basic functions are called polarization functions.

Generally, because of the variational principle, the more basis functions one uses, the more accurate the results will become. Above all, the total energy will get lower. On the other hand, the demand for computer resources grows with the number of basis functions, which ultimately results in the requirement that a reasonable compromise between accuracy and computational effort is sought. What exactly is meant by "reasonable" depends on both the system and the property being investigated.

When performing a calculation, one will (have to) choose a basis set. Normally, you will not create the basis functions by yourself, but you will use ones that have been developed and suggested by others. There are many of them, but here we will briefly discuss just a few of the most common basis sets.

We will look at an example, ferrocene, $Fe(C_5H_5)_2$. Ferrocene consists of H, C, and Fe atoms. For the isolated atoms, $1s$ orbitals for H, $1s$, $2s$, and $2p$ orbitals for C, and $1s$, $2s$, $2p$, $3s$, $3p$, $4s$, and $3d$ orbitals for Fe are occupied.

The smallest possible basis set is obtained by using one basis function per occupied orbital of the isolated atoms, i. e., the $1s$ orbitals for H, $1s$, $2s$, and $2p$ orbitals for C, and $1s$, $2s$, $2p$, $3s$, $3p$, $4s$, and $3d$ orbitals for Fe. This corresponds to a so-called minimal basis set. Ideally, one would use an STO [equation (14.23)] for each of these orbitals, but because not all the integrals that are needed then can be calculated analytically, STOs are hardly used. One way around this problem is to represent each STO in terms of a preselected linear combination of GTOs [equation (14.24)]. Then all the integrals that actually need to be calculated for the STOs can be calculated using the GTOs. Such a method is called STO-nG, where n indicates the number of GTOs used for the linear combination. A typical choice is STO-3G. Such a function, which consists of a preselected linear combination of preselected GTOs, is also called a Contracted Gaussian (CGTO).

With a minimal basis set, accurate results are rarely obtained. In order to increase the flexibility (i. e., to give the calculation more degrees of freedom for the description of the electronic orbitals), several CGTOs per atomic orbital can be used. One then speaks of double zeta, triple zeta, quadruple zeta, ... basis sets, if two, three, four, ... CGTOs are used per atomic orbital.

One frequently uses a mixture of such basis sets. This applies, e. g., for a basis set like 6-31G. The "-" separates core and valence orbits so that 6-31G means that one CGTO consisting of six GTOs per core orbital is used, while two CGTOs per valence orbital are used. The two CGTOs consist of three and one GTOs, respectively. For our example, ferrocene, the 1s orbitals of C and the 1s, 2s, 2p, 3s, and 3p orbitals of Fe are each described by means of a CGTO (consisting each of 6 GTOs), while the 1s orbitals of H, the 2s and 2p orbitals of C and the 4s and 3d orbitals of Fe are each described using two CGTOs. For example, for the 2s and 2p orbitals of C, a CGTO consisting of three Gaussians and a CGTO consisting of only one Gaussian are used, where the Gaussians for the s and p functions use the same α but have different coefficients. There are many related basis sets such as, e. g., 3-21G and 6-311G. 3-21G and 6-31G differ only in the number (and then in the values of α) of the Gaussians used, while for 6-311G not two but three CGTOs are used for the valence orbitals (indicated through the three integers after the "-").

Both polarization functions and diffuse functions can be added. Polarization functions are often indicated by "*" so that 6-31G** means that d functions are used for heavier atoms such as C (first "*") and p functions for H (second "*"). Alternatively, this could have been written as 6-31G(d,p). More explicit is a notation such as 6-31G(3df,2p), which means that three d and one f polarization function for C and two p polarization functions for H are used. Diffuse functions are similarly indicated by "+." Thus, 6-31G++ means that diffuse functions are used on both heavier atoms and on H. Where exactly the G as well as the + and * symbols are placed can vary.

Occasionally, results are reported at the so-called Hartree–Fock limit, i. e., the results you get when you solve the Hartree–Fock equations exactly. Actually, this is not possible in practice, but one can estimate the results by performing Hartree–Fock–Roothaan calculations with increasing basis set size and then extrapolating to the limit of an infinitely large basis set.

As an example of the results obtained with the different basis sets, which we have now discussed, Table 14.1 shows energies for the reaction $H^+ + H_2O \rightarrow H_3O^+$. It can be seen that the energies of the individual reactants and products depend very strongly on the basis set, but that the reaction energy (the protonation energy) is significantly less dependent on it. It can also be seen that the total energy becomes lower as the size of the basis set increases (as a result of the variational principle), but that this does not apply to the energy differences (reaction energies).

Table 14.1: The energy for the reaction $H^+ + H_2O \rightarrow H_3O^+$ (the protonation energy) as calculated using different basis sets. 1 Hartree = 27.21 eV; 1 eV = 23.06 kcal/mol.

Basis Set	$E_{tot}(H_2O)$ Hartree	$E_{tot}(H_3O^+)$ Hartree	Protonation energy Hartree	Protonation energy kcal/mol
STO-3G	−75.3133	−75.6817	−0.3684	−231.2
STO-6G	−76.0366	−76.4015	−0.3649	−229.0
6-31G	−76.3852	−76.6721	−0.2869	−180.1
6-31++G	−76.4000	−76.6753	−0.2753	−172.7
6-31G**	−76.4197	−76.7056	−0.2859	−179.4
6-31++G**	−76.4341	−76.7078	−0.2738	−171.8

Another set of commonly used basis functions are the so-called Correlation Consistent basis sets, which are all referred to as cc-pVXZ. XZ is DZ, TZ, QZ, 5Z, or 6Z and these indicate whether the basis set is of double-zeta, triple-zeta, quadruple-zeta, ... type. This designation refers to whether 2 (double-zeta), 3 (triple-zeta), 4 (quadruple-zeta) basis functions are used per atomic valence orbital. These basis sets also contain polarization functions and can additionally be extended through diffuse functions. As polarization functions, e. g., p functions for H and d functions for C are used with cc-pVDZ, while d functions for H and also f functions for C are used with cc-pVTZ. In the case that diffuse functions are to be included, the basis set names get the prefix "aug": aug-cc-pVXZ.

Of course, (many) other basis sets (with different names and nomenclatures) have been proposed, which are also used. But here they will not be discussed further.

More precisely, because of the variational principle, the more basis functions are used, the lower becomes the calculated ground state energy. Specifically, this means that if one has chosen N_b basis functions, $\{\chi_p\}$, the quantity

$$\tilde{E} = \frac{\langle \phi | \hat{H} | \phi \rangle}{\langle \phi | \phi \rangle} \tag{14.26}$$

with

$$\phi = \sum_{p=1}^{N_b} c_p \chi_p, \tag{14.27}$$

that has been minimized by varying the coefficients $\{c_p\}$, then the smallest value of $\tilde{E}$ will become smaller or stay unchanged if the functions in equation (14.27) are augmented though another function without changing the others. In practice, the functions change as the size of the basis set increases, but it is almost always true that this makes $\tilde{E}$ lower. Ultimately, this means that the smallest value of $\tilde{E}$ becomes smaller (and thus represents a closer approximation to the ground state energy), the more basis functions are used. One could then be tempted to use a very large number of basis functions N_b. However, the disadvantage of this is that the computational effort thereby also increases: typically with N_b^3 for a Hartree–Fock–Roothaan calculation. Of practical reasons, one therefore attempts to keep N_b reasonably small.

One way to reduce N_b is the above-mentioned contraction. From basis functions located on the same atom, certain linear combinations are constructed,

$$u_{\vec{R},a,n,l,m,k}(\vec{r}) = \sum_\alpha c_{\vec{R},a,n,l,m,k} \chi_{\vec{R},a,n,l,m}(\vec{r}), \tag{14.28}$$

e. g., are formed from more Gauss $2p_z$ functions, with different decay constants α on a certain atom. Subsequently, coefficients for these contracted basis functions are determined using a variational approach. If the number of contracted basis functions, $\{u_{\vec{R},a,n,l,m,k}(\vec{r})\}$, is smaller than that of the primitive basis functions, $\{\chi_{\vec{R},a,n,l,m}(\vec{r})\}$ this reduces the computational effort.

Contraction is mainly used for basis functions that are very localized to the area around the atomic nucleus. The changes in the electron density between an isolated atom and an atom when being a part of a chemical compound are observed mainly in the outer regions of the atoms, so that the assumption that the orbitals hardly change in the vicinity of the atomic nuclei is a very good approximation.

In order to obtain basis sets that are realistic, one can proceed as sketched in Figures 14.1 and 14.2 for the example of the H atom. It is here exploited that the isolated atom is spherically symmetric, so that accurate numerical calculations can be performed relatively easily. Subsequently, the obtained numerical atomic wave functions can be described in terms of various functions, e. g., Gaussians, and these optimized functions can ultimately be used as atom-centered basis functions for molecular calculations.

14.3 Semiempirical and ab initio methods

Having chosen the basis functions, it is hopefully possible to calculate all integrals of equation (14.21). These integrals are integrals for the operator of the kinetic energy, for the electrostatic potential of the nuclei, and for the electron–electron interactions.

Explicitly, these integrals are of the forms

$$-\frac{\hbar^2}{2m}\int \chi_p^*(\vec{r})\nabla^2\chi_q(\vec{r})\,d\vec{r},$$

$$-\int \chi_p^*(\vec{r})\frac{Ze^2}{4\pi\epsilon_0|\vec{r}-\vec{R}|}\chi_q(\vec{r})\,d\vec{r},$$

$$\int\int \chi_p^*(\vec{r}_1)\chi_r^*(\vec{r}_2)\frac{e^2}{4\pi\epsilon_0|\vec{r}_1-\vec{r}_2|}\chi_q(\vec{r}_1)\chi_s(\vec{r}_2)\,d\vec{r}_1\,d\vec{r}_2. \qquad (14.29)$$

The basis functions χ_p, χ_q, χ_r, and χ_s can be centered on different atoms, and in the second expression $\vec{R}$ does not have to be equal to the coordinates of one of the two nuclei, where χ_p or χ_q are centered.

Such methods, in which all integrals in equation (14.29) are explicitly evaluated are called ab initio methods. These differ from the so-called semiempirical methods, for which the matrix elements in equation (14.29), are not calculated directly. Indeed, it may then not even be necessary to specify the analytic expressions of the basis functions. Instead, the matrix elements are treated as parameters whose values can be determined from calculations or experiments for smaller molecules. The values are then set so that certain properties for a set of test systems can be reproduced. Subsequently, these parameter values are used to determine similar properties of related [but often (much) larger] molecules.

14.4 Hückel theory

The Hückel theory (after Erich Hückel) is a particularly simple, semiempirical method. This theory focuses on the π electrons of conjugated systems. As Figure 14.3 shows through an example, the π bonds of such systems are not very strong. While three of the four valence electrons of a carbon atom in a conjugated system are localized in strong σ bonds between neighboring atoms, the π orbital of the last valence electron is perpendicular to the plane of the atomic nuclei. Therefore, no strong covalent bonds can be formed with these orbitals. That these bonds are not strong can also be formulated by saying that their energies are not low. These orbitals therefore form the occupied orbitals with the highest orbital energies. Conversely, the antibonding π orbitals are also not particularly strong antibonding, so that the energetic lowest unoccupied orbitals are those formed from the π orbitals. This means that for excitation processes that take place at energies that are not too high, only the π electrons are relevant and, therefore, it would be helpful to treat only them. That is what the Hückel theory offers.

In the original Hückel theory, only the π electrons of the carbon atoms are treated. Thus, the orbitals of equation (14.16) are limited to the π orbitals, and the set of basis functions $\{\chi_p\}$ consists of exactly one π orbital per carbon atom. It is assumed that

Energy

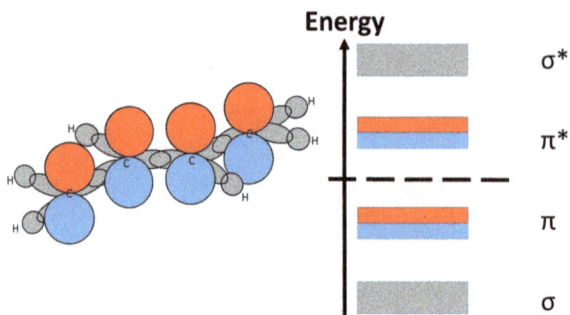

Figure 14.3: The fundamental idea behind the Hückel theory as illustrated through butadiene, C_4H_6. The gray orbitals symbolize localized, energetically low σ bonds between neighboring atoms, while the red/blue orbitals represent the less localized, energetically higher π orbitals. The occupied σ and unoccupied σ^* orbitals are energetically far from the Fermi energy separating the occupied and unoccupied orbitals, while the analoguous π and π^* orbitals are closer to the Fermi energy, as shown in the right part. Here, the dashed line marks the Fermi energy.

these are orthonormal

$$\langle \chi_p | \chi_m \rangle = \delta_{p,m}. \tag{14.30}$$

Furthermore, the Fock matrix elements are parameterized according to

$$\langle \chi_p | \hat{F} | \chi_m \rangle = \begin{cases} \alpha & m = p \\ \beta & \chi_p \text{ and } \chi_m \text{ at neighboring atoms} \\ 0 & \text{otherwise.} \end{cases} \tag{14.31}$$

This means that the Fock matrix elements have the same values, regardless of which molecule we are looking at. The $\pi - \pi$ matrix elements are always the same, both for small, simple molecules and for large, complex molecules. Furthermore, $\beta < 0$. Finally, it is possible to choose $\alpha = 0$ and $|\beta| = 1$ without loss of generality, whereby energies are expressed in so-called Hückel units. This will, however, not be done here.

As an example, consider the ethene molecule. For this we have two carbon atoms and the corresponding two π orbitals. The secular equation is then

$$\begin{pmatrix} \alpha & \beta \\ \beta & \alpha \end{pmatrix} \begin{pmatrix} c_1 \\ c_2 \end{pmatrix} = \epsilon \begin{pmatrix} 1 & 0 \\ 0 & 1 \end{pmatrix} \begin{pmatrix} c_1 \\ c_2 \end{pmatrix}. \tag{14.32}$$

Only for

$$\epsilon = \alpha \pm \beta \tag{14.33}$$

this equation has nontrivial solutions. These solutions are shown in Figure 14.4. Similarly, Figure 14.5 shows the solutions that are found for the butadiene molecule.

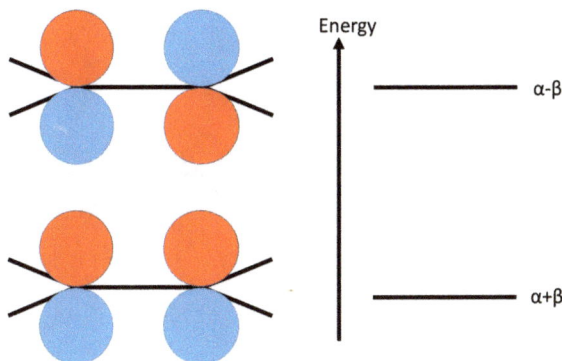

Figure 14.4: The Hückel theory applied to the ethene molecule. On the left are the two orbitals sketched, and on the right their energies.

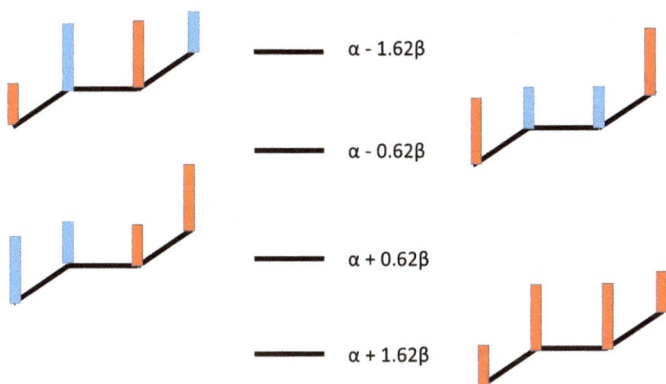

Figure 14.5: The Hückel theory applied to the butadiene molecule. For each orbital, the heights of the bars indicate the (real) coefficients to the atom-centered π functions, and the colors represent the sign of the coefficients.

The Hückel theory can also be used to find an explanation for the $4n$ and the $4n + 2$ rules (also called antiaromatics and aromatics rules). It can be shown that for cyclic polyenes one can proceed as follows, if one uses the Hückel theory. You position the molecule inside a circle so that one atom is at the bottom. This is shown in Figure 14.6 for benzene and cyclobutadiene. The circle should have the diameter $4|\beta|$. The projections of the positions of the atoms on the y axis then give the orbital energies relative to α. For polyenes with $4n + 2$ atoms, it is easy to identify a large energy gap between occupied and unoccupied orbitals, while for polyenes with $4n$ atoms, four orbitals (when the spin is taken into account) with the energy α occur, of which not all are occupied. Therefore, the polyenes with $4n + 2$ atoms are particularly stable, while those with $4n$ atoms are particularly unstable. This corresponds to the $4n + 2$ and $4n$ rules. Figure 14.7 shows the orbitals and their energies for the particularly stable benzene,

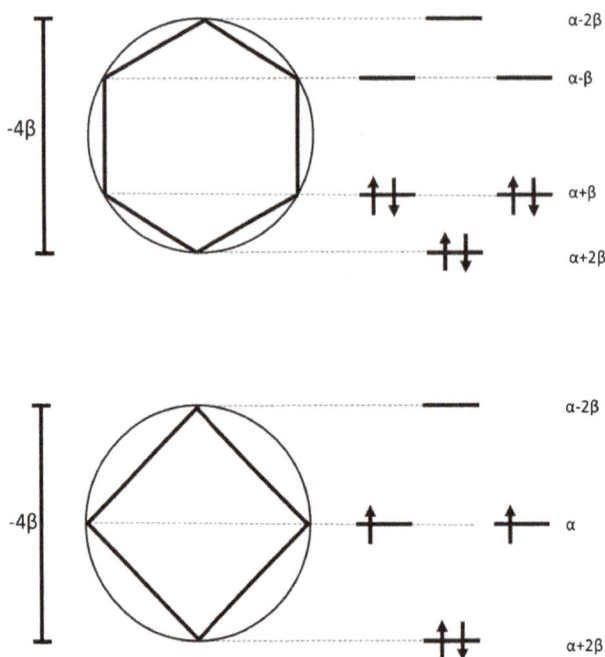

Figure 14.6: The Hückel theory applied to (top) the benzene molecule and (bottom) the cyclobutadiene molecule.

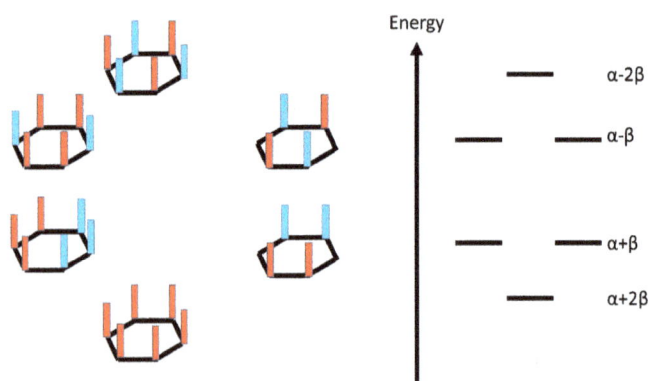

Figure 14.7: The Hückel theory applied to the benzene molecule. For each orbital, the bars indicate the (real) coefficients to the atom-centered π functions, and the colors represent the sign of the coefficients.

while Figure 14.8 shows the orbital energies for cyclic polyenes as a function of the total number of carbon atoms.

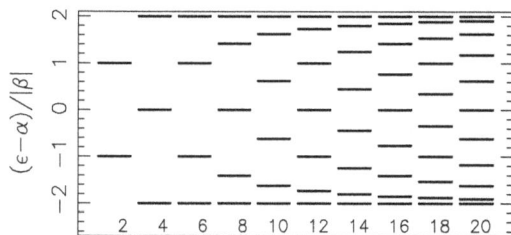

Figure 14.8: The Hückel theory applied to cyclic polyenes. The energies ϵ are relative to α and in units of $|\beta|$. The numbers under each set of energy values indicate the number of C atoms in each polyene.

14.5 Correlation

In Section 12.6, we discussed correlation effects for the H_2 molecule. We saw that the Hartree–Fock approximation leads to increasingly unreliable results for larger nucleus-nucleus distances. This failure of the Hartree–Fock approximation was attributed to correlation effects. In this case, they are purely static (as opposed to dynamic) properties of the system: the relative motion of the electrons among each other is not the cause of the correlation effects. Therefore, this is called static correlation.

However, there are also dynamical correlation effects that also are not covered by the Hartree–Fock approximation. These are shown schematically in Figure 14.9. We imagine that two electrons occupy two spatially separated orbitals, which we have labeled A and B for simplicity. Within the Hartree–Fock approximation, the electron in orbital B experiences an average potential from the electron in orbital A, as shown at the far left of Figure 14.9. But the electron in orbital A moves back and forth, and only on average does the distribution of the electron look like what orbital A represents. But if the electron is at the position represented by the black dot at some point in time (shown in the middle of Figure 14.9), the electron in orbital B will respond to it, and one can imagine that in this case the orbital B would look like the one in the middle of Figure 14.9. Later, the electron of orbital A may be where the point is in the right part of Figure 14.9, and the orbital B will look like shown there, too. On average, the orbitals look like what is shown in the left part of the figure, but only on average. What is missing is the dynamical correlation.

We have repeatedly emphasized that the orbital picture is abandoned through correlation effects. We will explain this with a simple model. We consider two electrons whose total wave function can be written as a linear combination of two Slater determinants,

$$\Psi_e(\vec{r}_1, \vec{r}_2) = c_A \begin{vmatrix} \psi_1(\vec{r}_1) & \psi_2(\vec{r}_1) \\ \psi_1(\vec{r}_2) & \psi_2(\vec{r}_2) \end{vmatrix} + c_B \begin{vmatrix} \psi_3(\vec{r}_1) & \psi_4(\vec{r}_1) \\ \psi_3(\vec{r}_2) & \psi_4(\vec{r}_2) \end{vmatrix}, \tag{14.34}$$

where we for simplicity have ignored spin. We will assume that

$$\langle \psi_1 | \psi_2 \rangle = \langle \psi_3 | \psi_4 \rangle = 0. \tag{14.35}$$

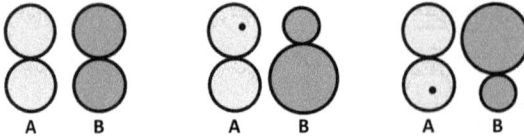

Figure 14.9: Schematic representation of the dynamical correlation. Each of the three scenarios shows the interaction between two electrons that are occupying two spatially separated orbitals, A and B. The left part shows the average, Hartree–Fock picture, whereas the other two show examples of the interactions at different times. For further details, see the text.

For the wave function of equation (14.34), the electron density is then

$$
\begin{aligned}
\rho(\vec{r}) = {} & 2\{|c_A|^2[|\psi_1(\vec{r})|^2 + |\psi_2(\vec{r})|^2] + |c_B|^2[|\psi_3(\vec{r})|^2 + |\psi_4(\vec{r})|^2] \\
& + c_A^* c_B[\psi_1^*(\vec{r})\psi_3(\vec{r})\langle\psi_2|\psi_4\rangle + \psi_2^*(\vec{r})\psi_4(\vec{r})\langle\psi_1|\psi_3\rangle \\
& - \psi_1^*(\vec{r})\psi_4(\vec{r})\langle\psi_2|\psi_3\rangle - \psi_2^*(\vec{r})\psi_3(\vec{r})\langle\psi_1|\psi_4\rangle] \\
& + c_B^* c_A[\psi_3^*(\vec{r})\psi_1(\vec{r})\langle\psi_4|\psi_2\rangle + \psi_4^*(\vec{r})\psi_2(\vec{r})\langle\psi_3|\psi_1\rangle \\
& - \psi_4^*(\vec{r})\psi_1(\vec{r})\langle\psi_3|\psi_2\rangle - \psi_3^*(\vec{r})\psi_2(\vec{r})\langle\psi_4|\psi_1\rangle]\}.
\end{aligned}
\tag{14.36}
$$

We now consider several different cases.

The orbital picture (the Hartree–Fock approximation) corresponds to, e. g.,

$$
c_A = \frac{1}{\sqrt{2}}
$$
$$
c_B = 0,
\tag{14.37}
$$

so that

$$
\rho(\vec{r}) = |\psi_1(\vec{r})|^2 + |\psi_2(\vec{r})|^2.
\tag{14.38}
$$

Thus, in each orbital we have exactly one electron.

On the other hand, if the two Slater determinants in equation (14.34) differ in all orbitals, we have in addition to equatin (14.35) also

$$
\langle\psi_1|\psi_3\rangle = \langle\psi_1|\psi_4\rangle = \langle\psi_2|\psi_3\rangle = \langle\psi_2|\psi_4\rangle = 0,
\tag{14.39}
$$

and then the electron density becomes

$$
\rho(\vec{r}) = 2\{|c_A|^2[|\psi_1(\vec{r})|^2 + |\psi_2(\vec{r})|^2] + |c_B|^2[|\psi_3(\vec{r})|^2 + |\psi_4(\vec{r})|^2]\},
\tag{14.40}
$$

so the two electrons are distributed over four orbitals, and the orbital picture is no longer valid.

It becomes even more complex when the two Slater determinants in equation (14.34) differ in only one orbital. Then the expression for the electron density also includes products from two different orbitals, which can hardly be reconciled with the orbital picture.

In the following, we will briefly discuss how correlation effects can be taken into account.

14.6 CI and CC

Within the Hartree–Fock approximation, the electronic wave function is approximated through a single-particle wave function, i. e.,

$$\Psi_e(\vec{x}_1, \vec{x}_2, \ldots, \vec{x}_N) \simeq \frac{1}{\sqrt{N!}} \begin{vmatrix} \psi_1(\vec{x}_1) & \psi_2(\vec{x}_1) & \cdots & \psi_N(\vec{x}_1) \\ \psi_1(\vec{x}_2) & \psi_2(\vec{x}_2) & \cdots & \psi_N(\vec{x}_2) \\ \vdots & \vdots & \ddots & \vdots \\ \psi_1(\vec{x}_N) & \psi_2(\vec{x}_N) & \cdots & \psi_N(\vec{x}_N) \end{vmatrix} \equiv \Phi_0. \tag{14.41}$$

We have explicitly made use of equation (14.22) and accordingly occupied the N orbitals of the N lowest orbital energies. Furthermore, we have denoted this configuration as Φ_0.

In some cases, however, we have seen that such a wave function is not sufficiently accurate to meet all the demands we can make. For the H_2 molecule, we have seen that with this wave function we get a description that is not realistic for larger nucleus-nucleus distances. We were able to correct this by using not one but more (here, two) Slater determinants/configurations.

Even single, isolated molecules often require wave functions that cannot be written as a single Slater determinant. For example, consider a carbon atom, for which four of the six electrons occupy the $1s$ and $2s$ orbitals. The last two electrons would then occupy two of the three p_x, p_y, and p_z orbitals, and have the same spin. For example, assume that the p_x and p_y orbital are occupied individually. Then the total electron density according to the Hartree–Fock approximation would be

$$\rho(\vec{r}) = 2|\psi_{1s}(\vec{r})|^2 + 2|\psi_{2s}(\vec{r})|^2 + |\psi_{2p_x}(\vec{r})|^2 + |\psi_{2p_y}(\vec{r})|^2 \tag{14.42}$$

when using a notation that should be understandable. This density is not spherically symmetric (see Section 11.7), and no matter how we linearly combine the p orbitals, we do not succeed in obtaining a spherically symmetric density within the Hartree–Fock approximation. Only when we write the total wave function as a time-dependent linear combination of several configurations, with the individual configurations differing in the p orbitals (i. e., using one configuration with the p_x and p_y orbitals singly occupied, one configuration with the p_x and p_z orbitals singly populated, as well as one configuration with the p_y and p_z orbitals singly occupied), a spherically symmetric density can be obtained, as we have discussed already in Section 11.7. So again, we have to leave the Hartree–Fock approximation to obtain physically/chemically meaningful results.

To formulate it more precisely: we realize that by solving the Hartree–Fock–Roothaan equations, we get a total of N_b orbitals. Of these, we use only the lowest energetic N to generate the wave function of equation (14.41). But it is possible to generate

$$\binom{N_b}{N} = \frac{N_b!}{N!(N_b - N)!} \tag{14.43}$$

different Slater determinants/configurations from the N_b orbitals. For those, we will use a notation like

$$\Phi_{ijk\cdots}^{abc\cdots} \tag{14.44}$$

which means that in Φ_0 [equation (14.41)] the occupied orbitals $\psi_i, \psi_j, \psi_k, \ldots$ are replaced through the unoccupied orbitals $\psi_a, \psi_b, \psi_c, \ldots$.

An improved electronic wave function [compared to that of equation (14.41)] is then (see Figure 14.10)

$$\Psi_e = C_0\Phi_0 + \sum_i \sum_a C_i^a \Phi_i^a + \sum_{ij} \sum_{ab} C_{ij}^{ab} \Phi_{ij}^{ab} + \sum_{ijk} \sum_{abc} C_{ijk}^{abc} \Phi_{ijk}^{abc} + \cdots \tag{14.45}$$

where the summations over $i, j, k, \ldots$ are over the occupied orbitals in equation (14.41), while the summations over $a, b, c, \ldots$, are over the unoccupied orbitals in equation (14.41). As mentioned above, Φ_0 is the ground state configuration where the N energetically lowest orbitals enter. Φ_i^a describes then that one electron is excited from the occupied orbital i to the unoccupied orbital a, and is therefore referred to as a single excitation. Similarly, Φ_{ij}^{ab}, Φ_{ijk}^{abs}, $\ldots$ are referred to as double, triple, $\ldots$ excitations.

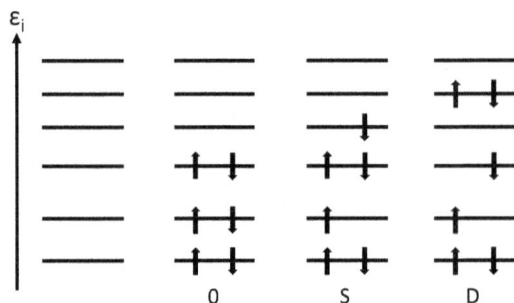

Figure 14.10: Schematic representation of different configurations for a system with 6 electrons. The orbital energies as obtained with a basis set with 6 functions (times 2, when including spin) from a Hartree–Fock–Roothaan calculation are shown to the left. Thereby, the orbitals are assumed being occupied as in the configuration marked 0, which corresponds the ground-state configuration. With the CI and CC methods, these orbitals are kept unchanged but single, double, ... excitations (exemplified in the figure through the configurations marked S and D for single and double excitations, respectively) are included in the electronic wave function. With the MCSCF (multiple-configuration self-consistent-field) method the changes in the orbitals and their energies due to the improved description of the electronic wave function are taken into account.

Because of the variational theorem, this wave function has a lower electronic energy than that of equation (14.41). Through this approximation, we have introduced a so-called CI approximation (CI = Configuration Interaction).

Because, for example,

$$\Phi_{ij}^{ab} = \Phi_{ji}^{ba} = -\Phi_{ij}^{ba} = -\Phi_{ji}^{ab}, \tag{14.46}$$

it is not necessary to include all terms in equation (14.45).

In practice, this sum is reduced to include terms in which only a smaller number of electrons are excited. If only the first term on the right-hand side in equation (14.45) is kept, we have "only" the Hartree–Fock approximation. It can be shown that, if the second term is taken into account, there is no improvement over the Hartree–Fock approximation (because, then, but only then, the coefficients will obey $C_i^a = 0$). Only when the sum in equation (14.45) is aborted after the third term (or later) one obtains an improved description of the system compared to the Hartree–Fock approximation (then, in general, both $C_i^a \neq 0$ and $C_{ij}^{ab} \neq 0$). This approximation corresponds to the so-called CISD (Confguration Interaction with Single and Double excitations) approximation.

When truncating the CI expansion after a few terms, a fundamental problem shows up, i. e., the so-called size consistency problem. This implies that when calculating the energy of a system of N identical, noninteracting molecules, one does not get the same energy by doing the same calculation for a single molecule and simply multiplying the energy of one molecule by N. The problem can be solved by using a special variant of the CI method, the so-called Coupled Cluster Method (CC Method). In practice, therefore, almost exclusively the CC method is used.

In equation (14.45), many of the configurations can also be ignored from the beginning because they have states with, e. g., an unrealistically high spin.

Finally, it should be mentioned that with wave functions of equations (14.45), we leave the orbital picture. It is not possible to assign individual orbitals to the electrons. But, as we saw for H_2, such wave functions can give accurate results.

Figure 14.11 shows a schematic representation of a CI description of the He atom, and Figure 14.12 shows a CI description of the He_2 molecule. In both cases, only single and double excitations are considered.

The inclusion of correlation effects cannot be done without a higher computational effort. While the computational requirements for a Hartree–Fock calculation scale with the number of basis functions N_b as N_b^3, it scales as N_b^7 for a CISD case. For an MP2 case, which we will describe below, it scales with N_b^5.

14.7 MCSCF and CASSCF

The basic assumption behind the CI and CC methods is that the Hartree–Fock approximation provides a good starting point for adding the correlation effects that are assumed to be small. Alternatively expressed, in the expression of equation (14.45), $|C_0|$ is much larger than the absolute values of all other coefficients. As a consequence, the

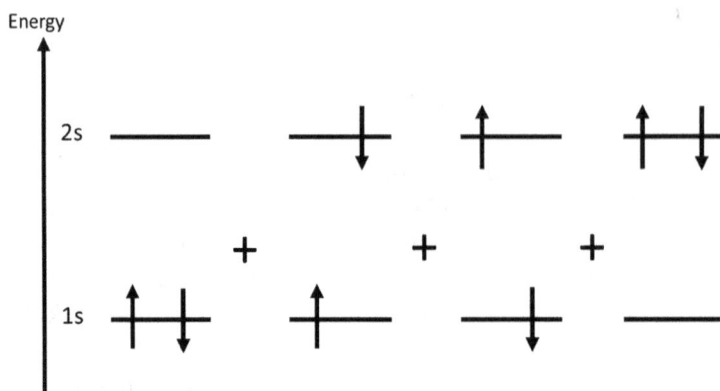

Figure 14.11: A CI description of the singlet state of the He atom, considering only the electronic 1s and 2s orbitals.

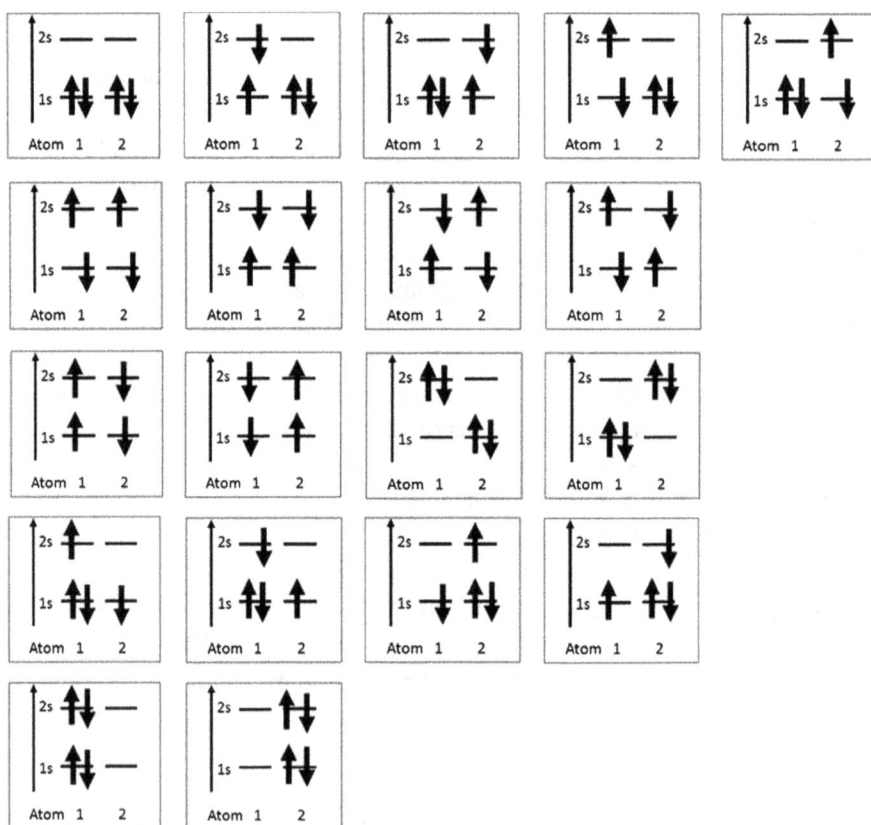

Figure 14.12: Configurations considered in a CISD description of the singlet state of the He₂ molecule, considering only the electronic 1s and 2s orbitals. The two lowest rows correspond to ionic configurations.

changes in the electronic distribution when passing from the Hartree–Fock approximation to the CI or CC wave function are so small that the orbitals calculated with the Hartree–Fock approximation provide a good basis for adding correlation effects; cf. Figure 11.7.

It may, however, happen that more configurations have significant contributions to the electronic wave function. This was, e. g., the case for the H_2 molecule for larger interatomic distances as we saw in Section 12.6. In that case, other methods can be more appropriate. To those belong the MCSCF (Multiple-Configuration Self-Consistent-Field) and CASSCF (Complete-Active-Space Self-Consistent-Field) methods that shall be discussed very briefly here.

The starting point for both methods is not the Hartree–Fock approximation (i. e., with a single Slater determinant) but with a linear combination of more Slater determinants (configurations). Which configurations are included, is chosen manually and is, accordingly, not a trivial task. In order to choose those configurations, a Hartree–Fock calculation can provide helpful information. The MCSCF and CASSCF methods differ in details about how the configurations are chosen.

Subsequently, a self-consistent calculation is performed, similar to the case for the Hartree–Fock approximation, but using that the electronic distribution is given by the full set of configurations used in setting up the electronic wave function. Thereby, the orbital energies may change compared to the results of the Hartree–Fock approximation. This aspect is briefly described in Figure 11.7.

Per construction, MCSCF and CASSCF calculations are computationally much more demanding than Hartree–Fock calculations, but the reward is that much more accurate results are obtained. When correlation effects are important, Hartree–Fock calculations may very well be too inaccurate, and MCSCF and CASSCF calculations can then provide a useful alternative. As we shall see in the next section, this is in particularly then the case, when we have several occupied and unoccupied orbitals that are energetically close to each other and to the Fermi energy that separates occupied and unoccupied orbitals.

14.8 MP

The CI wave function of equation (14.45) is an improvement over the Hartree–Fock approximation of equation (14.2). The differences between the two are by definition called correlation effects. Often, the correlation effects are small. Then it is possible to consider these with the help of perturbation theory. This is the content of the perturbation theory of Christian Møller and Milton Plesset (MP).

In order to use perturbation theory, it is necessary to identify the difference between the Hartree–Fock equations and the Schrödinger equation. The Hartree–Fock approximation solves single-electron equations,

$$\hat{F}\psi_i(\vec{x}) = \epsilon_i \psi_i(\vec{x}) \tag{14.47}$$

while the electronic Schrödinger equation is an equation for all (N) electrons,

$$\hat{H}_e\Psi_e(\vec{x}_1, \vec{x}_2, \ldots, \vec{x}_N) = E_e\Psi_e(\vec{x}_1, \vec{x}_2, \ldots, \vec{x}_N). \tag{14.48}$$

To be able to compare the two, we first form a new operator for all N electrons from the Fock operator. This is done simply by forming the sum of $\hat{F}$ for each electron

$$\hat{G}' = \sum_{i=1}^{N} \hat{F}(i). \tag{14.49}$$

If the individual ψ_l satisfy equation (14.47), a Slater determinant with N of the one-electron orbitals is an eigenfunction for $\hat{G}'$,

$$\hat{G}' \frac{1}{\sqrt{N!}} \begin{vmatrix} \psi_{i_1}(\vec{x}_1) & \psi_{i_2}(\vec{x}_1) & \cdots & \psi_{i_N}(\vec{x}_1) \\ \psi_{i_1}(\vec{x}_2) & \psi_{i_2}(\vec{x}_2) & \cdots & \psi_{i_N}(\vec{x}_2) \\ \vdots & \vdots & \ddots & \vdots \\ \psi_{i_1}(\vec{x}_N) & \psi_{i_2}(\vec{x}_N) & \cdots & \psi_{i_N}(\vec{x}_N) \end{vmatrix}$$

$$= (\epsilon_{i_1} + \epsilon_{i_2} + \cdots + \epsilon_{i_N}) \frac{1}{\sqrt{N!}} \begin{vmatrix} \psi_{i_1}(\vec{x}_1) & \psi_{i_2}(\vec{x}_1) & \cdots & \psi_{i_N}(\vec{x}_1) \\ \psi_{i_1}(\vec{x}_2) & \psi_{i_2}(\vec{x}_2) & \cdots & \psi_{i_N}(\vec{x}_2) \\ \vdots & \vdots & \ddots & \vdots \\ \psi_{i_1}(\vec{x}_N) & \psi_{i_2}(\vec{x}_N) & \cdots & \psi_{i_N}(\vec{x}_N) \end{vmatrix}. \tag{14.50}$$

This applies to all configurations and not only to the ground state configuration.

The eigenfunctions do not change if we modify $\hat{G}'$ by an additive constant although the eigenvalue must then be modified by this constant,

$$\hat{G} = \hat{G}' - E'. \tag{14.51}$$

We choose

$$E' = \frac{1}{2} \sum_{i,j=1}^{N} [\langle \phi_i\phi_j|\hat{h}_2|\phi_i\phi_j\rangle - \langle \phi_j\phi_i|\hat{h}_2|\phi_i\phi_j\rangle]. \tag{14.52}$$

The important point is that we may write

$$\hat{H}_e = \hat{G} + \Delta\hat{H} \tag{14.53}$$

where

$$\Delta\hat{H} = \frac{1}{2} \sum_{i\neq j=1}^{N} \hat{h}_2(i,j) - \sum_{i,j=1}^{N} [\hat{J}_j(i) - \hat{K}_j(i)] + \frac{1}{2} \sum_{i,j=1}^{N} [\langle \phi_i\phi_j|\hat{h}_2|\phi_i\phi_j\rangle - \langle \phi_j\phi_i|\hat{h}_2|\phi_i\phi_j\rangle] \tag{14.54}$$

indeed is a small perturbation (this will not be shown here). For this, we can then use the perturbation theory.

It then turns out that first the second-order correction to the energy is nonzero. This term equals

$$\sum_{i,j}\sum_{a,b}\frac{\langle\Phi_0|\Delta\hat{H}|\Phi_{i,j}^{a,b}\rangle\langle\Phi_{i,j}^{a,b}|\Delta\hat{H}|\Phi_0\rangle}{E_0-E_{ij}^{ab}}, \tag{14.55}$$

with

$$E_0 = \sum_{i=1}^{N}\epsilon_i - E' = E_{\text{HF}} \tag{14.56}$$

and

$$E_{ij}^{ab} = E_{\text{HF}} + \epsilon_a + \epsilon_b - \epsilon_i - \epsilon_j, \tag{14.57}$$

so that

$$E_0 - E_{ij}^{ab} = -\epsilon_a - \epsilon_b + \epsilon_i + \epsilon_j. \tag{14.58}$$

As above, the orbitals i and j are occupied and the orbitals a and b are unoccupied (within the Hartree–Fock approximation).

Equation (14.58) shows that correlation effects can be important if the energy differences between occupied and unoccupied orbitals are small. We have already seen this when we treated the H_2 molecule: the bond lengths for which the Hartree–Fock method failed mostly were indeed those for which the orbital energies of the (occupied) bonding and (unoccupied) antibonding orbitals are close. This happened for larger bond lengths. This is exemplarily shown in Figure 14.13 for the H_2^+ molecular ion.

Furthermore, since

$$\epsilon_i, \epsilon_j < \epsilon_a, \epsilon_b, \tag{14.59}$$

the denominator in the individual terms in equation (14.55) is real and negative while the numerator is real and positive. Thereby we realize that the correction in equation (14.55) is real and negative, i. e., that correlation effects lead to a reduction of the total energy. Nothing else should be expected: With a multi-determinant wave function, we have obtained a better approximation to the correct electronic wave function and the variational principle tells us that this will lead to a lowering of the energy.

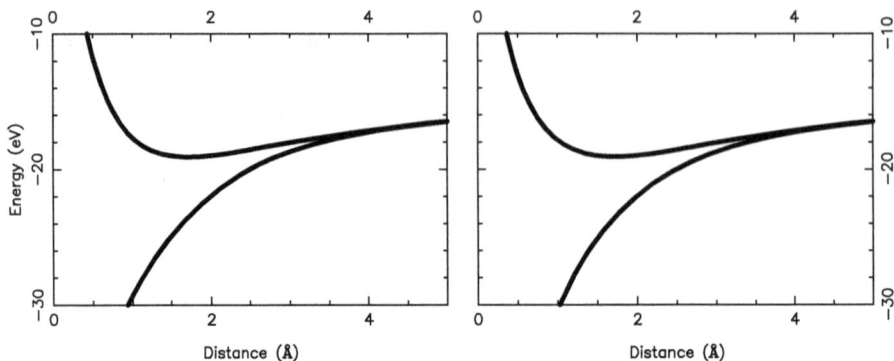

Figure 14.13: The energies of the bonding and antibonding orbital of the H_2^+ molecular ion as a function of the interatomic distance. The two diagrams correspond to the two cases in Figure 12.4.

14.9 DFT

An entirely different approach to calculate electronic properties of molecules or solids is based on the so-called Density Functional Theory (DFT). Although the first attempts to develop such a method were presented shortly after the introduction of the Schrödinger equation, a mathematically well-founded theory was presented only in 1964 and 1965 in two works by Walter Kohn, Pierre Hohenberg, and Lu Jeu Sham. In 1998, Walter Kohn received a part of the Nobel Prize in Chemistry for this work. Here, we will briefly introduce the basics of the theory, but do so without too many mathematical details.

In the first work, Hohenberg and Kohn presented two theorems. First, they could prove that, knowing the electron density $\rho(\vec{r})$ in position space of the ground state of any system, one can, in principle, calculate all the ground state properties of this system. This includes the electronic energy E_e. How to do this exactly is still unknown, but first of all you have a proof of existence. One takes the electron density $\rho(\vec{r})$, manipulates it in some way, possibly sets up some equations whose solutions one manipulates further, and so on. Exactly what you have to do is unknown, and may even be extremely complex. The only thing that is known is that you can calculate the electronic energy from the electron density. In other words, the electronic energy is a functional of the electron density,

$$E_e = E_e[\rho(\vec{r})]. \tag{14.60}$$

In the spirit of Section 3.1, a functional manipulates functions and returns numbers.

In the second theorem, Hohenberg and Kohn showed that assuming that the functional E_e is known, the lowest electronic energy is obtained by using the correct electron density, as long as the number of electrons is not changed. If the correct ground state electron density is called $\rho_0(\vec{r})$,

$$E_e[\rho(\vec{r})] \geq E_e[\rho_0(\vec{r})] \tag{14.61}$$

when

$$\int \rho(\vec{r})d\vec{r} = \int \rho_0(\vec{r})\, d\vec{r}. \tag{14.62}$$

Equation (14.61) together with equation (14.62) represents a variational theorem, which makes it possible to determine the electronic energy with arbitrary accuracy in principle. Compared to the variational theorem, which we have so far treated and used, the variational theorem of this chapter is much simpler: one needs not to determine the complete wave function, which depends on the coordinates of all electrons, but only the electron density in three-dimensional position space.

For a practical application, however, we face an initially insurmountable problem: we do not know what the functional $E_e[\rho(\vec{r})]$ looks like. It is possible to approximate this, but this often introduces so large inaccuracies that the results are no longer reliable. Kohn and Sham therefore proposed a different approach.

They introduced a nonexisting model system that should have the same energy and density as the electrons of the true system (see Figure 14.14). This means that E_e and $\rho(\vec{r})$ are identical for the two systems. But the particles of the model system are like electrons without charge, so they do not interact with each other. In order to achieve that these so-called quasiparticles, as they are often called, have the correct values of E_e and $\rho(\vec{r})$, they move in an initially unspecified external potential $V_{\text{eff}}(\vec{r})$, which is so constructed that the corresponding values for E_e and $\rho(\vec{r})$ are obtained.

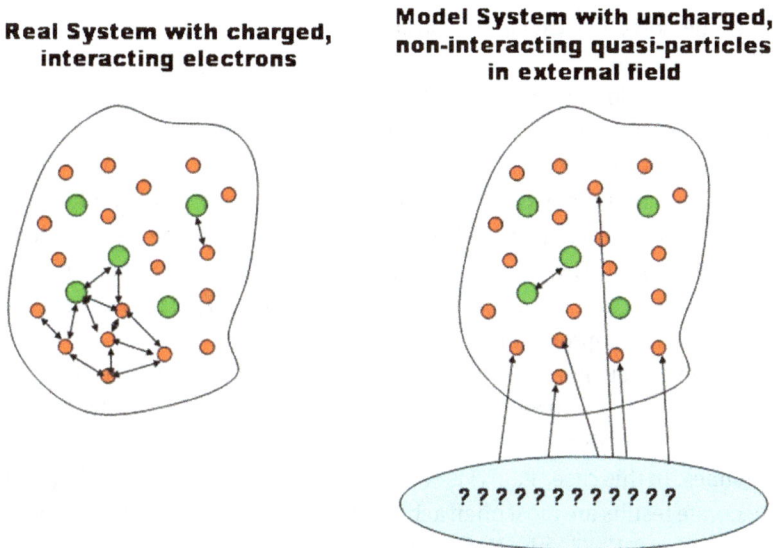

Figure 14.14: A schematic representation of the idea behind the Kohn–Sham method. Here, red circles mark electrons in the left part and quasiparticles in the right part, whereas the green circles mark the nuclei. The double arrows symbolize some of the interactions between the particles.

Because the quasiparticles do not interact with each other, the Hartree–Fock approximation is exact for this system. Furthermore, we can set up the corresponding Hartree–Fock equations,

$$\hat{h}_{\text{eff}}\psi_i^{\text{KS}}(\vec{r}) = \epsilon_i^{\text{KS}}\psi_i^{\text{KS}}(\vec{r}),$$
(14.63)

with the index KS indicating that they are the Kohn–Sham quasiparticles.

The one-particle operator $\hat{h}_{\text{eff}}$ is particularly simple,

$$\hat{h}_{\text{eff}} = -\frac{\hbar^2}{2m}\nabla^2 + V_{\text{eff}}(\vec{r}),$$
(14.64)

and by construction we know that

$$\rho(\vec{r}) = \sum_{i=1}^{N} |\psi_i^{\text{KS}}(\vec{r})|^2.$$
(14.65)

The problem that we do not know $E_e[\rho(\vec{r})]$ has been reformulated here: Now we do not know $V_{\text{eff}}(\vec{r})$.

Kohn and Sham were able to show that we already know the larger, most often dominating, parts of $V_{\text{eff}}(\vec{r})$. One part of $V_{\text{eff}}(\vec{r})$ consists of the external potential (which in most cases is the electrostatic potential due to the nuclei). Furthermore, we know another part: the electrostatic potential V_C, which is generated by the electron density $\rho(\vec{r})$; see equation (14.10). Then we can write

$$V_{\text{eff}}(\vec{r}) = V_{\text{ext}}(\vec{r}) + V_C(\vec{r}) + V_{\text{xc}}(\vec{r}).$$
(14.66)

The last term is the so-called exchange-correlation potential (short xc potential), where now this is unknown.

The problem that we do not know $E_e[\rho(\vec{r})]$ was initially only reformulated: we do not know $V_{\text{xc}}(\vec{r})$. But for most systems, $V_{\text{xc}}(\vec{r})$ is small, so we can approximate this term without introducing very large inaccuracies. And that is exactly what is done.

$V_{\text{xc}}(\vec{r}_1)$ is a number that depends on the position, i. e., $\vec{r}_1$, that we are looking at. According to density functional theory, this number may depend on the electron density in the whole space $\rho(\vec{r})$. The simplest way is to assume that $\vec{r}_1$ depends only on $\rho(\vec{r}_1)$, i. e., only on the electron density at exactly this one point, $\vec{r}_1$. This assumption leads to the Local Density Approximation, LDA.

It has been found over time that this approximation is often sufficiently accurate for solids but not for molecules. Therefore, improvements have been introduced, among which the so-called Generalized Gradient Approximation, GGA, is one of the most important ones. In this case, $V_{\text{xc}}(\vec{r}_1)$ is written as a function of $\rho(\vec{r}_1)$, $|\vec{\nabla}\rho(\vec{r}_1)|$, and $\nabla^2\rho(\vec{r}_1)$. Very accurate results are most often achieved with such methods. In even more complex approximations, $V_{\text{xc}}(\vec{r}_1)$ also depends on the Kohn–Sham orbitals, $\{\psi_i^{\text{KS}}(\vec{r}_1)\}$, which leads to so-called meta-GGA approximations.

It makes sense to compare the Hartree–Fock and Kohn–Sham approaches. One then finds the following:

  – The one-particle equations are very similar. For the Hartree–Fock method one solves the Hartree–Fock equations,

$$\left[-\frac{\hbar^2}{2m}\nabla^2 + V_{\text{ext}}(\vec{r}) + V_{\text{C}}(\vec{r}) - \sum_{j=1}^{N} \hat{K}_j \right] \psi_i^{\text{HF}}(\vec{r}) = \epsilon_i^{\text{HF}} \psi_i^{\text{HF}}(\vec{r}) \tag{14.67}$$

(with $\hat{K}_j$ being the exchange operator for the jth orbital), while the Kohn–Sham equations are

$$\left[-\frac{\hbar^2}{2m}\nabla^2 + V_{\text{ext}}(\vec{r}) + V_{\text{C}}(\vec{r}) + V_{\text{xc}}(\vec{r}) \right] \psi_i^{\text{KS}}(\vec{r}) = \epsilon_i^{\text{KS}} \psi_i^{\text{KS}}(\vec{r}) \tag{14.68}$$

Due to the similarity of the two types of equations, the same methods are used to solve them.

  – Among other things, this means that basis sets are used in both methods.

  – Furthermore, both methods require that the equations be solved self-consistently.

  – Although the Kohn–Sham equations are actually "only" for a nonexisting model system, the similarity of the Hartree–Fock and Kohn–Sham equations suggests that the solutions to the Kohn–Sham wavefunctions and orbital energies are very close to those of the electrons and, accordingly, can be interpreted as were they solutions to the Hartree–Fock equations. This is done very often (with success).

  – By definition, the Hartree–Fock equations do not contain any correlation effects in contrast to the Kohn–Sham equations.

  – Because V_{xc} is approximated, one currently solves approximate equations using the Kohn–Sham method, making it difficult to systematically improve the solutions. This is in contrast to the Hartree–Fock procedure.

Experience has shown that the results of the Hartree–Fock method and the Kohn–Sham method often compensate each other: the truth is often somewhere in between. For this reason, mixed methods are used where a fraction of exchange interactions are treated using Hartree–Fock methods, while the rest of exchange interactions and all correlation effects are treated using density functional methods. A very popular variant of such methods is the so-called B3LYP method. B3LYP stands for the following: Axel Becke (giving the B) has proposed such a hybrid method (hybrid, because it mixes density-functional theory and Hartree–Fock theory), which includes three parameters (explaining the 3) that describe how the two methods are combined. One needs an approximate DFT function, for which one often chooses the one by Chengtee Lee, Weitao Yang, and Robert G. Parr (hence LYP).

14.10 Crystals

Crystals are, in a sense, nothing other than (very) large molecules. Because of the size, however, the theoretical treatment of these would require very much computer time.

Instead, one exploits with enormous advantage the symmetry of the system. The system is assumed to be infinite and periodic. This means that one identifies a unit that is repeated periodically. By translating the system the vector

$$\vec{t}_{n_a,n_b,n_c} = n_a\vec{a} + n_b\vec{b} + n_c\vec{c}, \tag{14.69}$$

the system is mapped onto itself. In this equation, $\vec{a}$, $\vec{b}$, and $\vec{c}$ are the lattice vectors that describe the translational symmetry of the crystal, and n_a, n_b, and n_c are integers.

In a practical calculation, one will not treat the infinitely large system. Instead, one considers a smaller system with so-called cyclic boundary conditions. The smaller system is the unit that is repeated periodically as described through equation (14.69). We will illustrate this through a one-dimensional example (because it is simpler). Thus, as a model for the infinite, periodic system, we consider a large ring. In Figure 14.15, this ring has N units, and ultimately, we let $N \to \infty$. For the ring, there is a symmetry element that is important for our discussion: the rotation about $\frac{2\pi}{N}$ around the axis perpendicular to the plane of the ring. This rotational symmetry commutes with the Hamilton operator and with the Fock or Kohn–Sham operator of the system. This means that we can classify the electronic orbitals by means of their symmetry properties. We will not show this here, but that means we can assign a number k to each orbital. For systems that are periodic in two or three dimensions, this number becomes a vector, $\vec{k}$.

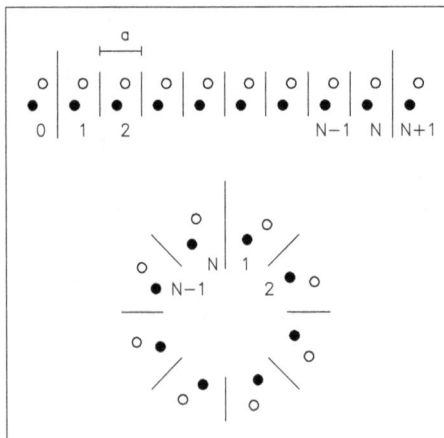

Figure 14.15: The lower part shows a ring as a finite model for a very large chain. In the upper part, it is shown that this ring is equivalent to the infinite system with cyclic boundary conditions.

For our ring, k describes how the orbitals change when we move from one cell to another, thereby moving the lattice vector $\vec{a}$,

$$\psi(k, \vec{r} + \vec{a}) = e^{ika} \psi(k, \vec{r}). \tag{14.70}$$

Here, a is the length of $\vec{a}$. Such a function, which obeys the symmetry properties of the infinite periodic system, is called a Bloch function (after Felix Bloch).

Repeating this N times, repeated use of equation (14.70) gives

$$\psi(k, \vec{r} + N\vec{a}) = [e^{ika}]^N \psi(k, \vec{r}) = e^{ikNa} \psi(k, \vec{r}) \equiv \psi(k, \vec{r}). \tag{14.71}$$

The final identity comes from the fact that we have returned to the starting point through the N rotations.

Equation (14.71) immediately gives the possible values of k,

$$kNa = 2n\pi, \tag{14.72}$$

with an integer n. This means that

$$k = 0, \pm\frac{2\pi}{aN}, \pm\frac{4\pi}{aN}, \dots, \begin{cases} \pm\frac{(N-3)\pi}{aN}, \pm\frac{(N-1)\pi}{aN} & \text{for } N \text{ odd} \\ \pm\frac{(N-2)\pi}{aN}, \frac{\pi}{a} & \text{for } N \text{ even.} \end{cases} \tag{14.73}$$

Other k values for which $|n| > N/2$, are equivalent to those listed here and are therefore rarely considered.

The condition in equation (14.71) is completely equivalent to the fact that we consider for the infinite, periodic chain as in the upper part of Figure 14.15 only those wave functions that satisfy the cyclic boundary conditions

$$\psi(k, \vec{r} + N\vec{a}) = \psi(k, \vec{r}). \tag{14.74}$$

The region with the length $L = Na$ is called the Born von Kármán zone (after Max Born and Theodore von Kármán).

In the limit $N \rightarrow \infty$, we have infinitely many k values that satisfy

$$-\frac{\pi}{a} < k \leq \frac{\pi}{a} \tag{14.75}$$

This region is called the first Brillouin zone (after Léon Brillouin). Although one must actually consider a whole continuous set of infinitely many k points, in practice a finite, not very large value of N_k k points is sufficient.

For each k, one will be able to calculate the orbital energies, which are then functions of k. In the limit $N_k \rightarrow \infty$, these then form no longer discrete energy levels, but a continuum with a finite width. This can easily be demonstrated with the help of the Hückel theory, as shown in Figure 14.16. Because k is a good quantum number, the

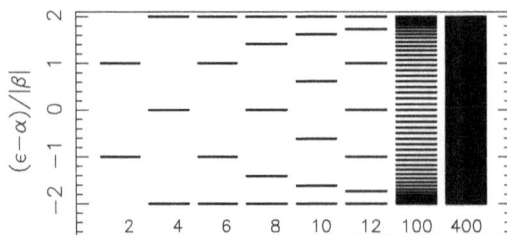

Figure 14.16: The formation of a continuum of orbital energies when a ring becomes very large. The results have been found using the Hückel theory. The energies ϵ are given relative to α and in units of $|\beta|$. The numbers under each set of energy values indicate the number of C atoms in each cyclic polyene.

orbital energies can also be represented as a function of k. This leads to the so-called band structures.

The width of the energy bands depends on the strength of the interactions between orbitals of different unit cells. These in turn depend on the distances between the atoms of different unit cells, or, in other words, on the lattice constant (or, in several dimensions, on the lattice constants). Figure 14.17 shows how from atomic energy levels (to the right in the figure) wider energy bands are formed as the interatomic distances become smaller. The figure shows the energy bands that consist of either atomic s and p orbitals or atomic s and d orbitals, and it can be seen how the resulting bands overlap at sufficiently small interatomic distances. If you have more different types of atomic orbitals, the whole picture becomes more complex, but in all cases you have several bands that can energetically overlap. The electrons occupy the lowest energy bands, with one electron per spin and unit cell fitting into each band. If a situation arises in which some bands are not completely filled, you have a metal, otherwise a semiconductor or an insulator. For the last two cases, the distinction between the two cases depends on the size of the energy gap between the occupied and unoccupied orbitals: for insulators, this is larger, which corresponds to above about 3 eV.

14.11 Electronic excitations

The energies necessary to excite electrons of for instance a molecule are relevant for many issues. Thus, they are responsible for the optical properties, including color, of the substance, as we briefly discussed in Section 4.7 using a simple model. Furthermore, these excitation energies can be used spectroscopically to characterize the compounds. In the last case, however, the information obtained experimentally is indirect: from the electronic excitation energies the hope is to be able to identify the underlying structure of the molecule.

Orbital energy

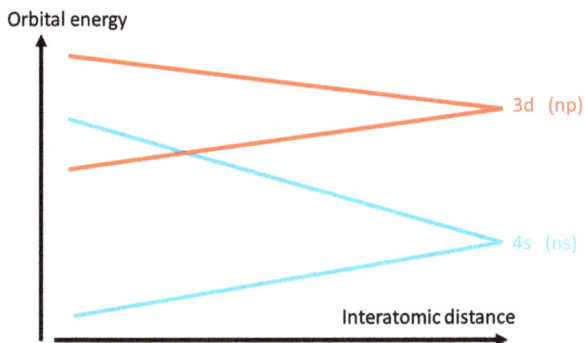

3d (np)

4s (ns)

Interatomic distance

Figure 14.17: The development of broad bands as the interatomic distances become smaller. As an example, bands of the 4s and 3d orbitals of transition metals, or the ns and np orbital of, e. g., Si, Ge, P, etc. are sketched.

For these reasons, there is a significant interest in theoretically determining the excitation energies, or more generally the electronic excitation spectra. More or less accurate methods for this purpose will be briefly discussed here.

When a molecule is experiencing an electric field, the molecule consisting of electrons and nuclei can be excited. Fermi's golden rule, equation (9.73), describes the probability that this happens in the case of a resonance. This corresponds to the case that the energy of the electromagnetic radiation, $\hbar\omega$, corresponds to an energy difference between two states of the undisturbed molecule.

$$\hbar\omega = E_f - E_i. \tag{14.76}$$

Here, E_i is the initial energy of the system and E_f is the final energy.

The finite probability that the system is excited can be determined either by the excitation rate W from equation (9.73) or, equivalently, by the so-called oscillator strength, f (which, like W, also depends on the excitation energy, the molecule, etc.). An electronic excitation spectrum will therefore show excitations for energies that satisfy equation (14.76) and the intensities are given through the oscillator strengths. An optimal theoretical method will be able to determine both excitation energies, or alternatively the possible energies of the system, as well as the oscillator strengths. This is only possible for particularly simple systems, so instead different approximations are used, which are discussed here.

Using the orbital picture (Hartree–Fock approximation), we can imagine that in the ground state of the molecule, the electrons are distributed as in the left part of Figure 14.18. By ionization or excitation, the distribution of the electrons changes, as shown in the middle part of Figure 14.18 in the case of ionization and in Figure 14.19 in the case of excitation. Ignoring electronic relaxation effects, it is assumed that the electronic orbitals and their energies remain unchanged after ionization or excitation

Energy

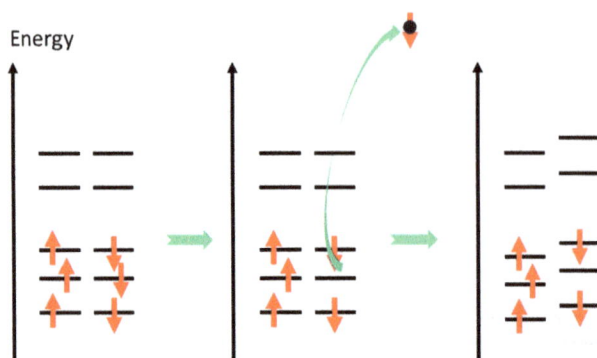

Figure 14.18: Schematic representation of the ionization of a molecule with six electrons.

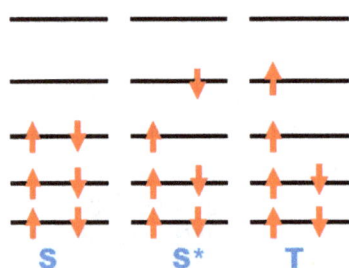

Figure 14.19: Schematic representation of an electronic excitation of a molecule with six electrons. The left part shows the distribution of the electrons in the singlet ground state, in the middle, the distribution in an excited singlet state and in the right part in an excited triplet state.

(middle part of Figure 14.18). Then we can estimate the excitation energies using equation (14.76), and ionization energies with the help of Koopmans' theorem. For the excitation energies, we have then

$$\hbar\omega \simeq \epsilon_j - \epsilon_i, \tag{14.77}$$

with ϵ_j and ϵ_i being the energy of an unoccupied and an occupied orbital before the excitation. But, as the right-hand part in Figure 14.18 indicates, because of the changed distribution of the electrons, the orbitals and their energies change, and at first we have not taken these relaxation effects into account.

It should be noted here that Koopmans' theorem, strictly speaking, is only valid within the Hartree–Fock approximation. When using a Kohn–Sham method instead, we face the fundamental problem that the orbitals and their energies are calculated for the nonexisting system of quasi-particles. A pragmatic solution is at first to ignore this problem. Experience has shown that for many systems the orbitals and their energies obtained by the two different methods are very similar. There is also a theorem related to Koopmans' theorem (Janak's theorem), which provides a more formal justification for treating the energies of the Kohn–Sham orbitals like those of the Hartree–Fock orbitals.

With this, we have a first approach for determining the electronic excitation energies. As mentioned, thereby electronic relaxation effects are ignored. This also applies to structural relaxation effects. To describe the latter, let us consider for simplicity a diatomic molecule. By electronic excitation, an electron may be excited from an orbital with a strong bonding character between the two atoms to an orbital of less bonding character. As a result, the bond between the two atoms becomes weaker, and the two atoms move slightly away from each other: the bond becomes longer. This change in the structure by the electronic excitation is a structural relaxation effect. Ignoring this one speaks of vertical excitations (the structure remains unchanged), otherwise of adiabatic excitations.

The method we have discussed, and with which vertical excitation energies can be estimated, is sometimes used as a first approximation. The method provides no information about the oscillator strengths. Furthermore, it has been found that the excitation energies determined by the Hartree–Fock method are larger than experimental values (often much larger), while with Kohn–Sham's method too small excitation energies are obtained.

It would be better to calculate directly the energy differences between excited states and the ground state. But there are some problems that we will sketch briefly with the help of Figure 14.20, even if structural relaxations are not considered.

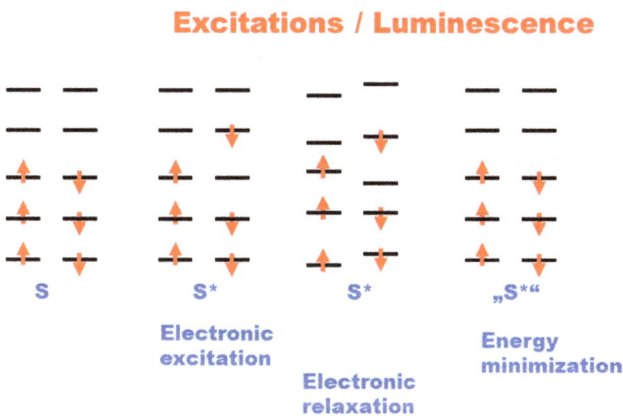

Figure 14.20: Schematic representation of the problems in the calculation of the energies of excited states.

We assume that the electronic ground state configuration, labeled S (for singlet), is as in the left part of Figure 14.20. Six electrons occupy the six energetically lowest orbitals. When an electron is excited, we will first consider the nonrelaxed orbitals and their energies, as shown in the second part of Figure 14.20. However, electronic relaxation effects will cause the orbitals and their energies to change as shown in the third part. We can try to determine these using the methods we have discussed in this manuscript. This means that we minimize the total electronic energy for a given structure. However, this leads to the fact that the electronically excited configuration S* will eventually relax to the ground state configuration S: this is ultimately the distribution of the electrons, which leads to the lowest energy. Figure 14.19 offers a solution to this problem. Instead of looking for the excited singlet state S*, one searches for the excited, energetically lowest triplet state T. The two S* and T states have total energies that differ only in exchange and correlation effects and are therefore often energetically close to one another.

However, there is an alternative method by which both excitation energies and oscillator strengths can be calculated for a given structure. The method, which is based on density functional theory, is called Time-Dependent Density Function Theory (TD-DFT), and there is also an analogue method based on the Hartree–Fock theory. The theories are mathematically not easy and will here therefore not be discussed any further. The methods are used as follows. First, the structure of the lowest total energy is determined. Subsequently, the electronic excitation energies and their oscillator strengths are calculated for this structure using the TD-DFT method. This means that only vertical excitation are treated.

Finally, we will discuss by way of example how luminescence can be treated theoretically. In luminescence, a molecule is first excited electronically, e. g., from the singlet ground state S to an excited singlet state S*; see the upper right part in Figure 14.21. The molecule has the lowest energy structure in state S. Subsequently, the structure of the molecule will change (structural relaxation) and assume the structure of the lowest energy in state S*. This relaxation is radiationless, so the energy released by the molecule is passed over to other molecules, for instance through collisions. In this structure, the molecule will relax to the state S and emit the energy difference in the form of radiation—this is the luminescence that can be observed. Eventually, the molecule relaxes again and ends up without radiation in the lowest-energy structure in state S.

We shall refer to the energy of the molecule in state X and with the structure of the energy minimum in state Y as $E(X, Y)$. Then the excitation energy is

$$E_{\text{excite}} = E(S^*, S) - E(S, S), \tag{14.78}$$

whereas the energy of the emitted radiation is

$$E_{\text{emit}} = E(S^*, S^*) - E(S, S^*). \tag{14.79}$$

As indicated in Figure 14.21, it is always true that the two so-called reorganization energies,

$$\lambda_1 = E(S^*, S) - E(S^*, S^*)$$

Excitations / Luminescence

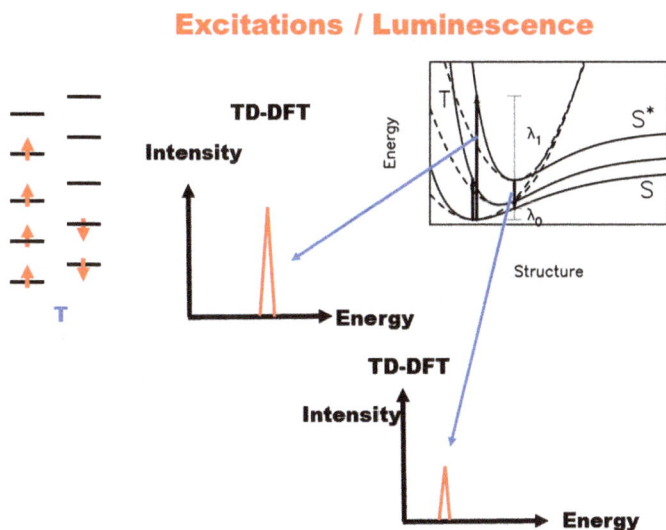

Figure 14.21: Schematic representation of luminescence. The energy differences in the upper right part are greatly exaggerated.

$$\lambda_2 = E(S, S^*) - E(S, S), \tag{14.80}$$

are positive and, therefore, also that

$$E_{\text{emit}} < E_{\text{excite}}. \tag{14.81}$$

With the help of, e. g., TD-DFT the two energies of equation (14.80) can be calculated, if the structures of the minima in states S and S^* are both known. In addition, the oscillator strengths of the transitions can also be calculated so that spectra can be obtained at the end, as sketched in the lower part of Figure 14.21. The main problem, however, is that it may be difficult to determine the structure of the energy minimum in the S^* state for the same reason we discussed above (see Figure 14.19). One possibility is then to replace the excited state S^* by a triplet state T; see the upper left part in Figure 14.21. In this illustration, one spin of the excited state is reversed. Compared to the "true" excited singlet state S^*, the total energy differs only in contributions from exchange and correlation interactions, which are often small, so the error introduced by this approach becomes small, i. e., much smaller than indicated in Figure 14.21.

14.12 Solvent effects

In the methods we have discussed so far, we have treated isolated molecules in the gas phase, i. e., molecules that feel no influence from any surrounding medium. However, the molecules are often generated and studied experimentally in a liquid medium, and it cannot be excluded that this medium has an influence on the properties of the molecule. In fact, there are, e. g., tautomers for which one tautomer is most stable in the gas phase while another is the more stable in the solution. Therefore, it is highly relevant to describe (briefly) methods for the treatment of solvent effects.

Basically, a distinction is made between explicit and implicit methods, which are shown schematically in Figure 14.22. With the explicit methods, individual solvent molecules are taken directly into account, whereas in the case of the implicit methods, the effects of the solvent are treated rather as those of a homogeneous medium.

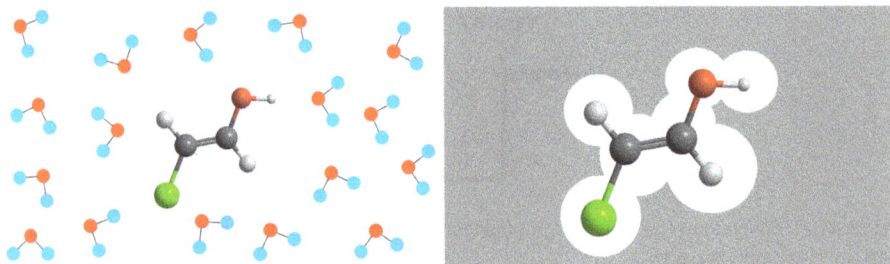

Figure 14.22: Schematic representation of an explicit method (left part) and an implicit method (right part) for the treatment of solvent effects.

The conceptually simplest explicit method is the supermolecular procedure, which is not necessarily the computationally most simple. In this case, it is exploited that the electronic effects are short-ranged, so that a perturbation in one place has no noticeable effects at another place, as long as the two places are not very close to each other. Accordingly, one can use a standard procedure, such as the ones we have discussed in this chapter, with a system consisting of the dissolved molecule and a few solvent molecules from the closest environment of the molecule of interest.

In principle, this method is very accurate, but the computational effort is very large if more than a few solvent molecules are to be considered.

On the other hand, this process involves a lot of computational effort to calculate the properties of solvent molecules that are not of direct interest. Instead, the solvent molecules should deliver "only" an external potential, in which the dissolved molecule is found. If we consider water as the simplest example as the solvent, the effects of the single water molecule in the simplest approximation can be considered as those of three point charges, $-2q$ on O and $+q$ on each of the two H atoms leading to an electrostatic potential. If one places a very large amount of water molecules in a random arrangement, but in such a way that they do not get too close to each other or to the dissolved molecule, and so that the density of the water molecules corresponds to the correct density of water, one can use this distribution to determine an electrostatic potential in which the dissolved molecule is found. Subsequently, the properties of the dissolved molecule can be calculated. Since it may happen that the properties may depend on the randomly generated arrangement of the water molecules, the process can be repeated for different arrangements, from which average values can be determined. Also this approach belongs to the class of explicit methods because the properties of the individual solvent molecules are included "somehow," in this case through electrostatic potentials from point charges.

Because point charges have very long-ranged effects, this method has the advantage that even a very large number of solvent molecules can be taken into consideration. On the other hand, there is no interaction at all between solvent molecules and dissolved molecules, which may be a disadvantage. Among those interactions not treated correctly with this approach are hydrogen bonds, that often play an important role in solutions. Finally, it should be noted that the simple model for water is not very accurate and can of course be improved, but without much consequences for the overall approach.

With the implicit methods, the solvent is considered to be a polarizable, homogenous medium (polarizable continuum). The dissolved molecule creates a cavity in it, as shown in Figure 14.22. If the dissolved molecule possesses a surplus of positive (or negative) charge locally, the surrounding medium will react in such a way that it will create an excess of positive (or negative) charge in the vicinity of the parts of the dissolved molecule with negative (or positive) charges, simultaneously ensuring that the medium as a whole remains neutral. The charge distribution of the continuum generates an electrostatic potential that is felt by the dissolved molecule, whose electron distribution will respond to this, which in turn can lead to a new charge distribution in the continuum. The process must therefore be carried out in a self-consistent manner: the electron distribution of the dissolved molecule leads to a charge distribution in the continuum, which leads to exactly that external potential for which the electron distribution of the dissolved molecule is found.

Thus, in this approach, there is a coupling between solvent and dissolved molecule, although again, e. g., hydrogen bonds are not treated, and also details of the solvent molecules in the nearest vicinity of the solute are not treated. Furthermore, the method suffers from the fact that the results depend on the shape of the cavity, which should not be. Nevertheless, this process is probably the method currently most frequently used to consider solvent effects. Furthermore, a combined approach has slowly become established: a supermolecule consisting of the solute molecule and few solvent molecules (e. g., those expected to have hydrogen bonds with the solute molecule) is treated as a whole together with the polarizable continuum approach.

14.13 Macromolecules and enzymes

For large, complex molecules, e. g., biomolecules a most important issue is that of their structure, whereas their composition is known. For these, so-called force field methods (Molecular Mechanics, MM) can often be used with advantage. With these, it is assumed that the total energy of the system to a good approximation is made up of contributions from individual atoms, pairs of atoms, etc. Accordingly, it is written as

$$E_{MM} = \sum_i E^{(1)}_{X_i} + \sum_{i>j} E^{(2)}_{X_i,X_j}(\vec{R}_i, \vec{R}_j) + \sum_{i>j>k} E^{(3)}_{X_i,X_j,X_k}(\vec{R}_i, \vec{R}_j, \vec{R}_k) + \cdots. \tag{14.82}$$

Here, the contribution of a single atom, $E^{(1)}_{X_i}$, depends only on its type, X_i, but does not depend on the environment in which this atom is located. Furthermore, the two-body contribution $E^{(2)}_{X_i,X_j}(\vec{R}_i, \vec{R}_j)$ depends only on the types of the two atoms and their relative arrangement, given by their coordinates $\vec{R}_i$ and $\vec{R}_j$. The same applies to the other interactions.

One often assumes simple functional dependencies, so that, for example, harmonic potentials, electrostatic interactions, Lennard–Jones interactions, or zero are assumed for the pair interactions:

$$E^{(2)}_{X_i,X_j}(\vec{R}_i, \vec{R}_j) = \begin{cases} \frac{1}{2}k_{X_i,X_j}(|\vec{R}_i - \vec{R}_j| - R_{0,X_i,X_j})^2 \\ \frac{q_{X_i}q_{X_j}}{|\vec{R}_i - \vec{R}_j|} \\ \epsilon_{X_i,X_j}[(\frac{\sigma_{X_i,X_j}}{|\vec{R}_i - \vec{R}_j|})^{12} - (\frac{\sigma_{X_i,X_j}}{|\vec{R}_i - \vec{R}_j|})^6] \\ 0. \end{cases} \tag{14.83}$$

The values of the parameters that are used in the expressions for the energy contributions can be determined by means of calculations for smaller model systems and then transferred to the large systems without change. It is assumed that one needs little more than 3- or 4-atom interactions, and that they are equal to zero if the interacting atoms are too far apart.

Such a (MM) method is fast in terms of computer resources and can therefore also be used for large systems. The result of such calculations is, above all, the structure of the (macro)molecule.

One problem is that electronic interactions are not taken into account. Such are important when the molecule is, e. g., active as an enzyme, and thus able to catalytically break chemical bonds of smaller systems or produce by-products. However, this catalytic activity is usually spatially localized to a very small part of the enzyme, a so-called active center. In order to be able to treat such systems, so-called QM/MM methods are used, i. e., Quantum Mechanics/Molecular Mechanics methods.

Using the quantum mechanical methods we have discussed in this book, chemical bonds, including chemical reactions, can be treated. Such methods are applied to the (small) part of the enzyme where the active site is located and where the reactions catalytically affected by the enzyme take place, while the rest of the enzyme is treated using MM methods. The MM treatment ensures that the active center is in the correct structural environment, while the formation and breaking of chemical bonds through the active site are properly treated by the QM method.

Exactly how such a method is realized, will not be discussed here. But it is not that easy. The development of such methods was honored in 2013 with the Nobel Prize in Chemistry to Martin Karplus, Michael Levitt, and Arieh Warshel.

14.14 Electromagnetic fields

The responses of matter to electromagnetic fields is of immense practical importance. First, they form the basis for spectroscopy and, furthermore, they can also be applied technologically. In the latter case, it is also used that there are nonlinear effects, e. g., that the application of one electromagnetic field with one frequency ω can cause emitted light with a frequency of 2ω or 3ω to be emitted (this is part of the field of nonlinear optics). Furthermore, e. g., static electric fields can lead to structural changes, which corresponds to a piezoelectric effect.

In order to investigate such effects theoretically, the fields must be directly included in the Hamilton operator. An electromagnetic field exerts a force on an electron. This force can be expressed by means of potentials, whereby only a vector potential can be applied for a magnetic field, whereas a scalar potential and/or a vector potential can be used for the electric field. With V_{EM} and $\vec{A}_{EM}$ being the scalar and the vector potential, respectively, the field strengths of the electric and magnetic field, respectively, can be calculated from

$$\vec{\mathcal{E}}(\vec{r}, t) = -\vec{\nabla} V_{EM}(\vec{r}, t) - \frac{1}{c}\frac{\partial}{\partial t}\vec{A}_{EM}(\vec{r}, t)$$
$$\vec{B}(\vec{r}, t) = \vec{\nabla} \times \vec{A}_{EM}(\vec{r}, t). \tag{14.84}$$

Here, c is the speed of light. Especially for the electric field, the separation into a scalar and a vector potential is not unique, and instead one can choose different so-called gauges.

We will now concentrate on static (i. e., time-independent) and homogeneous (i. e., position-independent) fields. Then we may choose

$$V_{EM}(\vec{r}) = -\vec{\mathcal{E}} \cdot \vec{r}$$
$$\vec{A}_{EM}(\vec{r}) = \frac{1}{2}\vec{B} \times (\vec{r} - \vec{R}_G). \tag{14.85}$$

$\vec{R}_G$ is unphysical and corresponds to the arbitrary origin of the magnetic vector potential. Ideally, the results should not depend on it.

The presence of the electrostatic field leads to an additional term in the electronic Hamiltonian,

$$\hat{H}_{e,el} = \sum_{n=1}^{N} \hat{h}_{el}(\vec{r}_n) = \sum_{n=1}^{N} e\vec{\mathcal{E}} \cdot \vec{r}_n, \tag{14.86}$$

where the sum runs over all the electrons of the system. There is a similar term for the nuclei,

$$\hat{H}_{n,el} = -\sum_{k=1}^{M} Z_k e \vec{\mathcal{E}} \cdot \vec{R}_k. \tag{14.87}$$

In principle, we could also have an origin for the scalar potential for the electrostatic field, but if the system is neutral, the results are independent of this origin.

In the presence of a magnetostatic potential, the expression for the kinetic energy in the electronic Hamiltonian changes to

$$\hat{H}_{e,km} = \sum_{n=1}^{N} \hat{h}_{km}(\vec{r}_n) = \sum_{n=1}^{N} \frac{1}{2m} \left[-i\hbar \vec{\nabla}_n + \frac{e}{2c} \vec{B} \times (\vec{r}_n - \vec{R}_G) \right]^2. \tag{14.88}$$

Here, $-i\hbar\vec{\nabla}_n$ is the momentum operator of the nth electron.

There are several suggestions to suppress the dependence of the results on $\vec{R}_G$. When the electronic orbitals are expanded in a basis set of atom-centered functions,

$$\psi_l(\vec{x}) = \sum_{p,\alpha} \chi_{p,\alpha}(\vec{x}) c_{p,\alpha,l} \tag{14.89}$$

with the atom p at the point $\vec{R}_p$, and α all other dependencies (main and secondary quantum numbers, decay constants, etc.), it has become common to use so-called GIAOs (Gauge-Including Atomic Orbitals). The basis functions (the AOs) are then replaced by GIAOs,

$$\chi_{p,\alpha}(\vec{x}) \rightarrow \chi_{p,\alpha}(\vec{x}) \exp\left[i\frac{e}{2c\hbar} \vec{B} \times (\vec{R}_G - \vec{R}_p) \cdot \vec{r} \right]. \tag{14.90}$$

With those, the results become independent of $\vec{R}_G$.

14.15 Experimental quantities

Many quantities measured in experiment can also be theoretically determined. In particular, the dependences of the total energy on the structure, the spin of the nuclei, and the components of electric and/or magnetic field vectors are important, i. e., quantities like

$$\frac{\partial E^{n_R + n_E + n_B + n_\Sigma}}{\partial R^{n_R} \partial \mathcal{E}^{n_E} \partial B^{n_B} \partial \Sigma^{n_\Sigma}}. \tag{14.91}$$

This notation implies that the experimentally relevant quantities are determined by the n_R-, n_E-, n_B-, and n_Σ-fold derivatives of the total energy E with respect to the nuclear coordinates, the vector components of the electric field, the vector components of the magnetic field, and the components of the nuclear spin.

In Table 14.2, we have collected some of the derivatives of equation (14.91) together with the quantities for which they are relevant. Not all of these have been explicitly dealt with in this text, but some of them will be in Chapter 15, so that a compilation here is useful. We briefly mention that infrared and Raman spectroscopy are used to determine vibrational spectra, and that the electric dipole moment as well as the polarizability and hyperpolarizabilities are relevant for linear and nonlinear optics.

Table 14.2: Some experimentally determinable quantities obtained by means of derivatives of the total energies of the type of equation (14.91).

n_R	n_E	n_B	n_Σ	Relevance
0	0	0	0	Total energy
1	0	0	0	Forces: structural optimization
0	1	0	0	Electric dipole moment
0	0	1	0	Magnetic dipole moment
0	0	0	1	Hyperfine structure
2	0	0	0	Harmonic vibrational frequencies and modes
0	2	0	0	Electric polarizability
0	0	2	0	Magnetic susceptibility
0	0	0	2	Coupling of spins of different nuclei
1	1	0	0	Infrared intensities
0	1	1	0	Circular dichroism
3	0	0	0	Anharmonic corrections to vibrational frequencies
0	3	0	0	First electrical hyperpolarizability
1	2	0	0	Raman intensities
4	0	0	0	Anharmonic corrections to vibrational frequencies
0	4	0	0	Second electrical hyperpolarizability

14.16 Problems with answers

1. **Problem:** Consider a symmetric ABA molecule, and suppose that the molecular orbitals can be described by one atomic orbital per atom:

$$\psi_i = c_{i,A_1}\chi_{A_1} + c_{i,B}\chi_B + c_{i,A_2}\chi_{A_2}. \tag{14.92}$$

Here, χ_{A_1}, χ_B, and χ_{A_2} are the atomic orbital centered on one A atom, on the (middle) B atom, and on the other A atom, respectively. The atomic orbitals are orthonormal and, furthermore,

$$\langle\chi_{A_1}|\hat{F}|\chi_{A_1}\rangle = \langle\chi_{A_2}|\hat{F}|\chi_{A_2}\rangle = \epsilon_A$$
$$\langle\chi_{A_1}|\hat{F}|\chi_{A_2}\rangle = 0$$
$$\langle\chi_{A_1}|\hat{F}|\chi_B\rangle = \langle\chi_{A_2}|\hat{F}|\chi_B\rangle = t. \tag{14.93}$$

Set up the secular equation for this basis set.

Next, consider the symmetry-adapted basis functions

$$\chi_1 = \frac{1}{\sqrt{2}}(\chi_{A_1} + \chi_{A_2})$$

$$\chi_2 = \chi_B$$

$$\chi_3 = \frac{1}{\sqrt{2}}(\chi_{A_1} - \chi_{A_2}) \tag{14.94}$$

and set up the secular equation for this basis set.

Finally, determine the orbital energies of this system.

Answer: In the original basis the secular equation is

$$\begin{pmatrix} \epsilon_A & t & 0 \\ t & \epsilon_B & t \\ 0 & t & \epsilon_A \end{pmatrix} \begin{pmatrix} c_{i,A_1} \\ c_{i,B} \\ c_{i,A_2} \end{pmatrix} = \epsilon_i \begin{pmatrix} 1 & 0 & 0 \\ 0 & 1 & 0 \\ 0 & 0 & 1 \end{pmatrix} \begin{pmatrix} c_{i,A_1} \\ c_{i,B} \\ c_{i,A_2} \end{pmatrix}. \tag{14.95}$$

For the symmetry-adapted basis set, we first determine the overlap matrix elements:

$$\langle \chi_1 | \chi_1 \rangle = \frac{1}{2}(\langle \chi_{A_1} | \chi_{A_1} \rangle + \langle \chi_{A_1} | \chi_{A_2} \rangle + \langle \chi_{A_2} | \chi_{A_1} \rangle + \langle \chi_{A_2} | \chi_{A_2} \rangle) = 1$$

$$\langle \chi_1 | \chi_2 \rangle = \frac{1}{\sqrt{2}}(\langle \chi_{A_1} | \chi_B \rangle + \langle \chi_{A_2} | \chi_B \rangle) = 0$$

$$\langle \chi_1 | \chi_3 \rangle = \frac{1}{2}(\langle \chi_{A_1} | \chi_{A_1} \rangle - \langle \chi_{A_1} | \chi_{A_2} \rangle + \langle \chi_{A_2} | \chi_{A_1} \rangle - \langle \chi_{A_2} | \chi_{A_2} \rangle) = 0$$

$$\langle \chi_2 | \chi_1 \rangle = \frac{1}{\sqrt{2}}(\langle \chi_B | \chi_{A_1} \rangle + \langle \chi_B | \chi_{A_2} \rangle) = 0$$

$$\langle \chi_2 | \chi_2 \rangle = \langle \chi_B | \chi_B \rangle = 1$$

$$\langle \chi_2 | \chi_3 \rangle = \frac{1}{\sqrt{2}}(\langle \chi_B | \chi_{A_1} \rangle - \langle \chi_B | \chi_{A_2} \rangle) = 0$$

$$\langle \chi_3 | \chi_1 \rangle = \frac{1}{2}(\langle \chi_{A_1} | \chi_{A_1} \rangle - \langle \chi_{A_1} | \chi_{A_2} \rangle + \langle \chi_{A_2} | \chi_{A_1} \rangle - \langle \chi_{A_2} | \chi_{A_2} \rangle) = 0$$

$$\langle \chi_3 | \chi_2 \rangle = \frac{1}{\sqrt{2}}(\langle \chi_{A_1} | \chi_B \rangle - \langle \chi_{A_2} | \chi_B \rangle) = 0$$

$$\langle \chi_3 | \chi_3 \rangle = \frac{1}{2}(\langle \chi_{A_1} | \chi_{A_1} \rangle - \langle \chi_{A_1} | \chi_{A_2} \rangle - \langle \chi_{A_2} | \chi_{A_1} \rangle + \langle \chi_{A_2} | \chi_{A_2} \rangle) = 1. \tag{14.96}$$

Similarly, the Fock-matrix elements in this basis become:

$$\langle \chi_1 | \hat{F} | \chi_1 \rangle = \epsilon_A$$

$$\langle \chi_1 | \hat{F} | \chi_2 \rangle = \sqrt{2}t$$

$$\langle \chi_1 | \hat{F} | \chi_3 \rangle = 0$$

$$\langle \chi_2 | \hat{F} | \chi_1 \rangle = \sqrt{2}t$$

$$\langle \chi_2 | \hat{F} | \chi_2 \rangle = \epsilon_B$$

$$\langle \chi_2 | \hat{F} | \chi_3 \rangle = 0$$
$$\langle \chi_3 | \hat{F} | \chi_1 \rangle = 0$$
$$\langle \chi_3 | \hat{F} | \chi_2 \rangle = 0$$
$$\langle \chi_3 | \hat{F} | \chi_3 \rangle = \epsilon_A. \tag{14.97}$$

Then the secular equation for this new basis set becomes

$$\begin{pmatrix} \epsilon_A & \sqrt{2}t & 0 \\ \sqrt{2}t & \epsilon_B & 0 \\ 0 & 0 & \epsilon_A \end{pmatrix} \begin{pmatrix} c_{i,1} \\ c_{i,2} \\ c_{i,3} \end{pmatrix} = \epsilon_i \begin{pmatrix} 1 & 0 & 0 \\ 0 & 1 & 0 \\ 0 & 0 & 1 \end{pmatrix} \begin{pmatrix} c_{i,1} \\ c_{i,2} \\ c_{i,3} \end{pmatrix}. \tag{14.98}$$

The possible values of ϵ_i are accordingly

$$\epsilon_i = \frac{\epsilon_A + \epsilon_B}{2} \pm \left[\left(\frac{\epsilon_A - \epsilon_B}{2} \right)^2 + 2t^2 \right]^{1/2}$$
$$\epsilon_i = \epsilon_A. \tag{14.99}$$

2. **Problem:** Use the Hückel theory for the linear $C_n H_{n+2}$ molecule for $n = 3$ and $n = 4$.

Answer: For $n = 3$, the secular equation becomes

$$\begin{pmatrix} \alpha & \beta & 0 \\ \beta & \alpha & \beta \\ 0 & \beta & \alpha \end{pmatrix} \begin{pmatrix} c_a \\ c_b \\ c_c \end{pmatrix} = \epsilon \begin{pmatrix} 1 & 0 & 0 \\ 0 & 1 & 0 \\ 0 & 0 & 1 \end{pmatrix} \begin{pmatrix} c_a \\ c_b \\ c_c \end{pmatrix}. \tag{14.100}$$

We proceed as in problem 1. With χ_a, χ_b, and χ_c equal to the three atom-centered π functions of the molecule, we create symmetry-adapted functions. This gives us two symmetric and one antisymmetric function:

$$\chi_1 = \frac{1}{\sqrt{2}} (\chi_a + \chi_c)$$
$$\chi_2 = \chi_b$$
$$\chi_3 = \frac{1}{\sqrt{2}} (\chi_a - \chi_c). \tag{14.101}$$

For this basis set, the secular equation becomes

$$\begin{pmatrix} \alpha & \sqrt{2}\beta & 0 \\ \sqrt{2}\beta & \alpha & 0 \\ 0 & 0 & \alpha \end{pmatrix} \begin{pmatrix} c_1 \\ c_2 \\ c_3 \end{pmatrix} = \epsilon \begin{pmatrix} 1 & 0 & 0 \\ 0 & 1 & 0 \\ 0 & 0 & 1 \end{pmatrix} \begin{pmatrix} c_1 \\ c_2 \\ c_3 \end{pmatrix}. \tag{14.102}$$

The solutions are

$$\epsilon_i = \alpha - \sqrt{2}\beta \qquad \vec{c}_i = \begin{pmatrix} -1/2 \\ 1/\sqrt{2} \\ -1/2 \end{pmatrix}$$

$$\epsilon_i = \alpha + \sqrt{2}\beta \qquad \vec{c}_i = \begin{pmatrix} 1/2 \\ 1/\sqrt{2} \\ 1/2 \end{pmatrix}$$

$$\epsilon_i = \alpha \qquad \vec{c}_i = \begin{pmatrix} 1/\sqrt{2} \\ 0 \\ -1/\sqrt{2} \end{pmatrix}. \qquad (14.103)$$

Here, $\vec{c}_i$ is the vector with the coefficients that multiply the original basis (χ_a, χ_b, χ_c). For $n = 4$, the secular equation is

$$\begin{pmatrix} \alpha & \beta & 0 & 0 \\ \beta & \alpha & \beta & 0 \\ 0 & \beta & \alpha & \beta \\ 0 & 0 & \beta & \alpha \end{pmatrix}\begin{pmatrix} c_a \\ c_b \\ c_c \\ c_d \end{pmatrix} = \epsilon \begin{pmatrix} 1 & 0 & 0 & 0 \\ 0 & 1 & 0 & 0 \\ 0 & 0 & 1 & 0 \\ 0 & 0 & 0 & 1 \end{pmatrix}\begin{pmatrix} c_a \\ c_b \\ c_c \\ c_d \end{pmatrix}. \qquad (14.104)$$

Again, we create symmetry-adjusted functions. This gives us two symmetric and two antisymmetric functions:

$$\chi_1 = \frac{1}{\sqrt{2}}(\chi_a + \chi_d)$$

$$\chi_2 = \frac{1}{\sqrt{2}}(\chi_b + \chi_c)$$

$$\chi_3 = \frac{1}{\sqrt{2}}(\chi_a - \chi_d)$$

$$\chi_4 = \frac{1}{\sqrt{2}}(\chi_b - \chi_c). \qquad (14.105)$$

For this basis set, the secular equation becomes

$$\begin{pmatrix} \alpha & \beta & 0 & 0 \\ \beta & \alpha+\beta & 0 & 0 \\ 0 & 0 & \alpha & \beta \\ 0 & 0 & \beta & \alpha-\beta \end{pmatrix}\begin{pmatrix} c_1 \\ c_2 \\ c_3 \\ c_4 \end{pmatrix} = \epsilon \begin{pmatrix} 1 & 0 & 0 & 0 \\ 0 & 1 & 0 & 0 \\ 0 & 0 & 1 & 0 \\ 0 & 0 & 0 & 1 \end{pmatrix}\begin{pmatrix} c_1 \\ c_2 \\ c_3 \\ c_4 \end{pmatrix}. \qquad (14.106)$$

The eigenvalues are then

$$\epsilon_i = \alpha \pm \frac{\beta}{2}(1 \pm \sqrt{5}). \qquad (14.107)$$

We did not attempt to determine the associated eigenvectors.

3. **Problem:** Show that the total energy by applying GIAOs instead of AOs does not depend on the origin of the vector potential when a magnetic field is switched on. NB: This problem is not easy.
   Answer: We will show that

$$E_{e,km} = \sum_{n=1}^{N} \langle \psi_n | \hat{h}_{km}(\vec{r}) | \psi_n \rangle = \sum_{n=1}^{N} \frac{1}{2m} \left\langle \psi_n \left| \left[-i\hbar\vec{\nabla} + \frac{e}{2c}\vec{B} \times (\vec{r} - \vec{R}_G) \right]^2 \right| \psi_n \right\rangle \qquad (14.108)$$

is independent of $\vec{R}_G$ when GIAOs are used instead of AOs. For this purpose, we write

$$\psi_n(\vec{x}) = \sum_{p,a} \chi_{p,a}(\vec{x}) c_{p,a,n} \exp\left[is\frac{e}{2c\hbar}\vec{B} \times (\vec{R}_G - \vec{R}_p) \cdot \vec{r}\right] \qquad (14.109)$$

with $s = 0$ for AOs and $s = 1$ for GIAOs. It is not important for us that the c coefficients also depend on whether GIAOs or AOs are used.
We need

$$\frac{1}{2m}\left[-i\hbar\vec{\nabla} + \frac{e}{2c}\vec{B} \times (\vec{r} - \vec{R}_G)\right]^2 \exp\left[is\frac{e}{2c\hbar}\vec{B} \times (\vec{R}_G - \vec{R}_p) \cdot \vec{r}\right]\chi_{p,a}(\vec{x})$$

$$= \frac{1}{2m}\left[-i\hbar\vec{\nabla} + \frac{e}{2c}\vec{B} \times (\vec{r} - \vec{R}_G)\right]\left[-i\hbar\vec{\nabla} + \frac{e}{2c}\vec{B} \times (\vec{r} - \vec{R}_G)\right]$$

$$\cdot \exp\left[is\frac{e}{2c\hbar}\vec{B} \times (\vec{R}_G - \vec{R}_p) \cdot \vec{r}\right]\chi_{p,a}(\vec{x})$$

$$= \frac{1}{2m}\left[-i\hbar\vec{\nabla} + \frac{e}{2c}\vec{B} \times (\vec{r} - \vec{R}_G)\right]\left\{\left[s\frac{e}{2c}\vec{B} \times (\vec{R}_G - \vec{R}_p) + \frac{e}{2c}\vec{B} \times (\vec{r} - \vec{R}_p)\right]\chi_{p,a}(\vec{x})\right.$$

$$\left. - i\hbar\vec{\nabla}\chi_{p,a}(\vec{x})\right\}\exp\left[is\frac{e}{2c\hbar}\vec{B} \times (\vec{R}_G - \vec{R}_p) \cdot \vec{r}\right]$$

$$= \frac{1}{2m}\left\{\left[-i\frac{e\hbar}{2c}\vec{\nabla} \cdot (\vec{B} \times (\vec{r} - \vec{R}_G))\right.\right.$$

$$+ 2\frac{e}{2c}\vec{B} \times (\vec{r} - \vec{R}_G) \cdot s\frac{e}{2c}\vec{B} \times (\vec{R}_G - \vec{R}_p)$$

$$+ \left(\frac{e}{2c}\vec{B} \times (\vec{r} - \vec{R}_G)\right)^2 + \left(s\frac{e}{2c}\vec{B} \times (\vec{R}_G - \vec{R}_p)\right)^2\right]\chi_{p,a}(\vec{x})$$

$$+ 2\left[s\frac{e}{2c}\vec{B} \times (\vec{R}_G - \vec{R}_p) + \frac{e}{2c}\vec{B} \times (\vec{r} - \vec{R}_G)\right] \cdot (-i\hbar\vec{\nabla}\chi_{p,a}(\vec{x}))$$

$$\left. - \hbar^2\nabla^2\chi_{p,a}(\vec{x})\right\}\exp\left[is\frac{e}{2c\hbar}\vec{B} \times (\vec{R}_G - \vec{R}_p) \cdot \vec{r}\right]. \qquad (14.110)$$

It is relatively easy to show that

$$\vec{\nabla} \cdot (\vec{B} \times (\vec{r} - \vec{R}_G)) = 0. \qquad (14.111)$$

Then the expression in equation (14.110) for $s = 1$ becomes

$$\frac{1}{2m}\left\{\left(\frac{e}{2c}\vec{B} \times (\vec{r} - \vec{R}_p)\right)^2\chi_{p,a} + 2\frac{e}{2c}\vec{B} \times (\vec{r} - \vec{R}_p) \cdot (-i\hbar\vec{\nabla}\chi_{p,a}) - \hbar^2\nabla^2\chi_{p,a}\right\}$$

$$\cdot \exp\left[i\frac{e}{2c\hbar}\vec{B} \times (\vec{R}_G - \vec{R}_p) \cdot \vec{r}\right], \qquad (14.112)$$

while we for $s = 0$ have

$$\frac{1}{2m}\left\{\left(\frac{e}{2c}\vec{B} \times (\vec{r} - \vec{R}_G)\right)^2\chi_{p,a} + 2\frac{e}{2c}\vec{B} \times (\vec{r} - \vec{R}_G) \cdot (-i\hbar\vec{\nabla}\chi_{p,a}) - \hbar^2\nabla^2\chi_{p,a}\right\}. \qquad (14.113)$$

Finally, $E_{e,km}$ for $s = 1$ then becomes

$$E_{e,km} = \frac{1}{2m} \sum_n \sum_{p_1,p_2,a_1,a_2} c^*_{p_1,a_1,n} c_{p_2,a_2,n} \left\langle \chi_{p_1,a_1} \left| \left\{ \left(\frac{e}{2c} \vec{B} \times (\vec{r} - \vec{R}_{p_2}) \right)^2 \chi_{p_2,a_2} \right. \right. \right.$$
$$+ 2\frac{e}{2c} \vec{B} \times (\vec{r} - \vec{R}_{p_2}) \cdot (-i\hbar\vec{\nabla}\chi_{p_2,a_2})$$
$$\left. \left. - \hbar^2\nabla^2\chi_{p_2,a_2} \right\} \exp\left[i\frac{e}{2ch} \vec{B} \times (\vec{R}_{p_1} - \vec{R}_{p_2}) \cdot \vec{r} \right] \right\rangle, \tag{14.114}$$

while the expression for $s = 0$ becomes

$$E_{e,km} = \frac{1}{2m} \sum_n \sum_{p_1,p_2,a_1,a_2} c^*_{p_1,a_1,n} c_{p_2,a_2,n} \left\langle \chi_{p_1,a_1} \left| \left\{ \left(\frac{e}{2c} \vec{B} \times (\vec{r} - \vec{R}_G) \right)^2 \chi_{p_2,a_2} \right. \right. \right.$$
$$+ 2\frac{e}{2c} \vec{B} \times (\vec{r} - \vec{R}_G) \cdot (-i\hbar\vec{\nabla}\chi_{p_2,a_2})$$
$$\left. \left. - \hbar^2\nabla^2\chi_{p_2,a_2} \right\} \right\rangle. \tag{14.115}$$

It can be clearly seen that the expression in equation (14.114) is independent of $\vec{R}_G$, which is not the case for the expression in equation (14.115).

14.17 Problems

1. Use the Hückel theory for the cyclic C_nH_n molecule for $n = 3$ and $n = 4$. The secular equation does not need to be solved.
2. Explain "CI."
3. Explain briefly the basics of density functional theory.
4. Compare the configuration interaction method and the Møller–Plesset method.
5. Compare Kohn–Sham and Hartree–Fock approaches.
6. Describe the term "CISD."
7. What is meant by "polarization functions" and "diffuse functions"?
8. Explain the advantages and disadvantages of using STOs or GTOs in electronic structure calculations.
9. Compare semiempirical and ab initio methods.
10. Explain the terms "Born von Kármán Zone," "Unit Cell," and "Brillouin Zone."
11. Explain briefly various methods used to consider the effects of solvents.
12. Consider a symmetric ABA molecule, and suppose that its molecular orbitals can be described in terms of one atomic orbital per atom,

$$\psi_i = c_{i,A_1}\chi_{A_1} + c_{i,B}\chi_B + c_{i,A_2}\chi_{A_2}. \tag{14.116}$$

Here, χ_{A_1}, χ_B, and χ_{A_2} is the atomic orbital centered on one A atom, on the (middle) B atom, and on the other A atom. The atomic orbitals are orthonormal and, moreover,

$$\langle \chi_{A_1} | \hat{F} | \chi_{A_1} \rangle = \langle \chi_{A_2} | \hat{F} | \chi_{A_2} \rangle = \epsilon_A$$
$$\langle \chi_{A_1} | \hat{F} | \chi_{A_2} \rangle = 0$$
$$\langle \chi_{A_1} | \hat{F} | \chi_B \rangle = -\langle \chi_{A_2} | \hat{F} | \chi_B \rangle = t. \tag{14.117}$$

[Notice the difference to equation (14.93) in the final identity in equatin (14.117)].
Set up the secular equation for this basis set.
Subsequently, use the symmetry-adapted basis functions

$$\chi_1 = \frac{1}{\sqrt{2}} (\chi_{A_1} + \chi_{A_2})$$
$$\chi_2 = \chi_B$$
$$\chi_3 = \frac{1}{\sqrt{2}} (\chi_{A_1} - \chi_{A_2}) \tag{14.118}$$

and set up the secular equation for this basis set.
Finally, determine the orbital energies of this system.

13. Explain briefly what can be calculated using the TD-DFT method.
14. Explain briefly how fluorescence can be studied theoretically.
15. Describe briefly how static electromagnetic fields can be taken into account in an electronic structure calculation.
16. Describe the notation 6-311++G**.
17. Describe aug-cc-pVTZ.
18. Use the Hückel theory for the cyclic C_8H_8 molecule and determine graphically the orbital energies of the π orbitals of this molecule.
19. Compare CI and CASSCF.

15 Larger systems: applications

15.1 Introduction

In principle, one can use the theoretical methods we have discussed so far to calculate properties of molecules with the help of computers. Figure 15.1 shows a schematic representation of how such a calculation will proceed. This scheme will now be briefly explained. The numbers refer to the numbering in Figure 15.1.

1. First, the number and types of the atoms as well as the number of electrons are specified. This means nothing more than deciding which molecule shall be treated.
2. Then the structure is fixed. This corresponds to using the Born–Oppenheimer approximation.
3. It is decided whether a (computationally demanding) ab initio method or a (faster, but perhaps also less accurate) semiempirical method should be used. If a semiempirical method is chosen, go directly to point 6.
4. If an ab initio method is chosen, it must also be decided whether a method based on the density-functional theory or one based on the Hartree–Fock approach should be used.
5. The number and type of the basis functions are selected.
6. Regardless of whether a density-functional (Kohn–Sham) or a Hartree–Fock method is chosen, we need to solve matrix eigenvalue equations of the form

$$\underline{\underline{h}}\,\underline{c}_i = \epsilon_i \underline{\underline{S}}\,\underline{c}_i \tag{15.1}$$

Here, $\underline{\underline{h}}$ is a matrix that depends on the solutions, so that the equations must be solved self-consistently. $\underline{\underline{h}}$ is the matrix for either the Fock operator (if a Hartree–Fock method is used) or the equivalent Kohn–Sham operator (if a density-functional method is used). In both cases, the operator includes the electrostatic potential of the electrons, which depends on the density of the electrons, that is, on the solutions to equation (15.1). Furthermore, the operator includes exchange and, in the case of a density-functional method, correlation effects, which also depend on the density of the electrons. Finally, the operator also includes external potentials such as electrostatic potentials from the nuclei, external, static, electromagnetic fields, and potentials from solvent molecules. On the other hand, the overlap matrix $\underline{\underline{S}}$ is independent of the solutions.

From this equation, the orbital energies ϵ_i as well as the wave functions of the single orbitals, given by the expansion coefficients $\underline{c}_i$, are determined, and with the basis functions $\{\chi_m\}$ the orbitals are given as

$$\psi_i(\vec{r}) = \sum_{m=1}^{N_b} \chi_m(\vec{r}) c_{mi}. \tag{15.2}$$

$\underline{c}_i$ is a column vector with the N_b elements c_{mi} for the ith orbital.

https://doi.org/10.1515/9783110742206-015

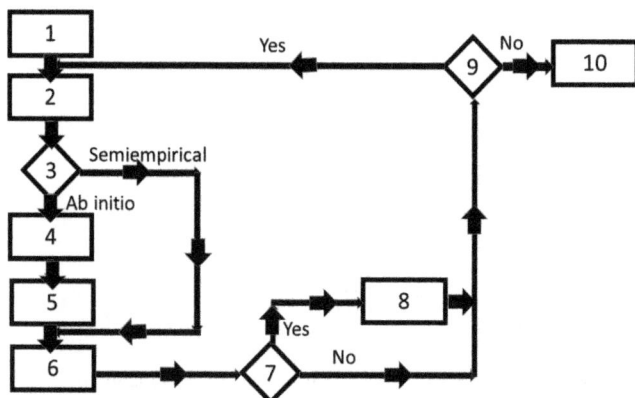

Figure 15.1: A flow chart for the theoretical calculation of properties of any system. For details, see the text.

7. With a Hartree–Fock method, correlation effects could be added at this place.
8. If this is the case, the correlation effects can be included by means of, e. g., Møller–Plesset perturbation theory or the CI or CC method.
9. Next, one has to decide whether another structure should be investigated. This could be the case if, e. g., the structure of the lowest total energy is to be determined.
10. If not, the desired properties will be calculated.

There are many limits to what is possible; see Figure 15.2. The basis sets cannot be made arbitrarily large without the required computation time becoming unacceptably large. Furthermore, very many different structures can rarely be treated, which can be a problem especially for larger molecules, where the computational time requirements are not exactly small anyway. If correlation effects have to be taken into account too, additional, large demands are placed on computer resources, so that limits are also set here. Finally, one will often treat isolated molecules in the gas phase, while the experiments may take place in solutions.

Overall, this means that computer calculations can never completely replace experimental studies, although they can provide very helpful, complementary information. As an example, Figure 15.3 shows schematically the limits with regard to accuracy and complexity that exist for the theoretical treatment of different systems. As an example, Figure 15.4 shows the limits for the treatment of catalytic processes and Figure 15.5 the limits for the systems that were treated theoretically in the AK Springborg in Saarbrücken.

Despite these limitations, theoretical calculations are often helpful especially for not too large molecular systems. In this chapter, we will present some examples of sys-

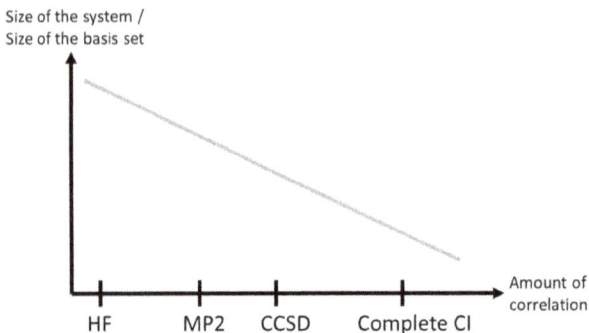

Figure 15.2: The limitations of the possibilities of computer calculations with wave function based methods such as HF, MP2, CCSD, and CI. The straight line symbolizes the limits to what is possible.

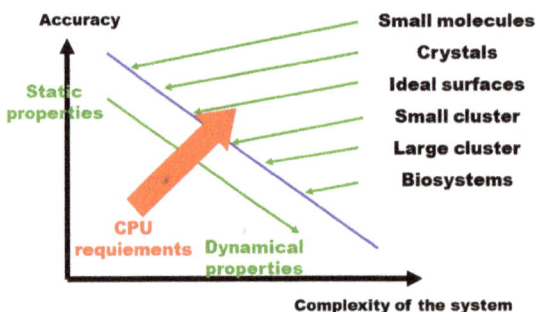

Figure 15.3: Accuracy and complexity of the system under study. The blue line represents schematically the limits of theoretical methods. Furthermore, the requirements for different systems are roughly indicated.

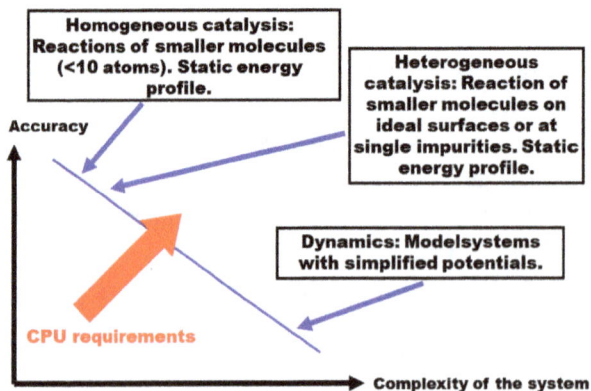

Figure 15.4: As Figure 15.3 but for catalysis as an example.

Our experience

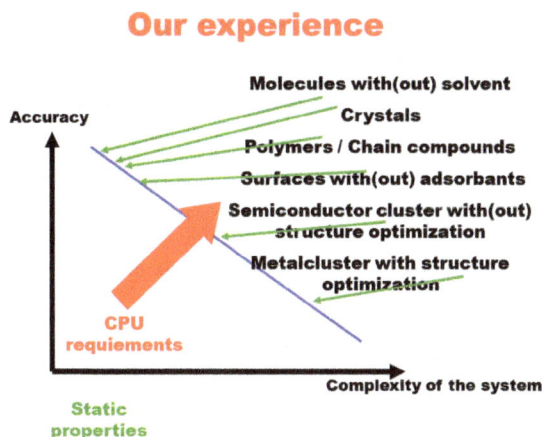

Figure 15.5: As Figure 15.3 but for systems that were treated theoretically in the AK Springborg in Saarbrücken.

tems and properties that can be handled with such computer calculations. Table 15.1 briefly summarizes details about the different methods used in this chapter.

15.2 Structure

With the Born–Oppenheimer approximation, the positions of the nuclei are fixed, and then the electronic properties are calculated for this structure. This also includes the electronic energy, and thus also the total energy E of the molecule for this structure. By varying the structure, it is possible to identify the structure of the lowest total energy (although this may not be easy).

The energy as a function of the structure, i. e., E as a function of the coordinates of the nuclei,

$$E = E(\vec{R}_1, \vec{R}_2, \ldots, \vec{R}_M) \equiv E(\vec{R}), \tag{15.3}$$

forms the so-called potential-energy surface (PES).

As we can see, E is a function of $3M$ coordinates, and finding a (local or global) minimum of this function can be difficult if you simply try to vary the coordinates of the nuclei by hand. It is helpful if the forces acting on the nuclei can also be determined. The force acting on the kth nucleus is given by

$$\vec{F}_k = -\vec{\nabla}_{\vec{R}_k} E = \left(-\frac{dE}{dR_{kx}}, -\frac{dE}{dR_{ky}}, -\frac{dE}{dR_{kz}} \right), \tag{15.4}$$

where

$$\vec{R}_k = (R_{kx}, R_{ky}, R_{kz}) \tag{15.5}$$

is the position vector of the kth nucleus.

Table 15.1: List of abbreviations for details of the methods used in this chapter.

Abbreviation	Meaning
MNDO	A semiempirical method
AM1	A semiempirical method
HF	The Hartree–Fock Method
MP2	The Møller–Plesset method, which uses perturbation theory to account for second-order correlation effects
CCSD(T)	A coupled-cluster method that accounts for single-, double-, and partly triple excitations
MCSCF	A method related to CI
Xα	An LDA method that takes into account only exchange but not correlation effects
LDA	A density-functional method according to which the potential $V_{xc}(\vec{r})$ in the point $\vec{r}$ depends only on $\rho(\vec{r})$ in the same point
GGA	A density-functional method, according to which the potential $V_{xc}(\vec{r})$ at the point $\vec{r}$ depends on $\rho(\vec{r})$, $|\vec{\nabla}\rho(\vec{r})|$, and $\nabla^2\rho(\vec{r})$ in the same point
BLYP	A GGA method
ACM	A hybrid method that combines HF, LDA, and GGA methods
B3LYP	A commonly used hybrid method combining HF, LDA, and GGA methods
3-21G	A small basis set, according to which a function consisting of three contracted Gaussian functions is used for core electrons, while two functions (consisting of two and one contracted Gaussian functions) are used for the valence electrons
6-31G*	Same as 3-21G, except that the contracted functions consist of more Gaussian functions, and that polarization functions are used for the heavier atoms (not H)
6-31G++	Same as 6-31G*, except that no polarization functions are used, but diffuse functions on all atoms (including H) are used

From equation (15.4), it can be seen that E gets smaller if you displace the kth nucleus slightly along $\vec{F}_k$. This can be done for all nuclei at the same time and can then make it possible to find a minimum of the total energy faster. So, for every nucleus, you change

$$\vec{R}_k \rightarrow \vec{R}_k + \tau\vec{F}_k, \tag{15.6}$$

until all forces $\{\vec{F}_k\}$ are very small. τ is a preselected constant. This is the so-called steepest-descent method.

It is possible to derive analytical expressions for the forces, so that their calculation is relatively easy. As a result, structural properties of molecules can be calculated "automatically" although most methods identify "only" the closest minimum on the PES and not necessarily the global total-energy minimum.

Alternatively, molecular dynamics may be used. The force $\vec{F}_k$ on the kth nucleus implies that this nucleus experiences an acceleration $\vec{a}_k$ (Newton's law)

$$\vec{F}_k = M_k \cdot \vec{a}_k \tag{15.7}$$

with M_k equal to the mass of the nucleus. We introduce a time coordinate t and consider the three times $t - \Delta t$, t and $t + \Delta t$, where Δt is a preselected small time interval.

With $\vec{v}_k$ as the velocity of the kth nucleus, we get by means of Tayler series

$$\vec{R}_k(t + \Delta t) = \vec{R}_k(t) + \vec{v}_k(t) \cdot \Delta t + \frac{1}{2}\vec{a}_k(t) \cdot (\Delta t)^2 + \cdots$$

$$\vec{R}_k(t - \Delta t) = \vec{R}_k(t) - \vec{v}_k(t) \cdot \Delta t + \frac{1}{2}\vec{a}_k(t) \cdot (\Delta t)^2 + \cdots . \tag{15.8}$$

When truncating this after the terms of second order in Δt, we get easily

$$\vec{R}_k(t + \Delta t) = 2\vec{R}_k(t) - \vec{R}_k(t - \Delta t) + \vec{a}_k(t) \cdot (\Delta t)^2$$

$$= 2\vec{R}_k(t) - \vec{R}_k(t - \Delta t) + \frac{1}{M_k}\vec{F}_k(t) \cdot (\Delta t)^2. \tag{15.9}$$

Thus, by knowing the coordinates at two times, we can automatically determine the coordinates at a later time. As a result, molecular dynamic calculations are also possible.

The procedure we have presented here is the so-called Verlet procedure. There are (many) other methods, but the principle remains the same: One calculates the temporal evolution of a structure by means of the calculated forces acting on the atomic nuclei.

The fact that different methods for calculating the total energy E can lead to different results is illustrated in Figure 15.6 and Table 15.2, where structural and energetic properties for the reaction of vinyl alcohol → acetaldehyde, CH_2CHOH → CH_3CHO, are shown. Here, results are shown that were obtained with semiempirical methods, with Hartree–Fock calculations with two different basis sets, from Møller–Plesset calculations, and from calculations with different density functional methods. Above all, the very different activation energies and reaction energies are important to mention. The scatter in the calculated values for these properties demonstrates how difficult it is to get accurate relative energies and, accordingly, that theoretical results must be considered with caution.

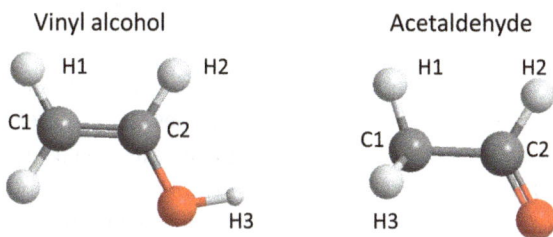

Figure 15.6: Structure of the reactant and product molecules of the reaction vinyl alcohol → acetaldehyde, CH_2CHOH → CH_3CHO.

Similar results are shown in Table 15.3. This table shows calculated C-C, C=C, C-N, and C-O bond lengths for some smaller, organic molecules. It can be seen that the Hartree–Fock values obtained with the smaller 3-21G basis set occasionally show significant

Table 15.2: Structural and energetic properties of the reaction vinyl alcohol → acetaldehyde, $CH_2CHOH \rightarrow CH_3CHO$. MNDO and AM1 mark two semiempirical methods, 3-21G designates Hartree–Fock calculations with a small basis set, while all other calculations were performed with the larger basis set 6-31G*. MP2 are calculations using the Møller–Plesset method, BLYP is a GGA, and ACM is a hybrid method. Exp. represents experimental results. E_A is the activation energy and ΔE is the reaction energy. Bond lengths and angles are given in Å and degrees, respectively, and energies in kcal/mol. A-B marks bond lengths between atom A and B, while A-B-C labels the bond angles between atoms A, B, and C. For the numbering of the atoms, see Figure 15.6.

Parameter	MNDO	AM1	3-21G	HF	MP2	BLYP	ACM	Exp.
Vinyl alcohol								
C2-O	1.357	1.372	1.377	1.347	1.368	1.377	1.357	1.369
C1-C2	1.350	1.336	1.314	1.318	1.337	1.345	1.334	1.335
O-H3	0.948	0.968	0.966	0.949	0.975	0.983	0.969	0.962
C2-H2	1.099	1.103	1.069	1.073	1.085	1.094	1.087	1.080
C1-C2-O	126.5	125.1	127.1	127.0	126.8	127.4	127.4	126.0
C2-O-H3	113.4	109.0	112.7	110.3	108.1	108.0	108.6	108.5
Transition state								
C2-O	1.280	1.296	1.282	1.252	1.295	1.302	1.283	
C1-C2	1.458	1.424	1.421	1.421	1.406	1.419	1.409	
C2-H2	1.090	1.096	1.072	1.081	1.092	1.102	1.094	
C1-H3	1.546	1.572	1.550	1.519	1.520	1.542	1.509	
O-H3	1.267	1.335	1.272	1.234	1.294	1.315	1.284	
C1-C2-O	103.6	107.0	108.3	109.2	110.9	111.2	110.6	
O-H3-C1	99.5	97.3	101.5	104.3	104.3	103.5	104.6	
Acetaldehyde								
C2-O	1.221	1.231	1.208	1.188	1.221	1.223	1.209	1.210
C1-C2	1.517	1.490	1.507	1.504	1.517	1.520	1.504	1.515
C2-H2	1.112	1.114	1.087	1.095	1.112	1.125	1.114	1.128
C1-C2-O	125.0	123.5	124.8	124.4	125.0	124.9	124.8	124.1
E_A	91.2	73.6	76.9	70.0	55.4	48.7	52.3	39.4
ΔE	−7.4	−8.0	−9.1	−17.8	−17.5	−16.1	−15.5	−9.8

deviations from the experimental values. Furthermore, it can be seen that most often accurate values for the C-C bond lengths are obtained, while the values for the purely covalent C=C bonds and for the partially ionic C-N and C-O bonds can be less accurate. Finally, it is observed that MP2 and B3LYP generally give accurate bond lengths.

Hydrogen bonds have significantly lower binding energies than covalent bonds. This also means that they are very sensitive to numerical inaccuracies. In particular, density-functional calculations with a local-density approximation, which tend to overestimate energies of covalent bonds, suggest the existence of bonds that have a larger covalent contribution than what is realistic. This is illustrated by the results for the intermolecular hydrogen bond between two water molecules (Figure 15.7) and the intramolecular hydrogen bonds of the enol tautomer of malonaldehyde (Figure 15.8). The bond lengths obtained (Tables 15.4 and 15.5) clearly show that the lengths of the

Table 15.3: Experimental and calculated bond lengths (in Å) for different molecules. HF refers to results of Hartree–Fock calculations, while MP2 gives results from Møller–Plesset calculations. B3LYP refers to results with the B3LYP hybrid method. In parentheses, the basis sets are given.

Molecule	Bond	Exp.	HF (3-21G)	HF (6-31G*)	MP2 (6-31G*)	B3LYP (6-31G*)
But-1-in-3-ene	C-C	1.431	1.432	1.439	1.429	1.424
Propyne		1.459	1.466	1.468	1.463	1.461
1,3-Butadiene		1.483	1.479	1.467	1.458	1.458
Propene		1.501	1.510	1.503	1.499	1.502
Cyclopropane		1.510	1.513	1.497	1.504	1.509
Propane		1.526	1.541	1.528	1.526	1.532
Cyclobutane		1.548	1.543	1.548	1.545	1.553
Cyclopropene	C=C	1.300	1.282	1.276	1.303	1.295
Allene		1.308	1.292	1.296	1.313	1.307
Propene		1.318	1.316	1.318	1.338	1.333
Cyclobutene		1.332	1.326	1.322	1.347	1.341
But-1-in-ene		1.341	1.320	1.322	1.344	1.341
1,3-Butadiene		1.345	1.320	1.323	1.344	1.340
Cyclopentadiene		1.345	1.329	1.329	1.354	1.349
Formamide	C-N	1.376	1.351	1.349	1.362	1.362
Methylisocyanide		1.424	1.432	1.421	1.426	1.420
Trimethylamine		1.451	1.471	1.445	1.455	1.456
Aziridine		1.475	1.490	1.448	1.474	1.474
Nitromethane		1.489	1.497	1.481	1.488	1.499
Formic acid	C-O	1.343	1.350	1.323	1.352	1.347
Furan		1.362	1.377	1.344	1.367	1.364
Dimethyl ether		1.410	1.435	1.392	1.416	1.410
Oxirane		1.436	1.470	1.401	1.439	1.430

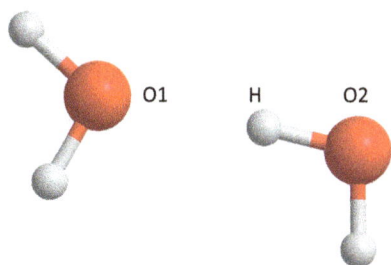

Figure 15.7: Structure of two water molecules connected by a hydrogen bond.

hydrogen bonds are significantly too small when calculated with LDA, and slightly too long with Hartree–Fock methods. For the LDA results in Table 15.5, it is even difficult to distinguish between the two bonds between H and its two neighboring O atoms.

Finally, we mention that in cases where little is known about the structure of a molecule, it can be advantageous to first perform semiempirical calculations that will

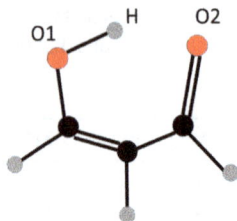

Figure 15.8: Structure of the enol tautomer of malonaldehyde.

Table 15.4: Computed and experimental bond lengths (in Å) of the system of Figure 15.7.

Method	O1–O2	O2–H
LDA	2.710	0.997
GGA	2.877	0.990
HF	2.886	0.948
MP2	2.910	0.976
Exp.	2.98	

Table 15.5: Computed and experimental bond lengths (in Å) of the system of Figure 15.8.

Method	O1–H	O2···H
LDA	1.204	1.220
GGA	1.042	1.568
HF	0.956	1.880
MP2	0.994	1.694
Exp.	0.969	1.680

provide a first, realistic estimate of the structure. This structure can subsequently be used as initial structure for more accurate, parameter-free calculations. A such approach will be able to reduce the computational demands, sometimes even significantly.

15.3 Vibrations

In Chapter 6, we discussed the vibrational properties of diatomic molecules. Thereby, the harmonic approximation was found to be very helpful. Even for larger molecules, this approach is helpful in calculating their vibrational properties. This means that $E(\vec{R})$ in equation (15.3) is approximated as follows:

$$E(\vec{R}) \simeq E(\vec{R}^e)$$
$$+ \frac{1}{2} \sum_{k_1,k_2=1}^{M} \sum_{\alpha_1,\alpha_2=x,y,z} \frac{\partial^2 E(\vec{R}^e)}{\partial R_{k_1,\alpha_1} \partial R_{k_2,\alpha_2}} (R_{k_1,\alpha_1} - R^e_{k_1,\alpha_1})(R_{k_2,\alpha_2} - R^e_{k_2,\alpha_2}). \quad (15.10)$$

This corresponds to a Taylor series up to second order around the equilibrium structure, which we have characterized through the upper index e. For this, the forces of equation (15.4) vanish so that the 1st order terms in the Taylor series disappear. That the forces have to disappear means that one must first have determined a structure of a minimum of the total energy.

Equation (15.10) defines a matrix,

$$\underline{\underline{H}} = \left(\frac{\partial^2 E(\vec{R}^e)}{\partial R_{k_1,\alpha_1} \partial R_{k_2,\alpha_2}} \right). \tag{15.11}$$

This $3M \times 3M$ matrix is the so-called Hessian. From this matrix, one can define the so-called dynamic matrix. This matrix is given by

$$\underline{\underline{D}} = \left(\frac{1}{\sqrt{M_{k_1} M_{k_2}}} \frac{\partial^2 E(\vec{R}^e)}{\partial R_{k_1,\alpha_1} \partial R_{k_2,\alpha_2}} \right), \tag{15.12}$$

where M_k is the mass of the kth atom. It can then be shown (which will not be done here) that the eigenvalues of this matrix are equal to the squares of the vibrational frequencies. Furthermore, the eigenvectors describe the vibrational patterns, i. e., the $\vec{u}_{nk}$ in equation (6.2).

In Table 15.6, we show calculated vibrational frequencies from Hartree–Fock calculations compared to experimental values. One recognizes a general problem here: the vibrational frequencies from Hartree–Fock calculations are fundamentally too large. It turns out, however, that this problem does not stem directly from the Hartree–Fock approximation, but rather from the harmonic approximation, that is, the approximation that the total energy as a function of the positions of the nuclei can be terminated after the second-order terms [equation (15.10)]. If one also considers higher-order terms (so-called anharmonic corrections), the results improve significantly, as shown in Table 15.7.

Table 15.6: Calculated and experimental vibrational frequencies in cm^{-1}. The theoretical values were obtained by means of HF calculations.

Molecule	Theory	Exp.	Molecule	Theory	Exp.
CH_3	3321	3184	NH_3	3985	3444
	3125	3002		3781	3336
	1470	1383		1814	1627
	776	580		597	950
CH_4	3372	3019	OH	3955	3735
	3226	2917	H_2O	4143	3756
	1718	1534		3987	3657
	1533	1306		1678	1595
NH_2	3676	3220	HF	4150	4138
	3554	3173	H_2	4644	4405
	1651	1499			

Table 15.7: Calculated and experimental vibrational frequencies in cm^{-1} for ethane, $(CH_3)_2$. Both the usual harmonic frequencies ("Harm") are given as well as those obtained by considering anharmonic corrections ("Anharm").

Mode	B3LYP Harm	B3LYP Anharm	HF Anharm	BLYP Harm	BLYP Anharm	Exp
1	3093	2953	2945	3020	2875	2978
2	3068	2932	2923	2994	2854	2955
3	3025	2870	2867	2958	2800	2920
4	3024	2868	2867	2956	2797	2915
5	1507	1462	1458	1473	1427	1472
6	1503	1456	1452	1469	1422	1468
7	1423	1391	1387	1385	1352	1388
8	1413	1379	1376	1381	1346	1379
9	1223	1191	1188	1191	1159	1190
10	995	972	969	958	934	995
11	827	821	823	809	802	822
12	305	273	267	297	265	289

Of mathematical reasons, from equation (15.10) we recognize also that $\vec{R}^e$ is only then a minimum of the total energy, if all eigenvalues of the Hessian are nonnegative. Six (or five if the molecule is linear) eigenvalues will be zero, and the associated eigenvectors of the Hessian will describe the translation and rotation of the total molecule. The same holds true for the eigenvalues and -vectors of the dynamic matrix. Therefore, it is customary to calculate the vibrational frequencies after the structure of a molecule has been optimized. It is then ascertained that all eigenvalues, except for the six (or five) that describe the translation and rotation of the molecule, all are positive, so that no imaginary vibrational frequencies are found. Such would imply that the structure being considered does not correspond to a minimum but rather to a saddle point on the PES. Such cases may, e. g., occur when one assumes too high a symmetry of the system. For instance, if you assume that H_2O is a linear H-O-H molecule, you will get a linear molecule at the end of the calculation, but then there will be imaginary vibrational frequencies and corresponding modes that indicate that the molecule wants to bend. Thereby, the symmetry will be reduced.

15.4 Total and relative energies

For the application of theoretical methods for the treatment of chemical issues, it is crucial that the energy changes associated with chemical reactions are accurately reproduced. In this section, we will discuss the extent to which this is the case for the various theoretical methods we have presented.

A first example has already been shown in Table 15.2, where the activation energy E_A and the reaction energy ΔE show large fluctuations depending on the theoretical approach. As a second example, we show in Table 15.8 dissociation energies for small, diatomic, homonuclear molecules, using various theoretical methods. The results show that the Hartree–Fock method results in too low binding energies, a result that is generally valid. It should also be noted that according to this method, F_2 is only metastable (the energy of the two noninteracting atoms is lower than that of the molecule), and that Be_2 is not stable at all. On the other hand, the local density approximation within the density functional theory tends to yield too large binding energies, which also corresponds to a general finding. We note that improved approaches within the density-functional theory (e. g., GGA) lead to accurate binding energies, which is also true for, e. g., MP2 calculations and other methods that add correlation effects to Hartree–Fock calculations.

Table 15.8: Experimental and calculated binding energies (in eV) for diatomic, homonuclear molecules. HF refers to results from Hartree–Fock calculations, while LDA denotes density functional calculations with a local approximation. Exp. represents the experimental values, and $X\alpha$ is an LDA approximation to the HF method. For Be_2, there is no stable structure with the HF approximation.

Molecule	Exp.	LDA	$X\alpha$	HF
H_2	4.75	4.91	3.59	3.64
Li_2	1.07	1.01	0.21	0.17
Be_2	0.10	0.50	0.43	
B_2	3.09	3.93	3.79	0.89
C_2	6.32	7.19	6.00	0.79
N_2	9.91	11.34	9.09	5.20
O_2	5.22	7.54	7.01	1.28
F_2	1.66	3.32	3.04	−1.37

Table 15.9 shows similar results for Cu_2 and Cr_2. Both elements are transition metals and accordingly have many $3d$ orbitals, which are almost all occupied for Cu, while they are only approximately half occupied for Cr_2. Because they are energetically high at the same time (near the Fermi energy, i. e., the energy separating occupied and unoccupied orbitals), we can use the Møller–Plesset theory to realize that correlation effects are probably important, especially for Cr_2. That this is true is clearly seen in Table 15.9.

Because of the (albeit fairly small) inaccuracies related to the total energies of various structures, the relative energy of different isomers can become very inaccurate. If each of the two energies E_1 and E_2 are underestimated or overestimated (see e. g., Table 15.2), the difference, $\Delta E = E_1 - E_2$, can be very inaccurate. Therefore, the accuracy found in Table 15.10 for the systems of Figure 15.9 is not at all self-evident.

For energetic properties, in recent years, the B3LYP method and the Møller–Plesset method have established themselves as accurate, reliable methods with the B3LYP method being computationally less intensive. The accuracy of the B3LYP

Table 15.9: Experimental and calculated properties for Cu_2 (upper-half) and Cr_2 (lower-half). HF refers to results from Hartree–Fock calculations, while LDA density denotes functional calculations with a local approximation. Exp. are the experimental values, and $X\alpha$ is an LDA approximation to the HF method. Finally, HF+Corr. denotes results using different methods based on HF calculations but also taking into account correlation effects. R_e is the bond length (in atomic units), D_e the bond energy (in eV), and ω_e the vibration frequency (in cm^{-1}).

Molecule	Method	R_e (a. u.)	ω_e (cm^{-1})	D_e (eV)
Cu_2	Exp.	4.195	265	1.97
	LDA	4.10–4.30	248–330	2.30–2.65
	$X\alpha$	4.12–4.20	286–290	2.10–2.16
	HF	4.58–4.61	198	0.51–0.56
	HF+Corr.	4.23–4.62	200–242	0.15–2.07
Cr_2	Exp.	3.17	470	1.56±0.3, 1.44±0.02
	LDA	3.17–3.21	441–470	1.80–2.80
	$X\alpha$	5.10–5.20	92–110	0.4–1.0
	HF	<1.5–2.95	7.50	
	HF+Corr.	3.04–6.14	70–396	0.1–1.86

Table 15.10: Calculated relative energies (in kcal/mol) for the structures in Figure 15.9. The MCSCF method is related to the CI method, so the results of this method can be considered the most accurate ones.

Method	I	II	III	IV	V
LDA	21	66	0	5	46
GGA	22	70	0	5	49
HF	20	89	0	7	
MP2	28	78	0	6	
MCSCF	35	86	0	7	66

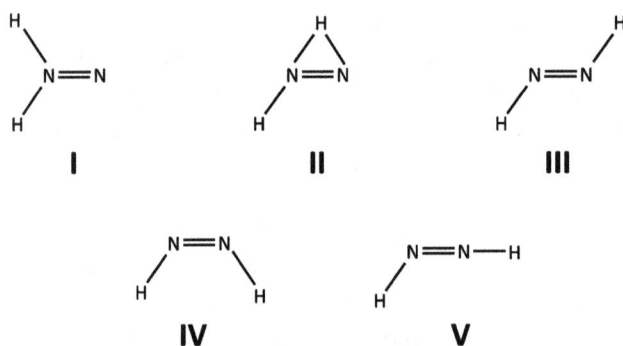

Figure 15.9: Different structures of N_2H_2, diazene.

method can be seen in Tables 15.11, 15.12, and 15.13. Table 15.11 presents results on atomization energies. In Table 15.12, the ionization potential of the molecule X is calculated as the energy of the reaction

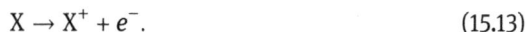

$$X \to X^+ + e^-. \tag{15.13}$$

Similarly, the proton affinity of a molecule X is the energy of the reaction

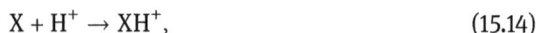

$$X + H^+ \to XH^+, \tag{15.14}$$

which is largely the energy of a bond between a molecule and a hydrogen nucleus.

Table 15.11: Calculated atomization energies (in kcal/mol) using a hybrid method in comparison to experimental values.

Molecule	Exp.	Hybrid	Molecule	Exp.	Hybrid
H_2	103.5	101.6	LiH	56.0	52.9
CH	79.9	79.9	$CH_2(^3B_1)$	179.6	184.1
$CH_2(^1A_1)$	170.6	168.2	CH_3	289.2	292.6
CH_4	392.5	393.5	NH	79.0	81.3
NH_2	170.0	173.1	NH_3	276.7	276.8
OH	101.3	101.9	H_2O	219.3	217.0
HF	135.2	133.3	Li_2	24.0	17.9
LiF	137.6	131.7	C_2H_2	388.9	389.0
C_2H_4	531.9	534.3	C_2H_6	666.3	668.7
CN	176.6	176.7	HCN	301.8	302.4
CO	256.2	253.4	HCO	270.3	273.7
H_2CO	357.2	357.9	CH_3OH	480.8	480.8
N_2	225.1	230.0	N_2H_4	405.4	407.2
NO	150.1	151.5	O_2	118.0	123.1
H_2O_2	252.3	249.8	F_2	36.9	35.6
CO_2	381.9	385.1	$SiH_2(^1A_1)$	144.4	142.8
$SiH_2(^3B_1)$	123.4	126.4	SiH_3	214.0	213.3
SiH_4	302.8	300.0	PH_2	144.7	146.8
PH_3	227.4	225.6	H_2S	173.2	172.7
HCl	102.2	102.0	Na_2	16.6	13.2
Si_2	74.0	76.3	P_2	116.1	112.2
S_2	100.7	105.8	Cl_2	57.2	58.6
NaCl	97.5	92.6	SiO	190.5	184.2
CS	169.5	166.9	SO	123.5	126.5
ClO	63.3	66.6	ClF	60.3	60.7
Si_2H_6	500.1	496.7	CH_3Cl	371.0	373.2
CH_3SH	445.1	446.2	HOCl	156.3	156.2
SO_2	254.0	251.4	BeH	46.9	54.5

Table 15.12: Calculated first ionization potentials (in eV) obtained using a hybrid method compared to experimental values.

Molecule	Exp.	Hybrid	Molecule	Exp.	Hybrid
H	13.60	13.71	He	24.59	24.71
Li	5.39	5.56	Be	9.32	9.02
B	8.30	8.71	C	11.26	11.58
N	14.54	14.78	O	13.61	13.95
F	17.42	17.58	Ne	21.56	21.60
Na	5.14	5.27	Mg	7.65	7.57
Al	5.98	6.12	Si	8.15	8.25
P	10.49	10.57	S	10.36	10.48
Cl	12.97	13.04	Ar	15.76	15.80
CH_4	12.62	12.47	NH_3	10.18	10.12
OH	13.01	13.09	H_2O	12.62	12.54
HF	16.04	15.99	SiH_4	11.00	10.85
PH	10.15	10.31	PH_2	9.82	10.03
PH_3	9.87	9.81	SH	10.37	10.43
$SH_2(^2B_1)$	10.47	10.42	$SH_2(^2A_1)$	12.78	12.64
HCl	12.75	12.74	C_2H_2	11.40	11.23
C_2H_4	10.51	10.36	CO	14.01	14.05
$N_2(^2\Sigma_g)$	15.58	15.77	$N_2(^2\Pi_u)$	16.70	16.65
O_2	12.07	12.46	P_2	10.53	10.41
S_2	9.36	9.58	Cl_2	11.50	11.35
ClF	12.66	12.55	CS	11.33	11.34

Table 15.13: Calculated proton affinities (in kcal/mol) obtained with a hybrid method compared to experimental values.

Molecule	Exp.	Hybrid
H_2	100.8	100.9
C_2H_2	152.3	157.0
NH_3	202.5	204.4
H_2O	165.1	165.7
SiH_4	154.0	153.9
PH_3	187.1	186.1
H_2S	168.8	168.9
HCl	133.6	134.6

Due to the often reliable results (but unfortunately with exceptions) of the B3LYP method, this, among a few others, has established itself as a standard approach, so that results are automatically considered credible with it. However, because there may be deviations, it is not advisable to have 100 % confidence in results from B3LYP calculations. Control of the credibility and accuracy of the results is always very important.

Also, for structures of different molecules consisting of the same atoms, but having different bonding patterns (tautomers and isomers), one can often obtain very accurate relative energies, although some methods give less accurate results. This is illustrated in Table 15.14. The small 3-21G basis set together with the Hartree–Fock method does not produce accurate results, but even with the larger 6-31G* basis set you will not always get accurate results. However, such ones are usually obtained using the MP2 and B3LYP methods, with neither method being systematically better than the other.

Table 15.14: Calculated and experimental energy differences (in kcal/mol) between different isomers. The numbers are the energy of structure 1 minus the energy of structure 2.

Structure 1	Structure 2	HF (3-21G)	HF (6-31G*)	MP2 (6-31G*)	B3LYP (6-31G*)	Exp.
C_2H_3N	C_2H_3N					
Acetonitrile	methyl isocyanide	88	100	121	113	88
C_2H_4O	C_2H_4O					
Acetaldehyde	Oxirane	142	130	113	117	113
$C_2H_4O_2$	$C_2H_4O_2$					
Acetic acid	dimethyl ether	25	29	38	21	50
C_3H_4	C_3H_4					
Propine	Allene	13	8	21	−13	4
Propine	Cyclopropene	167	109	96	92	92
C_3H_6	C_3H_6					
Propene	Cyclopropane	59	33	17	33	29
C_4H_6	C_4H_6					
1,3-Butadiene	2-Butyne	17	29	17	33	38
1,3-Butadiene	Cyclopropen	75	54	33	50	46
1,3-Butadiene	Bicyclo(1,1,0)Butane	192	126	88	117	109

As a further example, Table 15.15 shows results for the transition-metal complexes $M(CO)_6$ with M = Cr, Mo, or W. As previously mentioned, we can expect for such transition-metal-containing compounds that correlation effects are important and, accordingly, even MP2 calculations may not be accurate, whereas the more advanced CCSD(T) method (related to CC) provides reliable results, although at a considerably increased computational effort. Interestingly, the methods based on density functional theory are accurate. Ultimately, W is an atom for which relativistic effects play a not quite negligible role, as the table also shows.

If methods are used that provide exact relative energies for the problem you have, quite detailed results can be obtained. As an example, Figure 15.10 shows the variation of the total energy of an acrolein molecule when rotating around the central C–C bond.

Finally, we shall discuss a small problem that often is met. Imagine that we want to calculate the interaction energy between two systems (atoms, molecules, . . .), A and B.

Table 15.15: Calculated and experimental properties of M(CO)$_6$ molecules with M = Cr, Mo, and W. M–C indicates the bond length between the metal and a carbon atom, and the dissociation energy ΔE is for the reaction M(CO)$_6$ → M(CO)$_5$ + CO. The calculation with +R also includes relativistic effects, whereas CCSD(T) represents a method related to CC.

Method	Cr–C Å	Mo–C Å	W–C Å	ΔE kcal/mol
LDA	1.866	2.035		
GGA	1.910	2.077	2.116	38.8
GGA+R			2.049	43.7
B3LYP	1.921	2.068	2.078	44.8
HF	2.00			37.7
MP2	1.883	2.066	2.060	54.9
CCSD(T)	1.939			48.0
Exp.	1.918	2.063	2.058	46.0 ± 2

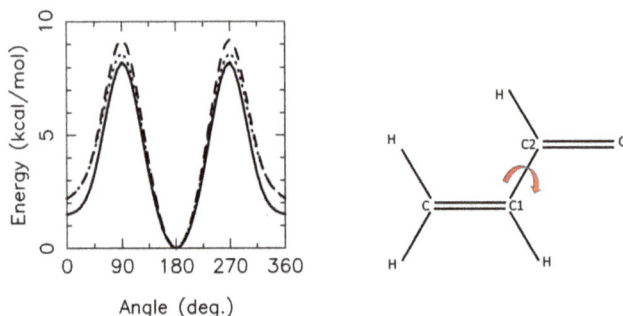

Figure 15.10: The variation in the total energy of an acrolein molecule (right half) when rotated around the central C1–C2 bond. The solid, dashed, and dotted curves show results from calculations using the MP2 method, an LDA method, and a GGA method, respectively.

We first make a calculation for the isolated A and B systems, using, e. g., for the study of A some basis set. If we added further basis functions to this calculation, the total energy of A would become lower—a simple consequence of the variational theorem. This is, e. g., the case when we add the basis functions that we subsequently use for the calculation for B, and when we center these basis functions where the B system is placed for the complete A–B system. Thus, simply by including the basis functions of the other system for each of the isolated A and B systems, the total energies of the individual A and B systems become lower. However, since we do not (normally) do so, we will erroneously predict a too large interaction energy between A and B by directly comparing the total energy of the AB system with those of the noninteracting A and B systems. The error is called Basis Set Superposition Error (BSSE). To correct the error, one proceeds just as indicated above: one carries out calculations for the isolated systems, which also include the basis functions of the other system, the other system being placed as in the A–B system. This is the so-called counterpoise method.

15.5 Dipole moment

For a set of point charges $\{q_i, i = 1, 2, \ldots, N\}$, which are somehow distributed in space, the dipole moment is defined as

$$\vec{\mu} = \sum_i q_i \vec{r}_i. \tag{15.15}$$

Here, $\vec{r}_i$ is the position vector of the ith charge.

By shifting the whole system by $\vec{r}_0$,

$$\vec{r}_i \to \vec{r}_i + \vec{r}_0, \tag{15.16}$$

the dipole moment changes as follows:

$$\vec{\mu} \to \sum_i q_i(\vec{r}_i + \vec{r}_0) = \left(\sum_i q_i \vec{r}_i \right) + \left(\sum_i q_i \vec{r}_0 \right) = \vec{\mu} + \left(\sum_i q_i \right) \vec{r}_0 = \vec{\mu} + Q\vec{r}_0, \tag{15.17}$$

with Q equal to the total charge of the system. From this, we learn that the dipole moment for a neutral system is independent of the choice of the origin of the coordinate system.

For a neutral molecule, we have nuclei, which can be thought of as point charges, and the electron density, which forms a continuous delocalized charge "cloud." The dipole moment of this system is then obtained by summing over the nuclei and integrating over the electron density,

$$\vec{\mu} = \sum_{k=1}^{M} Z_k e \vec{R}_k - e \int \rho(\vec{r}) \vec{r} d\vec{r}. \tag{15.18}$$

Using theoretical methods, we determine the electron density $\rho(\vec{r})$ by making the total energy of the system as low as possible. The potential felt by the electrons goes to $-\infty$ at the nuclear sites, i.e., it becomes very small. This means that the energy minimization process gives an electron density that is particularly well described at the nuclear sites. Far from the nuclei, the density may be much less accurate.

But this is the region where the electron density contributes most to the overall dipole moment because $|\vec{r}|$ there is largest, if one conveniently places the coordinate origin in the middle of the system (for a neutral system, this choice is irrelevant, as we have seen above). Therefore, the theoretical description of the dipole moment is often not as accurate as that of other properties. This can be recognized immediately in Table 15.16 (see also Figure 15.11). Here, we show results for the HF molecule, which

Table 15.16: Results of calculations for the HF molecule with different basis sets. E denotes the total energy (in Hartree), R_e the optimized bond length (in bohr), and μ the dipole moment (in units of elementary charge times bohr). In the first row, the notation describes the primitive basis set, in this case $6s3p/3s$, which means that 6 Gaussian functions of s type and 3 Gaussian functions of each p type (i. e., p_x, p_y, and p_z) centered on the fluorine atom were used, while 3 Gaussian functions of s type centered on the hydrogen atom were used. Subsequently, the basis set is contracted, so that from these Gaussian functions 2 linear combinations of s type centered on fluorine, one linear combination of each of the p_x, p_y, and p_z types also centered on fluorine, and finally 1 linear combination of s type functions centered on hydrogen were used. The same applies to the other rows, and it can be seen that with the larger basis sets d functions on F and p functions on H were also used.

Primitive basis set	Contracted basis set	E (a. u.)	R_e (a. u.)	μ (a. u.)
$6s3p/3s$	$2s1p/1s$	−98.572844	1.8055	0.49258
$12s6p/6s$	$2s1p/1s$	−99.501718	1.8028	0.51000
$8s4p/4s$	$3s2p/2s$	−99.887286	1.7410	0.89971
$10s4p/4s$	$3s2p/2s$	−99.983425	1.7386	0.90487
$9s5p/4s$	$3s2p/3s$	−100.018895	1.7467	0.95544
$9s5p/4s$	$3s2p/2s$	−100.020169	1.7475	0.96334
$9s5p/5s$	$3s2p/3s$	−100.020665	1.7376	0.96256
$9s5p/4s$	$4s3p/2s$	−100.022946	1.7390	0.93645
$11s6p/5s$	$4s2p/3s$	−100.026364	1.7422	0.91244
$9s5p/4s2p$	$3s2p/2s1p$	−100.034266	1.7257	0.87851
$10s6p/5s$	$5s3p/3s$	−100.036872	1.7380	0.93757
$10s6p/5s$	$5s4p/3s$	−100.037008	1.7371	0.93656
$9s5p/4s2p$	$4s3p/2s1p$	−100.040470	1.7046	0.83604
$11s6p/5s2p$	$4s2p/3s1p$	−100.044050	1.7168	0.84243
$10s6p/5s2p$	$5s4p/3s1p$	−100.044751	1.7206	0.81251
$9s5p2d/4s2p$	$3s2p1d/2s1p$	−100.049112	1.7053	0.74383
$9s5p2d/4s2p$	$4s3p1d/2s1p$	−100.049799	1.7046	0.74154
$11s6p2d/5s2p$	$4s2p1d/3s1p$	−100.057755	1.7036	0.69515
$10s6p1d/5s2p$	$5s4p1d/3s1p$	−100.059724	1.7078	0.74436
$10s6p2d/5s2p$	$5s3p1d/3s1p$	−100.062343	1.7027	0.74871

were obtained with basis sets of different sizes. It is seen that the total energy is steadily becoming smaller as a function of increasing size of the basis set (which can be understood as a logical consequence of the variational theorem), and a convergence behavior is recognizable, which is also true for the optimized bond length. On the other hand, the dipole moment shows quite a different behavior. Only with good will one may recognize a tendency toward convergence for the largest basis sets.

The problems in the calculation of the dipole moment can also be seen on Table 15.17. The fact that the calculated dipole moment is subject to some inaccuracy also means that its change, e. g., due to the vibrations or external, electromagnetic fields can be difficult to determine with theoretically. These properties are otherwise very important for spectroscopy and optical properties.

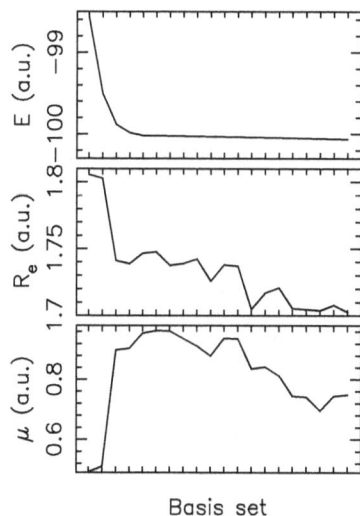

Figure 15.11: Graphical representation of the results from Table 15.16.

Table 15.17: Calculated dipole moment along the major axis of the molecules for different systems. The results are given in D (Debye).

Molecule	Basis set	HF	MP2	LDA	GGA	B3LYP
$NH_2(C_6H_4)NO_2$	6-31G	3.23	2.78	3.23	3.14	3.15
	6-31G++	3.19	2.81	3.30	3.23	3.20
$NH_2(C_6H_4)C_2H_2(C_6H_4)NO_2$	6-31G	3.88	3.16	4.29	4.29	4.01
$NH_2(C_2H_2)_6NO_2$	6-31G	5.87	4.03	6.91	6.73	6.40
$NH_2(C_2H_2)_{12}NO_2$	6-31G	6.73	4.18	10.03	9.73	8.50

15.6 Electron densities

The dipole moment gives limited information on the spatial distribution of the charge in the molecule, but only in the form of three (real) numbers. More information, of course, provides the electron density, $\rho(\vec{r})$. This gives a (real) number in each point in position space.

Representing the electron density is not easy: one cannot plot a fourth quantity as a function of three continuous variables. But there are some solutions to this problem. On the one hand, it is possible to graphically represent the electron density in some prechosen plane. For this, one can use either contour curves or a spatial representation. If one is only interested in the density along a straight line, it can be drawn relatively easily. We have already presented such figures in this manuscript when we have treated the electron density and/or orbitals in small molecules or in atoms.

Another approach is to plot the surface for which

$$\rho(\vec{r}) = \text{constant.} \tag{15.19}$$

The value of this constant has a strong impact on what the surface looks like. For an atom, the electron density as a function of the distance to the nucleus is largely monotonously decreasing. When the atom is incorporated into a molecule, the electron density changes only marginally: The effects of the chemical bonds on the chemical density are really very small. Therefore, the electron density in the vicinity of the individual nuclei also largely decays monotonously and is spherical symmetric. It follows that the largest electron density is found near the nuclei. Because of this, by choosing a large value for the constant in equation (15.19), you get surfaces that are localized around the atomic nuclei. With smaller values, one will increasingly obtain a surface from which one can extract chemical information. All this is illustrated in Figure 15.12.

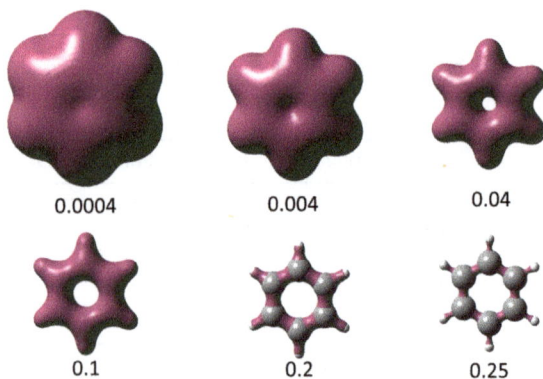

0.0004	0.004	0.04
0.1	0.2	0.25

Figure 15.12: Different surfaces for benzene with constant electron density. The various constant values are also indicated.

Once a reasonable procedure has been found, the plot of the electron density can be quite informative. As an example we show in Figure 15.13, the electron density for Si_7 together with a schematic representation of the bonding of this molecule. This molecule consists of a pentagon of five Si atoms and two Si atoms symmetrically placed above and below this pentagon. To what extent there is a bond between the last two atoms is not clear. In fact, the electron density in a plane with the two atoms and one of the five atoms of the pentagon (left part of Figure 15.13) shows that there is a smaller increase in the electron density between the two single atoms than between neighboring atoms of the pentagon (right part of Figure 15.13). This indicates that there is a weaker chemical bond—if any—between the two Si atoms above and below the pentagon than between the Si atoms of the pentagon.

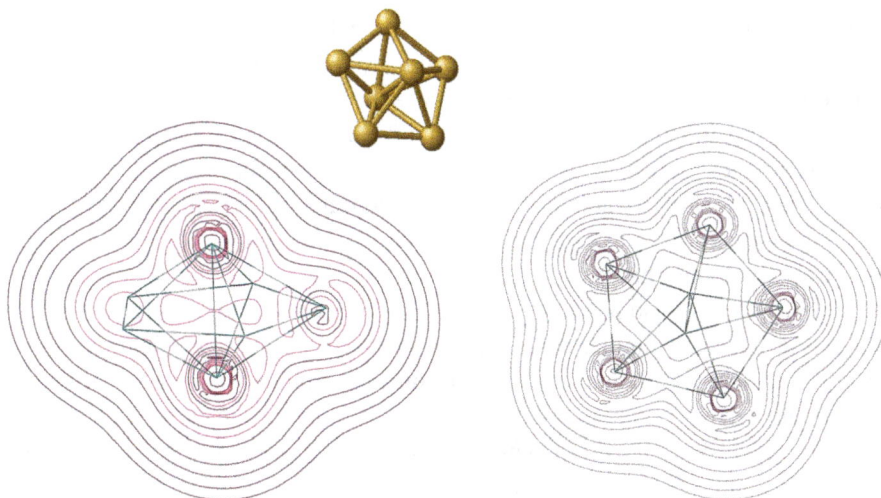

Figure 15.13: The electron density and structure of Si_7.

15.7 Atomic charges

As we have just seen, the electron density is a fairly delocalized cloud around the whole molecule. It is therefore not easy or clear how to divide this density into atomic components. This issue is shown schematically in Figure 15.14. In general, the electron density of a molecule differs only slightly from the superpositioned densities of the individual, isolated, spherical atoms. But since the atoms, as well as their electron densities, do not necessarily have the same size, several, different, meaningful divisions of the electron density can be proposed, as Figure 15.14 shows.

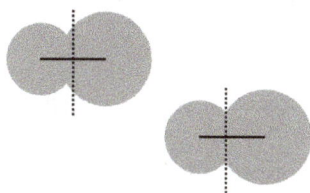

Figure 15.14: Two different attempts to divide the electron distribution of a diatomic molecule into atomic proportions. The left-hand part separates the density according to whether one point is closer to one or the other atomic nucleus, while the right-hand part uses a separation that is more constructed according to the shape of the electron distribution.

Experimentally, however, it is possible to estimate atomic charges. This is often done using ESCA (Electron Spectroscopy for Chemical Analysis). It is used that core electrons are strongly localized near the atomic nuclei, so that they differ only slightly from the orbitals of the isolated atoms. But they already sense how many electrons are in the next vicinity of the nucleus. If one measures the energies of the core electrons for neutral or charged, isolated atoms, one gets different energies. If one makes similar measurements for the atoms in any compound, one gets again different energies. By interpolating the energies obtained for the isolated, neutral, or charged atoms, one can estimate the number of electrons on the atom in the compound. From that, you can extract an atomic charge.

In principle, it is also possible to determine such core electron energies with calculations. However, a simpler method is the so-called Mulliken population analysis (according to Robert Mulliken). This is based on atom-centered basis functions, as we have introduced in Section 14.2. To illustrate the whole process in more detail, we first consider a single orbital of an AB molecule. The orbital is described by means of one basis function, χ_A, of atom A, and one basis function, χ_B, of atom B,

$$\psi = c_A \chi_A + c_B \chi_B. \tag{15.20}$$

To make things even simpler, we assume that all functions and coefficients are real. Because the wave function ψ is normalized, we have

$$1 = \langle \psi | \psi \rangle = c_A^2 \langle \chi_A | \chi_A \rangle + c_B^2 \langle \chi_B | \chi_B \rangle + 2 c_A c_B \langle \chi_A | \chi_B \rangle$$
$$\equiv n_A + n_B + n_{AB} = \left(n_A + \frac{1}{2} n_{AB} \right) + \left(n_B + \frac{1}{2} n_{AB} \right)$$
$$\equiv N_A + N_B. \tag{15.21}$$

Here, n_A and n_B are called the net populations of the two atoms, while n_{AB} is called the overlap population. Finally, N_A and N_B are called gross populations of the two atoms.

In the general case, we explicitly write down the different dependencies of the basis functions and write accordingly

$$\chi_j(\vec{x}) \equiv \chi_{p,l,m,a}(\vec{x}). \tag{15.22}$$

Here, p specifies the atom on which the function is centered, the angular dependence is given by (l, m), and a describes everything else. The latter could be spin-dependence, but also the decay constants for GTOs or STOs [see equations (14.23) and (14.24)] or the main quantum numbers. That the basis functions are assigned to the individual atoms, is what we will use to obtain atomic charges. We emphasize that the method is closely related to the definition of the basis functions and, therefore, that the results can not be considered as absolute values. Rather, they provide tendencies, especially when comparing different, related structures or molecules.

We use that the orbitals are normalized,

$$
1 = \langle \psi_k | \psi_k \rangle
$$

$$
= \left\langle \sum_{p_1,l_1,m_1,\alpha_1} c_{p_1,l_1,m_1,\alpha_1,k} X_{p_1,l_1,m_1,\alpha_1} \Bigg| \sum_{p_2,l_2,m_2,\alpha_2} c_{p_2,l_2,m_2,\alpha_2,k} X_{p_2,l_2,m_2,\alpha_2} \right\rangle
$$

$$
= \sum_{p_1,p_2} \left\{ \sum_{l_1,l_2} \sum_{m_1,m_2} \sum_{\alpha_1,\alpha_2} c^*_{p_1,l_1,m_1,\alpha_1,k} c_{p_2,l_2,m_2,\alpha_2,k} \langle X_{p_1,l_1,m_1,\alpha_1} | X_{p_2,l_2,m_2,\alpha_2} \rangle \right\}
$$

$$
= \sum_p n_{p,k} + \sum_{p_1 \neq p_2} n'_{p_1,p_2,k}, \tag{15.23}
$$

in which

$$
n_{p,k} = \sum_{l_1,l_2} \sum_{m_1,m_2} \sum_{\alpha_1,\alpha_2} c^*_{p,l_1,m_1,\alpha_1,k} c_{p,l_2,m_2,\alpha_2,k} \langle X_{p,l_1,m_1,\alpha_1} | X_{p,l_2,m_2,\alpha_2} \rangle \tag{15.24}
$$

defines the net population on the pth atom of the kth orbital. Furthermore,

$$
n'_{p_1,p_2,k} = \sum_{l_1,l_2} \sum_{m_1,m_2} \sum_{\alpha_1,\alpha_2} \tilde{c}^*_{p_1,l_1,m_1,\alpha_1,k} \tilde{c}_{p_2,l_2,m_2,\alpha_2,k} \langle X_{p_1,l_1,m_1,\alpha_1} | X_{p_2,l_2,m_2,\alpha_2} \rangle \tag{15.25}
$$

is the overlap population between atoms p_1 and p_2 for the kth orbital,

$$
n_{p_1,p_2,k} = n'_{p_1,p_2,k} + n'_{p_2,p_1,k}. \tag{15.26}
$$

The separation into atomic and overlapping populations corresponds to writing

$$
1 = \sum_p n_{p,k} + \sum_{p_1 \neq p_2} n'_{p_1,p_2,k} = \sum_p n_{p,k} + \sum_{p_1 > p_2} n_{p_1,p_2,k}. \tag{15.27}
$$

To remove the overlap populations, we write

$$
1 = \sum_p n_{p,k} + \frac{1}{2} \sum_{p_1 \neq p} n_{p_1,p,k} = \sum_p \left[n_{p,k} + \frac{1}{2} \sum_{p_1 \neq p} n_{p_1,p,k} \right] \equiv \sum_p N_{p,k} \tag{15.28}
$$

with

$$
N_{p,k} = n_{p,k} + \frac{1}{2} \sum_{p_1 \neq p} n_{p_1,p,k} \tag{15.29}
$$

equal to the so-called gross population of the pth atom and the kth orbital.

To get an atomic charge, all the gross populations of the occupied orbitals are added, multiplied by the charge of an electron $(-e)$ and to this we add the charge of the nucleus,

$$
Q_p = Z_p e - e \sum_k N_{p,k}. \tag{15.30}
$$

The Mulliken population analysis is not a method that can be used to determine exact atomic charges. It depends heavily on the choice of basis functions and, in addition, the separation of each overlap population into two equal parts is quite arbitrary. Nevertheless, the analysis provides chemical insight, especially when comparing related systems. On the other hand, this also means that there are many other concepts according to which the electron distribution can be decomposed into atomic components. But this will not be treated any further here.

15.8 Electrostatic potential

When two molecules start interacting, e. g., in the initial stage of a chemical reaction, they first experience the electrostatic potential of each other. This potential can therefore be used to obtain ideas about how the two molecules approach each other: regions of positive electrostatic potential are preferred by regions of the other molecule with a negative charge, and vice versa. Therefore, the electrostatic potential of a molecule is often studied.

The electrostatic potential at point $\vec{r}$ can be easily calculated from the distribution of nuclei and electrons in the molecule. It equals

$$V_{es}(\vec{r}) = \sum_k \frac{Z_k e}{4\pi\epsilon_0 |\vec{R}_k - \vec{r}|} - \int \frac{e\rho(\vec{r}_1)}{4\pi\epsilon_0 |\vec{r}_1 - \vec{r}|} d\vec{r}_1. \tag{15.31}$$

As an example, we show the electrostatic potential around an ethanol molecule, C_2H_5OH, in Figure 15.15. For comparison, we also show the electron density of this molecule, and as you can see, it can be of advantage to combine the two properties in a single graphical representation.

However, we should add that currently other approaches are considered more useful to study how molecules interact with each other and whether chemical reactions between them may take place. But this will not be discussed further in this manuscript.

15.9 Orbitals

With the orbital picture we have seen that the energies of the highest occupied orbital (HOMO) and of the lowest unoccupied orbital (LUMO) are related to the first ionization potential and the first electron affinity, respectively. This is the content of Koopmans' theorem. Electronic relaxation effects are not taken into account: it is assumed that the orbitals do not change when the number of electrons of the molecule changes.

Equivalently, it is assumed that an additional electron will occupy the LUMO and an electron will be removed from the HOMO. Spatial representations of these orbitals can therefore provide information on how the electron distribution will change when electrons are added or removed. As an example we show in Figure 15.16 the HOMO

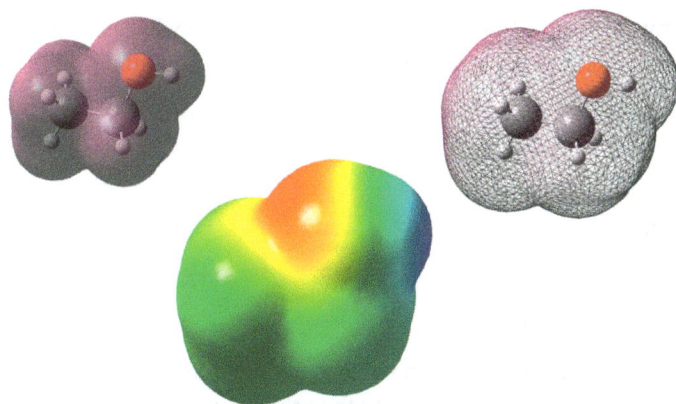

Figure 15.15: In the two upper figures, two surfaces are shown on which the electron density of ethanol has a constant value. At the bottom, such a surface is color-coded with values of the electrostatic potential. In this case, red marks a very negative potential value, and blue a very positive potential value.

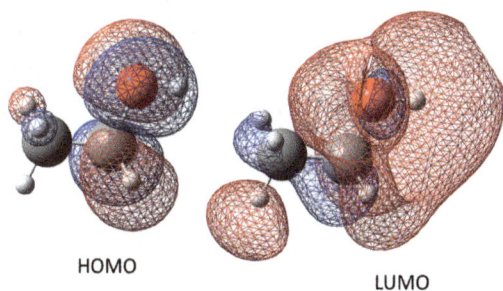

HOMO

LUMO

Figure 15.16: The HOMO (right) and LUMO (left) of ethanol, C_2H_5OH.

and LUMO orbitals of ethanol, C_2H_5OH. It can be seen that these orbitals are quite delocalized over the whole molecule.

In the theoretical treatment of the fundamentals of the orbital picture, we have also seen that the orbital picture is equivalent to the assumption of the validity of the Hartree–Fock approximation. In the general case, the Hartree–Fock equations are solved,

$$\hat{F}\tilde{\psi}_k = \sum_i \lambda_{ki}\tilde{\psi}_i. \tag{15.32}$$

The paramters $\{\lambda_{ki}\}$ are the Lagrange multipliers, who are introduced to guarantee that the orbitals are orthonormal. It was possible to choose them to be nonzero only for $k = i$, so that we get the usual Hartree–Fock equations,

$$\hat{F}\psi_k = \epsilon_k\psi_k. \tag{15.33}$$

We assume that the ϵ_k are ordered by increasing energy, $\epsilon_i \leq \epsilon_{i+1}$.

Considering the ground state of the molecule, the N energetically lowest of these orbitals are occupied. But it is easy to show that other solutions to equation (15.32)

$$\tilde{\psi}_l = \sum_{k=1}^{N} U_{lk}\psi_k \tag{15.34}$$

can be generated as long as

$$\sum_k U_{lk}^* U_{mk} = \delta_{l,m} \tag{15.35}$$

is fulfilled. Furthermore, the same expectation value is obtained for any experimentally measurable observable, regardless of whether the orbitals $\{\tilde{\psi}_k\}$ or $\{\psi_k\}$ are used. Accordingly, none of these sets of orbitals is more correct.

The experience has shown that the orbitals obtained by solving equation (15.33), (the so-called canonical orbitals) are often delocalized over the entire molecule. This may or may not be a consequence of symmetry properties. But such delocalized orbitals are rarely consistent with the frequently used picture of localized chemical bonds between neighboring atoms. However, through a transformation as in equation (15.34) it is possible to obtain more localized orbitals, that then correspond more closely to the usual concepts. An example of this is shown in Figure 15.17.

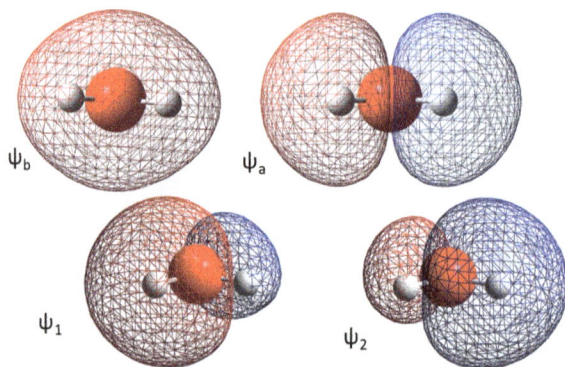

Figure 15.17: An example of the transformation of (top) delocalized molecular orbitals to (below) localized molecular orbitals. Shown are orbitals of a H_2O molecule.

In Figure 15.17, we have created two new orbitals from the delocalized orbitals ψ_b and ψ_a,

$$\psi_1 = \frac{1}{\sqrt{2}}(\psi_b + \psi_a)$$
$$\psi_2 = \frac{1}{\sqrt{2}}(\psi_b - \psi_a), \tag{15.36}$$

or

$$\psi_b = \frac{1}{\sqrt{2}}(\psi_1 + \psi_2)$$

$$\psi_a = \frac{1}{\sqrt{2}}(\psi_1 - \psi_2). \tag{15.37}$$

ψ_b and ψ_a are canonical orbitals, i. e., eigenfunctions to the Fock operator,

$$\hat{F}\psi_b = \epsilon_b \psi_b$$

$$\hat{F}\psi_a = \epsilon_a \psi_a, \tag{15.38}$$

but this does not apply to ψ_1 and ψ_2. For these, we have

$$\hat{F}\psi_1 = \hat{F}\left[\frac{1}{\sqrt{2}}(\psi_b + \psi_a)\right] = \frac{1}{\sqrt{2}}(\epsilon_b \psi_b + \epsilon_a \psi_a)$$

$$= \frac{1}{2}[(\epsilon_b(\psi_1 + \psi_2) + \epsilon_a(\psi_1 - \psi_2)] = \frac{\epsilon_b + \epsilon_a}{2}\psi_1 + \frac{\epsilon_b - \epsilon_a}{2}\psi_2 \tag{15.39}$$

and

$$\hat{F}\psi_2 = \hat{F}\left[\frac{1}{\sqrt{2}}(\psi_b - \psi_a)\right] = \frac{1}{\sqrt{2}}(\epsilon_b \psi_b - \epsilon_a \psi_a)$$

$$= \frac{1}{2}[(\epsilon_b(\psi_1 + \psi_2) - \epsilon_a(\psi_1 - \psi_2)] = \frac{\epsilon_b - \epsilon_a}{2}\psi_1 + \frac{\epsilon_b + \epsilon_a}{2}\psi_2. \tag{15.40}$$

In matrix form, we have

$$\hat{F}\begin{pmatrix} \psi_b \\ \psi_a \end{pmatrix} = \begin{pmatrix} \epsilon_b & 0 \\ 0 & \epsilon_a \end{pmatrix}\begin{pmatrix} \psi_b \\ \psi_a \end{pmatrix}$$

$$\hat{F}\begin{pmatrix} \psi_1 \\ \psi_2 \end{pmatrix} = \begin{pmatrix} \frac{1}{2}(\epsilon_b + \epsilon_a) & \frac{1}{2}(\epsilon_b - \epsilon_a) \\ \frac{1}{2}(\epsilon_b - \epsilon_a) & \frac{1}{2}(\epsilon_b + \epsilon_a) \end{pmatrix}\begin{pmatrix} \psi_1 \\ \psi_2 \end{pmatrix}, \tag{15.41}$$

which shows that in the second case, the matrix on the right-hand side is not diagonal (i. e., there are nonzero matrix elements outside the diagonal). On the other hand,

$$|\psi_1(\vec{r})|^2 + |\psi_2(\vec{r})|^2 = \frac{1}{2}|\psi_b(\vec{r}) + \psi_a(\vec{r})|^2 + \frac{1}{2}|\psi_b(\vec{r}) - \psi_a(\vec{r})|^2 = |\psi_b(\vec{r})|^2 + |\psi_a(\vec{r})|^2, \tag{15.42}$$

so that the electron density is independent of which set of orbitals that is used.

15.10 Orbital energies

With the Hartree–Fock approximation, we get the following expression for the (approximate) electronic energy:

$$E_e \approx \sum_{k=1}^{N} \epsilon_k - \frac{1}{2}\sum_{k,l=1}^{N}[\langle\psi_k\psi_l|\hat{h}_2|\psi_k\psi_l\rangle - \langle\psi_l\psi_k|\hat{h}_2|\psi_k\psi_l\rangle]. \tag{15.43}$$

This equation shows that a substantial fraction of the electronic energy is simply the sum of the energies of the occupied orbitals. Therefore, by analyzing the orbital energies as a function of the structure, one can obtain a simple method to rationalize the structural properties of a molecule. Schematic representations of orbital energies as a function of structure are called Walsh diagrams (after Arthur Donald Walsh).

We will illustrate this here through the example of the bond angle of the water molecule.

Figure 15.18 shows the orbitals and their energies as a function of the H–O–H bond angle in an H_2O molecule. The energetically lowest orbital has bonding interactions between all atoms. Therefore, if the bond angle is increased, the H–H bonding interaction (which is not a "true" chemical bond, although the atomic hydrogen 1s orbitals sense each other) loses strength. Accordingly, the orbital energy increases as a function of the bond angle.

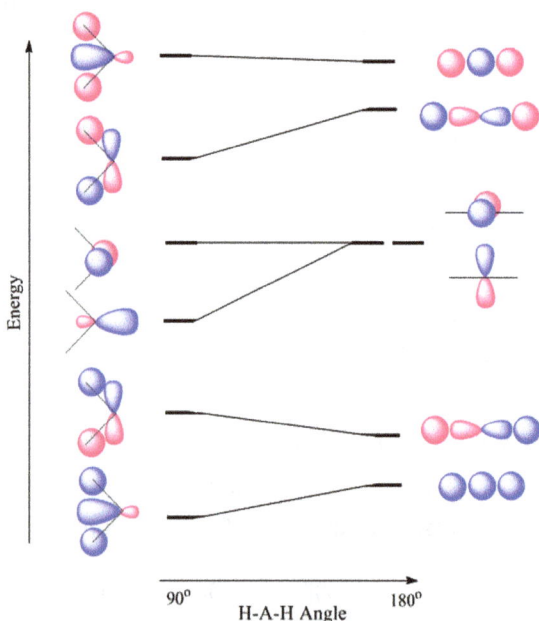

Figure 15.18: Qualitative representation of the orbital energies as a function of the H–O–H bond angle for H_2O. Only the four lowest orbitals are occupied. Copied on 09.02.17 from https://commons.wikimedia.org/wiki/File_AWalshdiagram.gif. From Chem540f09grp5 (Own work), via Wikimedia Commons.

The opposite is found for the energetically next-higher orbital. Here, the O–H bonding interactions are enhanced and the H–H antibonding interactions are reduced when the bond angle becomes larger. Thus, the energy of this orbital decreases as a function of the bond angle.

The next orbital becomes slowly a pure p function on the oxygen atom when the bond angle approaches 180°. This also means that the (weak) bonding interactions decrease (or rather disappear), so that the orbital energy of this orbital increases as a function of the bond angle.

Ultimately, the topmost occupied orbital is a pure p function on the oxygen atom. Therefore, its energy is largely independent of the bond angle.

Adding all the energies of the occupied orbitals, one obtains a function that has a minimum at a bond angle roughly between 100° and 120°. This value corresponds quite well to the experimental bond angle of the water molecule.

15.11 Problems with answers

1. **Problem:** Consider a diatomic, heteroatomic molecule for which an LCAO-MO description can be used, using only one AO per atom. The two molecular orbitals are (in increasing energetic order) $\psi_1 = c_{a1}\chi_a + c_{b1}\chi_b$ and $\psi_2 = c_{a2}\chi_a + c_{b2}\chi_b$. The two atomic orbitals χ_a and χ_b are normalized but not orthogonal. The molecule has two electrons. Determine the Mulliken net, overlap, and gross populations on the two atoms. Explain all the quantities you introduce.

Answer: We write the two orbitals as

$$\psi_j = c_{aj}\chi_a + c_{bj}\chi_b. \tag{15.44}$$

Each orbital is normalized,

$$1 = \langle \psi_j | \psi_j \rangle = |c_{aj}|^2 \langle \chi_a | \chi_a \rangle + |c_{bj}|^2 \langle \chi_b | \chi_b \rangle + c_{aj}^* c_{bj} \langle \chi_a | \chi_b \rangle + c_{bj}^* c_{aj} \langle \chi_b | \chi_a \rangle$$
$$= |c_{aj}|^2 + |c_{bj}|^2 + \left[c_{aj}^* c_{bj} S_{ab} + c_{bj}^* c_{aj} S_{ab}^* \right] = n_{aj} + n_{bj} + n_{abj}. \tag{15.45}$$

Here is

$$\langle \chi_a | \chi_a \rangle = 1$$
$$\langle \chi_b | \chi_b \rangle = 1$$
$$\langle \chi_a | \chi_b \rangle = S_{ab}$$
$$\langle \chi_b | \chi_a \rangle = S_{ba} = S_{ab}^*. \tag{15.46}$$

Then

$$n_{aj} = |c_{aj}|^2$$
$$n_{bj} = |c_{bj}|^2 \tag{15.47}$$

are the atomic net populations on the atoms a and b for the jth orbital, while the overlap population between the atoms a and b for the same orbital is equal to

$$n_{abj} = c_{aj}^* c_{bj} S_{ab} + c_{bj}^* c_{aj} S_{ab}^*. \tag{15.48}$$

Because we have two electrons in total, the energetically lowest orbital (with $j = 1$) is doubly occupied and that with $j = 2$ is empty. Then

$$n_a = 2n_{a1}$$
$$n_b = 2n_{b1}$$
$$n_{ab} = 2n_{ab1}$$
$$N_a = n_a + \frac{1}{2}n_{ab}$$
$$N_b = n_b + \frac{1}{2}n_{ab}. \tag{15.49}$$

2. **Problem:** More different theoretical methods were used to study the relative energies of three isomers, which we will denote I, II, and III. According to Hartree–Fock calculations with a small basis set, the total energies of the three isomers are -110, -112, and -108 eV, while Hartree–Fock calculations with a larger basis set gave total energies of -115, -116, and -112 eV. With CCSD calculations with the larger basis set, -122, -121, and -119 eV were found. LDA calculations with the larger basis set yielded -126, -128, and -125 eV, and similar GGA calculations found -123, -121, and -119 eV. What can be said about the relative energies?

Answer: For the wave function-based calculations (Hartree–Fock and CCSD), the variational principle applies without restrictions, so that improvements in the calculations automatically lead to more accurate results. The same is true only to a limited extent for density functional calculations, with first of all LDA calculations often leading to inaccurate total energies. More precise are GGA calculations, but above all the CCSD calculations can be considered as the most accurate. Therefore, the total energies equal to -122, -121, and -119 eV can be interpreted as the most accurate ones, from which we find relative energies of 0, 1, and 3 eV. It should be emphasized that, strictly speaking, these considerations that are based on the variational principle apply only for the total energies, but not for the relative energies.

15.12 Problems

1. Explain the term "Walsh diagram."
2. Sketch a Walsh diagram for H_2O.
3. Sketch a Walsh diagram for CO_2.
4. Why is it difficult to define atomic charges?
5. Explain the relationship between the potential energy surface and the geometry of a molecule.
6. Explain how to calculate the structure of a molecule.
7. Explain how to calculate the vibrational spectrum of a molecule.

8. Describe how the dipole moment of a molecule is calculated and why it can be difficult to obtain accurate values for this property.

9. Explain how to calculate the electrostatic potential of a molecule.

10. Explain briefly what happens to the orbital energies when generating localized orbitals from those that are obtained with the usual Hartree–Fock equations $\hat{F}\psi_k = \epsilon_k \psi_k$.

11. Explain briefly the term "BSSE."

16 Supporting information

16.1 Continuous probability distributions

As an example, we shall consider the game of throwing darts at a dartboard. A typical dartboard has a spherical region and you have to throw the darts so that they stay somewhere in this spherical region. This part consists of two small, concentric spheres in the middle. The remaining part is split into 20 radial sections, each separated into four parts of different sizes. Thus, when successfully throwing a dart, it will stay in one of those 82 parts. We will distinguish those parts through integers going from $i = 1$ to $i = 82$.

We will now assume that we throw a dart very many (N) times at the board and count the number of times, N_i, we hit the different parts of the boards. When N is very large, N_i becomes proportional to N and we can write

$$N_i = NP_i \tag{16.1}$$

with P_i being independent of N. P_i is the probability that we have hit the ith part of the dartboard. Of obvious reasons, we have

$$\sum_{i=1}^{82} P_i = 1. \tag{16.2}$$

This example provides an example of a discrete probability distribution. Discrete, because the variable, i, can take only discrete values. As is well known, we can use this probability in calculating various expectation values like, for instance, the average value of i (here, we do not discuss whether this is useful),

$$\langle i \rangle = \sum_{i=1}^{82} P_i \cdot i. \tag{16.3}$$

Such discrete distribution functions are well known, whereby most often the case of throwing a dice is used to illustrate the concept.

We shall now modify the experiment a little. Thus, we remove the lines that separate the spherical region into 82 parts and have, accordingly, only the spherical region with a radius we call R. We place a cartesian coordinate system so that it has the origin in the middle of the sphere. Again, throwing a dart successfully implies that it stays somewhere for which

$$x^2 + y^2 \leq R^2. \tag{16.4}$$

Again, we may throw the dart a very large number of times and consider the possibility of getting a well-defined (x, y) that satisfies equation (16.4). However, the probability of getting exactly this value is vanishingly small. Instead, we consider the probability

https://doi.org/10.1515/9783110742206-016

$P_c(x, y) \Delta x \Delta y$, which is the probability that the dart hits a point in the small area $[x - \Delta x/2; x + \Delta x/2] \times [y - \Delta y/2; y + \Delta y/2]$. We shall ultimately consider the limit $\Delta x \to dx$, $\Delta y \to dy$, i. e., the small area becomes infinitesimally small.

$P_c(x, y)$ is an example of a continuous probability distribution. When integrating over all possible outcomes of throwing the dart, we shall get 1, similar to equation (16.2),

$$\iint\limits_{x^2+y^2 \leq R^2} P_c(x, y) \, dx \, dy = 1. \tag{16.5}$$

Also in this case, we may calculate average values like, for instance,

$$\langle x \rangle = \iint\limits_{x^2+y^2 \leq R^2} P_c(x, y) x \, dx \, dy. \tag{16.6}$$

It may be useful to use not the cartesian (x, y) coordinates but instead the polar coordinates (r, θ). Then we have

$$P_p(r, \theta) \, dr \, d\theta = P_c(r \cos \theta, r \sin \theta) r \, dr \, d\theta \tag{16.7}$$

for which equation (16.5) becomes

$$\int\limits_0^R \int\limits_0^{2\pi} P_p(r, \theta) \, d\theta \, dr = 1. \tag{16.8}$$

One may for instance use $P_p(r, \theta)$ to calculate the probability distribution for the distances to the center, i. e., for r, irrespectively of θ. This becomes $\int_0^{2\pi} P_p(r, \theta) \, d\theta$.

Finally, we shall consider a simple example of a continuous probability distribution of one variable, s, i. e., $P(s)$. Two arbitrary examples of such distributions are shown in Fig. 16.1. The distribution functions satisfy

$$\int\limits_{-\infty}^{\infty} P(s) = 1. \tag{16.9}$$

From $P(s)$, one may calculate an average value (expectation value)

$$\langle s \rangle = \int\limits_{-\infty}^{\infty} sP(s). \tag{16.10}$$

This value is marked in the figure, too. Also the width (uncertainty) of the distribution can be calculated

$$\Delta s = [\langle (s - \langle s \rangle)^2 \rangle]^{1/2} = [\langle s^2 \rangle - \langle s \rangle^2]^{1/2}. \tag{16.11}$$

This is also shown in the two examples of Figure 16.1. When repeating a measurement of s many times, $\langle s \rangle$ will give an estimate of the average value that is obtained, whereas most of the measurements will lie in the range $[\langle s \rangle - \Delta s/2; \langle s \rangle + \Delta s/2]$, i. e., Δs describes the size of the interval inside which most measurements will lie.

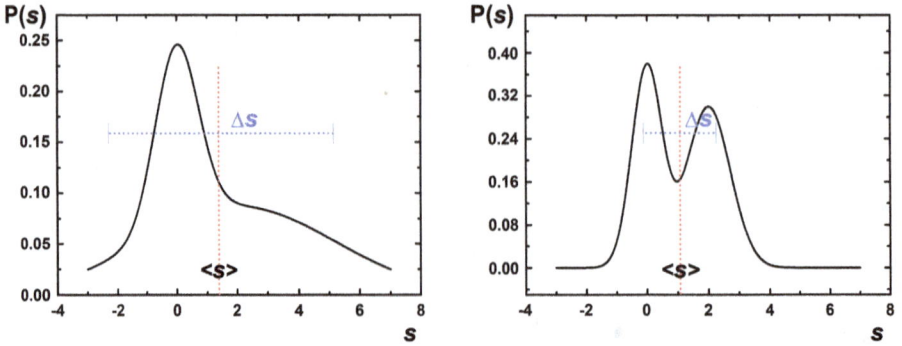

Figure 16.1: Two examples of continuous distribution functions of a single variable, s. In each figure, also ⟨s⟩ and Δs are marked through the vertical and horizontal lines, respectively.

16.2 Dirac's δ function

Dirac's δ function $\delta(x - x_0)$ in one dimension can be defined by requiring that

$$\int_{x_1}^{x_2} f(x)\delta(x - x_0) \, dx = f(x_0) \tag{16.12}$$

for all functions $f(x)$ when

$$x_1 < x_0 < x_2. \tag{16.13}$$

Otherwise

$$\int_{x_1}^{x_2} f(x)\delta(x - x_0) \, dx = 0 \tag{16.14}$$

when

$$x_1 < x_2 < x_0 \quad \text{or} \quad x_0 < x_1 < x_2. \tag{16.15}$$

It may be described as

$$\delta(x - x_0) = \begin{cases} \infty & x = x_0 \\ 0 & x \neq x_0 \end{cases} \tag{16.16}$$

It may also be considered as the limit of a Gaussian function for which the width vanishes,

$$\delta(x - x_0) = \lim_{a \to 0} \frac{1}{\sqrt{\pi a}} \exp\left(-\frac{(x - x_0)^2}{a}\right). \tag{16.17}$$

The generalization to for instance three dimensions is according to

$$\delta(\vec{r} - \vec{r}_0) = \delta(x - x_0)\delta(y - y_0)\delta(z - z_0) \tag{16.18}$$

with $\vec{r}_0 = (x_0, y_0, z_0)$.

16.3 Diagonalization

We consider a hermitian matrix $\underline{\underline{A}}$, i. e., a square matrix for which

$$A_{ij} = A_{ji}^{*}. \tag{16.19}$$

Diagonalizing this matrix means writing it as

$$\underline{\underline{A}} = \underline{\underline{U}}\,\underline{\underline{\Lambda}}\,\underline{\underline{U}}^{\dagger}. \tag{16.20}$$

This is equivalent to solving the eigenvalue equation

$$\underline{\underline{A}}\,\underline{c}_i = \lambda_i\,\underline{c}_i. \tag{16.21}$$

Here, $\underline{c}_i$ is a column vector with the eigenvector for the ith eigenvalue of $\underline{\underline{A}}$, λ_i. Below, we illustrate this through an example.

In equation (16.20), $\underline{\underline{\Lambda}}$ is a diagonal matrix containing the eigenvalues of $\underline{\underline{A}}$,

$$\Lambda_{ij} = \delta_{i,j}\lambda_i. \tag{16.22}$$

Furthermore, the ith column of $\underline{\underline{U}}$ is the (normalized) ith eigenvector of $\underline{\underline{A}}$. Finally, $\underline{\underline{U}}^{\dagger}$ is the hermitian conjugate of $\underline{\underline{U}}$, i. e.,

$$(U^{\dagger})_{ij} = U_{ji}^{*}. \tag{16.23}$$

$\underline{\underline{U}}$ is unitary,

$$\underline{\underline{U}}^{-1} = \underline{\underline{U}}^{\dagger}. \tag{16.24}$$

Solving equations (16.20) or (16.21) is called diagonalizing $\underline{\underline{A}}$.

We shall illustrate this through an example,

$$\underline{\underline{A}} = \begin{pmatrix} 3 & \sqrt{2} \\ \sqrt{2} & 2 \end{pmatrix}. \tag{16.25}$$

The eigenvalues of $\underline{\underline{A}}$ are found from

$$0 = \begin{vmatrix} 3-\lambda & \sqrt{2} \\ \sqrt{2} & 2-\lambda \end{vmatrix} = (3-\lambda)(2-\lambda) - 2 \tag{16.26}$$

which has the solutions

$$\lambda = \lambda_1 = 4$$
$$\lambda = \lambda_2 = 1. \tag{16.27}$$

For these two values of λ, the two equations

$$(3 - \lambda)x + \sqrt{2}y = 0$$
$$\sqrt{2}x + (2 - \lambda)y = 0 \tag{16.28}$$

are linear dependent (i. e., proportional to each other). Considering the second of those equations, we find

$$x = \frac{\lambda - 2}{\sqrt{2}}y. \tag{16.29}$$

Then the normalization of the eigenvectors (x, y) gives

$$1 = x^2 + y^2 = \left[\frac{(\lambda - 2)^2}{2} + 1\right]y^2 \tag{16.30}$$

which for the two values for λ of equation (16.27) gives

$$y_1 = \sqrt{\frac{1}{3}}$$
$$y_2 = \sqrt{\frac{2}{3}} \tag{16.31}$$

and subsequently from equation (16.29)

$$x_1 = \sqrt{\frac{2}{3}}$$
$$x_2 = -\sqrt{\frac{1}{3}} \tag{16.32}$$

$\underline{\underline{\Lambda}}$ is the diagonal matrix with the eigenvalues of equation (16.27),

$$\underline{\underline{\Lambda}} = \begin{pmatrix} 4 & 0 \\ 0 & 1 \end{pmatrix}. \tag{16.33}$$

Moreover, $\underline{\underline{U}}$ contains the normalized eigenvectors as columns,

$$\underline{\underline{U}} = \begin{pmatrix} \sqrt{\frac{2}{3}} & -\sqrt{\frac{1}{3}} \\ \sqrt{\frac{1}{3}} & \sqrt{\frac{2}{3}} \end{pmatrix}. \tag{16.34}$$

Then,

$$\underline{\underline{U}}^{\dagger} = \begin{pmatrix} \sqrt{\frac{2}{3}} & \sqrt{\frac{1}{3}} \\ -\sqrt{\frac{1}{3}} & \sqrt{\frac{2}{3}} \end{pmatrix}. \tag{16.35}$$

As a control, we may easily verify that

$$\underline{\underline{U}}^\dagger \cdot \underline{\underline{U}} = \underline{\underline{U}} \cdot \underline{\underline{U}}^\dagger = \begin{pmatrix} 1 & 0 \\ 0 & 1 \end{pmatrix}. \tag{16.36}$$

Finally, equation (16.20) takes the form

$$\begin{pmatrix} 3 & \sqrt{2} \\ \sqrt{2} & 2 \end{pmatrix} = \begin{pmatrix} \sqrt{\frac{2}{3}} & -\sqrt{\frac{1}{3}} \\ \sqrt{\frac{1}{3}} & \sqrt{\frac{2}{3}} \end{pmatrix} \begin{pmatrix} 4 & 0 \\ 0 & 1 \end{pmatrix} \begin{pmatrix} \sqrt{\frac{2}{3}} & \sqrt{\frac{1}{3}} \\ -\sqrt{\frac{1}{3}} & \sqrt{\frac{2}{3}} \end{pmatrix} \tag{16.37}$$

which without too much effort can be verified, also.

16.4 Boltzmann, Fermi–Dirac, and Bose–Einstein distributions

Within the field of statistical thermodynamics, properties of macroscopic samples are calculated by utilizing the fact that such samples contain a very large number of equivalent particles (electrons, atoms, molecules) so that statistical arguments can be applied. Here, we shall recall just some few of the foundations for these arguments.

The starting point is the assumption is that each particle can occupy one out of a set of different energy levels. The energies of those levels are given and the main goal is to determine the distribution of the particles in those levels, i. e., the number of particles n_i in the ith level with the (given) energy ϵ_i. The n_i are determined under various boundary conditions, leading to different distributions, but here we shall just consider some few cases.

At first, we restrict ourselves to the case that the total number of particles N and the total energy E are given

$$N = \sum_i n_i$$

$$E = \sum_i n_i \epsilon_i. \tag{16.38}$$

Subsequently, three different sets of rules can be applied to place the individual particles in the different energy levels. The Boltzmann distribution (after Ludwig Boltzmann) is obtained by assuming that the particles are distinguishable and that each n_i can take any value. This is often referred to as a classical distribution.

Both the Fermi–Dirac (after Enrico Fermi and Paul Adrien Maurice Dirac) and the Bose–Einstein (after Satyendra Nath Bose and Albert Einstein) distribution are quantum distributions. In both cases, it is assumed that the particles are indistinguishable, but whereas for the Bose–Einstein distribution n_i can take any value, for the Fermi–Dirac distribution, n_i can take only the values 0 and 1. Particles that satisfy the Bose–Einstein distribution are called bosons; those that follow the Fermi–Dirac distribution are called fermions. Electrons are fermions.

For the Boltzmann distribution, one finds

$$n_i = N \frac{\exp(-\frac{\epsilon_i}{kT})}{\sum_j \exp(-\frac{\epsilon_j}{kT})} \tag{16.39}$$

with k being Boltzmann's constant and T being the temperature. This expression shows that n_i decreases with increasing energy, ϵ_i.

17 Mathematical formulas

17.1 Trigonometric functions

$$\sin(x) = \frac{1}{2i}\left(e^{ix} - e^{-ix}\right)$$

$$\cos(x) = \frac{1}{2}\left(e^{ix} + e^{-ix}\right) \tag{17.1}$$

or

$$e^{is} = \cos(s) + i\sin(s) \tag{17.2}$$

(Formula of Euler)

$$\sin(\alpha)\sin(\beta) = \frac{1}{2}\left[\cos(\alpha - \beta) - \cos(\alpha + \beta)\right]$$

$$\sin(\alpha)\cos(\beta) = \frac{1}{2}\left[\sin(\alpha - \beta) + \sin(\alpha + \beta)\right]$$

$$\cos(\alpha)\cos(\beta) = \frac{1}{2}\left[\cos(\alpha - \beta) + \cos(\alpha + \beta)\right] \tag{17.3}$$

Furthermore,

$$\sin(\alpha - \beta) = \sin\alpha\cos\beta - \cos\alpha\sin\beta$$

$$\cos(\alpha - \beta) = \cos\alpha\cos\beta + \sin\alpha\sin\beta. \tag{17.4}$$

17.2 Spherical coordinates

Relations between spherical coordinates and cartesian coordinates:

$$x = r\sin\theta\cos\phi$$

$$y = r\sin\theta\sin\phi$$

$$z = r\cos\theta, \tag{17.5}$$

and vice versa

$$r = \left(x^2 + y^2 + z^2\right)^{1/2}$$

$$\theta = \mathrm{Arccos}\frac{z}{r}$$

$$\phi = \begin{cases} \mathrm{Arccos}\frac{x}{(x^2+y^2)^{1/2}} & y > 0 \\ 2\pi - \mathrm{Arccos}\frac{x}{(x^2+y^2)^{1/2}} & y < 0. \end{cases} \tag{17.6}$$

https://doi.org/10.1515/9783110742206-017

17.3 Laplace operator

In cartesian coordinates:

$$\nabla^2 = \Delta = \frac{\partial^2}{\partial x^2} + \frac{\partial^2}{\partial y^2} + \frac{\partial^2}{\partial z^2}. \tag{17.7}$$

In spherical coordinates,

$$\Delta\Psi = \nabla^2\Psi = \frac{1}{r}\frac{\partial^2}{\partial r^2}(r\Psi) + \frac{1}{r^2}\hat{\Lambda}^2\Psi = \frac{\partial^2}{\partial r^2}\Psi + \frac{2}{r}\frac{\partial}{\partial r}\Psi + \frac{1}{r^2}\hat{\Lambda}^2\Psi$$

$$\hat{\Lambda}^2\Psi = \frac{1}{\sin^2\theta}\left(\frac{\partial^2\Psi}{\partial\varphi^2}\right) + \frac{1}{\sin\theta}\frac{\partial}{\partial\theta}\left(\sin\theta\frac{\partial\Psi}{\partial\theta}\right) \tag{17.8}$$

17.4 Integrals

Integrals with trigonometric functions:

$$\int \sin^2(\alpha z)\,dz = \frac{-1}{4\alpha}\sin(2\alpha z) + \frac{z}{2}$$

$$\int \cos^2(\alpha z)\,dz = \frac{1}{4\alpha}\sin(2\alpha z) + \frac{z}{2}$$

$$\int \cos(\alpha z)\sin(\alpha z)\,dz = \frac{-1}{4\alpha}\cos(2\alpha z)$$

$$\int z\sin(\alpha z)\,dz = \frac{1}{\alpha^2}\sin(\alpha z) - \frac{1}{\alpha}z\cos(\alpha z)$$

$$\int z\cos(\alpha z)\,dz = \frac{1}{\alpha^2}\cos(\alpha z) + \frac{1}{\alpha}z\sin(\alpha z)$$

$$\int z^2\sin(\alpha z)\,dz = \frac{2}{\alpha^3}\cos(\alpha z) + \frac{2}{\alpha^2}z\sin(\alpha z) - \frac{1}{\alpha}z^2\cos(\alpha z)$$

$$\int z^2\cos(\alpha z)\,dz = -\frac{2}{\alpha^3}\sin(\alpha z) + \frac{2}{\alpha^2}z\cos(\alpha z) + \frac{1}{\alpha}z^2\sin(\alpha z)$$

$$\int z\sin^2(\alpha z)\,dz = \frac{z^2}{4} - \frac{z}{4\alpha}\sin(2\alpha z) - \frac{1}{8\alpha^2}\cos(2\alpha z)$$

$$\int z\cos^2(\alpha z)\,dz = \frac{z^2}{4} + \frac{z}{4\alpha}\sin(2\alpha z) + \frac{1}{8\alpha^2}\cos(2\alpha z)$$

$$\int z\cos(\alpha z)\sin(\alpha z)\,dz = \frac{-z}{4\alpha}\cos(2\alpha z) + \frac{1}{8\alpha^2}\sin(2\alpha z) \tag{17.9}$$

Integrals with exponential functions:

$$\int_0^\infty e^{-\beta s^2}\,ds = \frac{1}{2}\sqrt{\frac{\pi}{\beta}}$$

$$\int_0^\infty s e^{-\beta s^2}\,ds = \frac{1}{2\beta}$$

$$\int_0^\infty s^2 e^{-\beta s^2}\, ds = \frac{1}{4\beta}\sqrt{\frac{\pi}{\beta}}$$

$$\int_0^\infty s^3 e^{-\beta s^2}\, ds = \frac{1}{2\beta^2}$$

$$\int_0^\infty s^4 e^{-\beta s^2}\, ds = \frac{3}{8\beta^2}\sqrt{\frac{\pi}{\beta}}$$

$$\int_0^\infty s^n e^{-as}\, ds = n!/a^{n+1} \tag{17.10}$$

Index

www.ingramcontent.com/pod-product-compliance
Lightning Source LLC
Chambersburg PA
CBHW080909220326
41598CB00034B/5525